CISM International Centre for Mechanical Sciences

Courses and Lectures

Volume 602

For more than 40 years the book series edited by CISM, "International Centre for Mechanical Sciences: Courses and Lectures", has presented groundbreaking developments in mechanics and computational engineering methods. It covers such fields as solid and fluid mechanics, mechanics of materials, micro- and nanomechanics, biomechanics, and mechatronics. The papers are written by international authorities in the field. The books are at graduate level but may include some introductory material.

More information about this series at http://www.springer.com/series/76

Martin Kronbichler · Per-Olof Persson
Editors

Efficient High-Order Discretizations for Computational Fluid Dynamics

Springer

Editors
Martin Kronbichler
Institute for Computational Mechanics
Technical University Munich
Garching bei München, Germany

Per-Olof Persson
Department of Mathematics
University of California, Berkeley
Berkeley, CA, USA

ISSN 0254-1971 ISSN 2309-3706 (electronic)
CISM International Centre for Mechanical Sciences
ISBN 978-3-030-60609-1 ISBN 978-3-030-60610-7 (eBook)
https://doi.org/10.1007/978-3-030-60610-7

This Springer imprint is published by the registered company Springer Nature Switzerland AG
The registered company address is: Gewerbestrasse 11, 6330 Cham, Switzerland

Preface

The chapters presented in this volume correspond to lectures given during the course "Efficient High-Order Discretizations for Computational Fluid Dynamics" that was held at the CISM in Udine, Italy, July 16–20, 2018.

High-order methods have been gaining a lot of momentum during the last decades and are being increasingly deployed in new-generation fluid dynamics solvers. As compared to low-order finite volume methods predominant in today's production codes, higher order discretizations significantly reduce dispersion and dissipation errors, the main source of error in long-time simulations of flows at higher Reynolds numbers. Furthermore, high-order methods have the potential to better use modern computer hardware by doing more arithmetic operations on cacheable data. Combined with a reduced number of unknowns, they promise to relax the memory transfer and communication limits in well-optimized low-order simulators. This combination makes previously intractable problems accessible to simulation in an increasingly wide range of applications. However, high-order methods are no silver bullet, requiring a careful selection of algorithmic ingredients and implementations.

The focus of this book is to give a comprehensive insight into the state of the art and selected research directions of a particular class of high-order methods for CFD, the Discontinuous Galerkin (DG) methods. DG methods inherit ingredients from finite element and finite volume methods. High-order shape functions are used on a mesh of elements, making the method accurate also on complicated geometries with unstructured and deformed meshes. At the same time, DG uses the finite volume mechanisms of data exchange between neighboring elements in terms of Riemann solvers from two discontinuous states. This weak imposition of continuity allows to take directionality of flow into account and introduce a controlled amount of numerical dissipation, depending on the application needs.

The carefully selected chapters of this book aim to cover the breadth of this research field. The volume starts with an introduction to the main ingredients of discontinuous Galerkin schemes by Martin Kronbichler. Besides the main stability and convergence properties for linear transport and a Poisson problem, the chapter contrasts the convergence rates for smooth analytic cases to the more practically relevant behavior of high-order methods in under-resolved situations. Each of the subsequent chapters starts with the basics of the respective algorithmic and computational context, before delving into cutting-edge research topics. The second chapter by Martin Kronbichler discusses the implementation of discontinuous Galerkin schemes on modern computer hardware, with the main focus on explicit time integration. The chapter by Gregor Gassner and co-authors introduces to a class of robust split-form discontinuous Galerkin methods, constructing algorithms that are provably stable also for variable or nonlinear coefficients and deformed elements where integrals cannot be computed exactly. The chapter by Stefano Rebay and co-authors approaches the solution of the implicit linear system in the context of compressible flows. The next chapter by Per-Olof Persson focuses on implicit solvers for various time-integration schemes as well as mixed implicit-explicit methods. Next, the chapter by Sonia Fernández-Méndez introduces the hybridizable discontinuous Galerkin method, which aims to increase efficiency through a reduced size of the associated linear systems. Finally, the chapter by Rainald Löhner gives a critical view on high-order methods from an engineering perspective, and points out challenges that still need to be resolved by the research community.

The content primarily focuses on doctoral students and postdoctoral researchers in engineering, applied mathematics, physics, and high-performance computing with a strong interest in the interdisciplinary aspects of computational fluid dynamics. However, the book is also well-suited for industrial researchers or practicing computational engineers who would like to gain a comprehensive overview of discontinuous Galerkin methods, modern algorithmic realizations, and high-performance implementations.

The editors would like to thank all contributors to this volume and lecturers of the corresponding CISM for presenting a wide topic of content, to the course participants who stimulated further discussions, and to all members of CISM and CISM Springer for their support and patience during the course and during the preparation of this volume.

Garching bei München, Germany — Martin Kronbichler
Berkeley, CA, USA — Per-Olof Persson

Contents

Chapter 1
The Discontinuous Galerkin Method: Derivation and Properties

Martin Kronbichler

Abstract This text introduces to the main ingredients of the discontinuous Galerkin method, combining the framework of high-order finite element methods with Riemann solvers for the information exchange between the elements. The concepts are explained by the example of linear transport and convergence is evaluated in one, two, and three space dimensions. Finally, the construction of schemes for second derivatives is explained and detailed for the symmetric interior penalty method.

Introduction

The discontinuous Galerkin (DG) method, originally introduced by Reed and Hill (1973) for studying neutron transport, has emerged as one of the most important discretization schemes for the partial differential equations (PDE) of computational fluid dynamics (CFD). Discontinuous Galerkin methods enable a high formal order of accuracy on complex domains that need to be captured by unstructured meshes. As compared to low-order finite volume methods predominant in today's CFD production codes, higher order discretizations significantly reduce dispersion and dissipation errors also in the pre-asymptotic regime, the main source of inaccuracy in long-time simulations of flows at higher Reynolds numbers or wave propagation problems. Therefore, previously intractable problems are becoming accessible to simulation in an increasingly wide range of applications. An overview of the history and the most important application areas can be found in Cockburn et al. (2000), Hesthaven and Warburton (2008) and references therein. For the specific application in fluid dynamics, the work by Wang et al. (2013) gives an overview of the main contributions and research directions. The success of DG methods compared to globally continuous approximations has spurred additional research along with a multitude of directions.

M. Kronbichler (✉)
Institute for Computational Mechanics, Technical University of Munich, Boltzmannstr. 15, 85748 Garching bei München, Germany
e-mail: kronbichler@lnm.mw.tum.de

M. Kronbichler and P.-O. Persson (eds.), *Efficient High-Order Discretizations for Computational Fluid Dynamics*, CISM International Centre for Mechanical Sciences 602, https://doi.org/10.1007/978-3-030-60610-7_1

A variant widely used in CFD are flux reconstruction methods, reviewed in Huynh et al. (2014).

Discontinuous Galerkin methods derive their high-order accuracy from the variational principles of finite and spectral elements, finding solutions as some best fit over a rich basis, typically polynomials, within each element of the mesh. Instead of strongly enforcing continuity over the element boundaries as in conventional finite element methods, the discontinuous Galerkin methods keep the approximation independent between the elements, leading to a discontinuity in the solution representation. This initially counterintuitive step allows to add physics-informed mechanisms, the numerical fluxes, for the inter-element coupling necessary to solve the global PDE. Numerical flux functions act point-wise along the element faces by the concept of Riemann solvers, taking two adjacent states and a normal vector to produce a result consistent with or approximating the underlying partial differential equation. As a consequence, DG schemes can be equipped with similar upwinding mechanisms as finite volume methods, introducing dissipation over the cell boundaries in a controlled way, without compromising the high-order accuracy. This principle provides robustness for transport-dominated processes of marginally resolved simulations, whereas the continuous finite element method needs to rely on additional stabilization mechanisms such as streamline-upwind diffusion.

This text derives high-order discontinuous Galerkin methods with the focus on problems in fluid dynamics, building on the analogy with finite element and spectral element methods but with additional terms at the places of discontinuity. Given the introductory nature, there is a close relation to the textbooks by Hesthaven and Warburton (2008) and the books on spectral elements by Deville et al. (2002), Karniadakis and Sherwin (2005), Kopriva (2009), but with an algorithmic focus. The steps to construct a DG scheme are introduced for a scalar conservation law in section "The Main Concepts". Section "Stability and Convergence" presents basic stability results and illustrates convergence on analytical test cases as well as more application-oriented metrics. The development of DG schemes for second derivatives of parabolic and elliptic PDEs is presented in section "DG Discretizations for Second Derivatives". Finally, section "Concluding Remarks" summarizes the methods and gives an outlook to the challenges in the implementation and practical applications in terms of suitable polynomial degrees.

The Main Concepts

The concepts of discontinuous Galerkin methods are introduced for a scalar conservation law in d dimensions (typically, $d = 1, 2, 3$), which seeks to find a scalar function $u(\boldsymbol{x}, t)$ dependent on the spatial variables $\boldsymbol{x} = (x_1, x_2, \ldots, x_d)$ and time t that fulfills the equation

$$
\begin{aligned}
&\frac{\partial u(\boldsymbol{x},t)}{\partial t} + \nabla \cdot \boldsymbol{f}(u(\boldsymbol{x},t),\boldsymbol{x},t) = b(\boldsymbol{x},t), && \boldsymbol{x} \in \Omega \subset \mathbb{R}^d, t \in (0,T), \qquad (1.1)\\
&u(\boldsymbol{x},t) = g(\boldsymbol{x},t), && \boldsymbol{x} \in \partial\Omega_{\text{in}}, t \in (0,T),\\
&u(\boldsymbol{x},0) = u_0(\boldsymbol{x}).
\end{aligned}
$$

The vector-valued function $\boldsymbol{f}$ specifies the physics of the problem and is called the flux. The right-hand side function b represents possible source terms. Here, ∇ denotes the gradient vector containing partial derivatives, $\nabla = [\partial/\partial x_1, \ldots, \partial/\partial x_d]$ and $\cdot$ the inner product between two vectors. Thus, the quantity $\nabla \cdot \boldsymbol{f} = \sum_{i=1}^{d} \partial f_i/\partial x_i$ denotes the divergence of the vector field $\boldsymbol{f}$. In order to ensure well-posedness of the partial differential equation, see, e.g., Gustafsson (2008) or Gustafsson et al. (2013), boundary conditions g need to be set on all inflow boundaries $\partial\Omega_{\text{in}}$ where the flux Jacobian points into the computational domain,

$$
\boldsymbol{n} \cdot \frac{\partial \boldsymbol{f}}{\partial u} < 0,
$$

where $\boldsymbol{n}$ denotes the unit outer normal on the boundary $\partial\Omega$. Note that the inflow boundary can change over time, depending on $\boldsymbol{f}$.

The starting point for the discontinuous Galerkin method is the weak form of the PDE, similarly to the framework of the continuous finite element method, but restricted to a suitable subdomain $\Omega' \subset \Omega$. Equation (1.1) is multiplied by a test function $v \in H^1(\Omega')$ with square-integrable first derivative and integrated over Ω',

$$
\int_{\Omega'} v \frac{\partial u}{\partial t}\,\mathrm{d}\boldsymbol{x} + \int_{\Omega'} v \nabla \cdot \boldsymbol{f}(u)\,\mathrm{d}\boldsymbol{x} = \int_{\Omega'} v b\,\mathrm{d}\boldsymbol{x}.
$$

We apply integration by parts to the term with $\nabla \cdot \boldsymbol{f}$, which yields

$$
\int_{\Omega'} v \frac{\partial u}{\partial t}\,\mathrm{d}\boldsymbol{x} - \int_{\Omega'} (\nabla v) \cdot \boldsymbol{f}(u)\,\mathrm{d}\boldsymbol{x} + \int_{\partial\Omega'} v \boldsymbol{n} \cdot \boldsymbol{f}(u)\,\mathrm{d}\boldsymbol{x} = \int_{\Omega'} v b\,\mathrm{d}\boldsymbol{x}. \qquad (1.2)
$$

To simplify the notation, spatial integrals over a volume are denoted by the bilinear form

$$
(a, b)_{\Omega'} = \int_{\Omega'} a \odot b\,\mathrm{d}\boldsymbol{x},
$$

where a and b can be scalar functions, vector-valued functions with d components each, or tensor-valued functions with $d_1 \times d_2$ components each for some number of components d_1 and d_2. The product $\odot$ denotes the inner product, summing over the product $a_i b_i$ of all components in a and b, with the special case of the product of a and b in the scalar case or $\cdot$ for the vector-valued case. Similarly, integrals over the boundary of a domain $\partial\Omega'$ are denoted by the bilinear form

$$
\langle a, b \rangle_{\partial\Omega'} = \int_{\partial\Omega'} a \odot b\,\mathrm{d}\boldsymbol{x}.
$$

Using this notation, the weak form (1.2) is written as

$$\left(v, \frac{\partial u}{\partial t}\right)_{\Omega'} - (\nabla v, \boldsymbol{f}(u))_{\Omega'} + \langle v\boldsymbol{n}, \boldsymbol{f}(u)\rangle_{\partial\Omega'} = (v, b)_{\Omega'} . \tag{1.3}$$

In the following subsections, the process of finding a DG approximation to the problem is presented step-by-step, following the algorithm structure of a finite element discretization.

Triangulation of the Computational Domain

The first step in a finite element method is to subdivide the computational domain into a set of disjoint subdomains of simple geometrical shapes, the elements,

$$\Omega_h = \bigcup_{e=1}^{n} \Omega_e,$$

with $\Omega_h \approx \Omega$. The two main classes of element shapes are triangular (2D) and tetrahedral (3D) elements on the one hand as well as quadrilateral (2D) and hexahedral (3D) elements on the other hand.

The main advantage of tetrahedral elements is the ease of mesh generation. Off-the-shelf mesh generators can create high-quality meshes with good aspect ratios and no distorted element shapes for almost any kind of geometry. This is in contrast to hexahedral meshing, which has remained a challenge up to date and is restricted to simpler shapes of the computational domain. At the same time, hexahedral or prismatic elements are the natural shape for boundary layers created by extrusion from a surface that is meshed with quadrilateral or triangular elements. Outside from mostly academic research where simple geometries suffice, hexahedral-dominated meshing, where most elements are hexahedra but with some fill-up of pyramids, wedges, or tetrahedra, or non-fitted meshes with a suitable mechanism to impose boundary conditions on immersed interfaces, are commonly used for algorithms which prefer hexahedral meshes.

Moreover, the common polynomial spaces used on triangles, nodal polynomials of *complete* degree k (see also section "Polynomial Approximation" below), have the advantage that linear basis functions with $k = 1$ exactly correspond to the span of $\{1, x, y\}$. This keeps the number of unknowns to reach a certain polynomial degree to a minimum, which is especially relevant in 3D. When compared to the common basis on hexahedra, the tensor product of one-dimensional polynomials with $(4 + 1)^3 = 125$ basis functions for degree $k = 4$, the polynomial space up to complete degree 4 only involves $5 \cdot 6 \cdot 7/6 = 35$ functions. Put differently, investing 120 basis functions per element of a complete polynomial space allows for a polynomial degree $k = 7$ and the associated higher convergence rates.

Intriguingly, this statement is misleading and consensus in most applications is that tensor product bases on hexahedral elements typically deliver more accurate results for a fixed number of unknowns. On the one hand, hexahedra span a considerably larger volume for a given edge length. The increased number of unknowns is thus not simply a waste of polynomials, but merely an aggregation of the unknowns that would be located on several tetrahedra. The additional mixed terms also improve the error constants and enable a better representation of anisotropic effects in the solution. While it has been believed that the additional coupling between unknowns in the final system matrix might eliminate the benefit of tensor product bases, see, e.g., Löhner (2011, 2013), Chang et al. (2018), fast sum factorization algorithms that have been established by the spectral element community, see Deville et al. (2002), Fischer et al. (2020) and references therein, have demonstrated the opposite for certain classes of solvers, involving less work per degree of freedom than equivalent complete polynomial bases.

The concepts presented in this chapter are general and do not depend on the element shape, even though all numerical results use intervals in 1D, quadrilateral elements in 2D, and hexahedral elements in 3D.

Mapping to Curved Elements

In a finite element method, approximating functions, typically polynomials, are defined on each mesh element. In order to simplify the implementation of polynomial spaces and the computation of the underlying integrals, it is common to define the concept of a reference element. In the case of quadrilateral elements, we denote by $\hat{\boldsymbol{x}}$ the reference coordinates on the reference element $\hat{\Omega} = [-1, 1]^d$. Reference positions are mapped in a one-to-one fashion to the actual element Ω_e by a mapping $\boldsymbol{\Phi}_e$, see Fig. 1.1 for an illustration. The most common definition of the mapping is to use a polynomial approximation using a set of nodes (or control points) $\boldsymbol{v}_i$ on the real geometry,

$$\Omega_e = \left\{ \boldsymbol{x} \in \mathbb{R}^d : \boldsymbol{x} = \boldsymbol{\Phi}_e(\hat{\boldsymbol{x}}) = \sum_i \boldsymbol{v}_i \varphi_i(\hat{\boldsymbol{x}}),\ \hat{\boldsymbol{x}} \in \hat{\Omega} = [-1, 1]^d \right\}. \tag{1.4}$$

Here, φ_i are the linear, quadratic, or higher order Lagrange polynomials on the respective element, depending on the degree of the geometry representation. For higher order basis functions for the solution a purely linear approximation of curved geometries often leads to a significant loss of accuracy, see, e.g., Bassi and Rebay (1997a), which is why higher order approaches are widespread. Figure 1.2 shows an illustration of the approximation of a part of the sphere with a bi-quadratic polynomial interpolation. Choosing the same polynomial degree k as for the numerical solution, also called an isoparametric geometry approximation, is a common approach. Alternatives are non-uniform rational B-spline representations of the geometry, used in

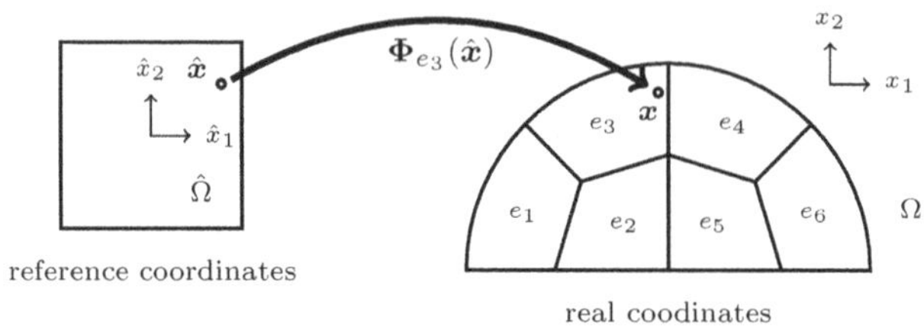

Fig. 1.1 Schematic representation of a mesh with 6 quadrilateral elements, 4 of which using curved edges, and the mapping from the reference element

Fig. 1.2 Sixth of a sphere (shaded in blue) approximated by a bi-quadratic interpolation through nine node points

the context of isogeometric analysis (Hughes et al. 2005) for both geometry and solution or for the geometry alone (Sevilla et al. 2008), or exact geometry manipulations (Heltai et al. 2019).

The most common workflow for high-order finite element and discontinuous Galerkin codes, see also Šolín et al. (2004) for a general description, is to create meshes by specialized mesh generator software. Most mesh generators only provide information about the vertices, which allows only for a linear representation of the boundary. In a second stage, additional intermediate points are placed between the node vertices to define the polynomial description (1.4), following the curved geometry in form of a manifold. The curved description can either be given analytically for basic geometric entities or via a computer-aided design (CAD) information. Often, the analytical curved description is only available at the boundary and not in the interior of the computational domain. Thus, a mechanism is needed to extend the curved description in a smooth way while retaining retain a smooth representation in (1.4), see Strang and Fix (1988, Sect. 3.3) for what smooth means in this context. The most widespread concept is transfinite interpolation (Gordon and Thiel 1982), which implements a linear blending of the curved description along the surrounding boundaries. Transfinite interpolation is closely related to the earlier proposed Coons patches (Coons 1967). For software concepts of geometry operations in a modern finite element implementation, we refer to Heltai et al. (2019).

Given a smooth and invertible mapping $\boldsymbol{\Phi}$, the integral of a function f over the element Ω_e can be transformed to an integral in reference coordinates on $\hat{\Omega}$ by the transformation formula

$$\int_{\Omega_e} f(\boldsymbol{x})\,\mathrm{d}\boldsymbol{x} = \int_{\hat{\Omega}} f\left(\boldsymbol{\Phi}_e(\hat{\boldsymbol{x}})\right) \left|\det\left(\boldsymbol{J}_e(\hat{\boldsymbol{x}})\right)\right| \,\mathrm{d}\hat{\boldsymbol{x}}, \tag{1.5}$$

using the relation $\boldsymbol{x} = \boldsymbol{\Phi}_e(\hat{\boldsymbol{x}})$. Here, the Jacobian matrix of the transformation is defined by

$$\boldsymbol{J}_e(\hat{\boldsymbol{x}}) = \hat{\nabla}\boldsymbol{\Phi}_e(\hat{\boldsymbol{x}}) = \left[\frac{\partial \Phi_{e,i}}{\partial x_j}\right]_{i,j}.$$

In the Jacobian matrix, the partial derivatives are aligned in columns and the different spatial components i in $\boldsymbol{\Phi}_e$ in rows. In most finite element codes, the orientation of the transformation $\boldsymbol{\Phi}_e$ is such that the determinant is always positive, allowing to skip the absolute value of the Jacobian determinant.

Furthermore, derivatives ∇ in terms of the physical coordinates $\boldsymbol{x}$ can be expressed in terms of the gradient with respect to reference coordinates $\hat{\boldsymbol{x}}$, denoted $\hat{\nabla}$, and a metric term $\boldsymbol{J}_e^{-T}$ derived from the chain rule using the inverse relation $\hat{\boldsymbol{x}}(\boldsymbol{x}) = \boldsymbol{\Phi}_e^{-1}(\boldsymbol{x})$, such that

$$\nabla\varphi = \boldsymbol{J}_e^{-T}\hat{\nabla}\varphi. \tag{1.6}$$

When defined from the polynomial representation (1.4) in three space dimensions, the naive inverse of the derivative of the Jacobian leads to schemes that do not exactly satisfy certain metric identities formulated in Kopriva (2006). There exist modifications for the inverse transformation, such as the so-called conservative curl form of the contravariant transformation scaled by the determinant of the Jacobian,

$$J_{ij}^{\mathrm{cov},(e)} = -\frac{1}{\det\left(\boldsymbol{J}^{(e)}\right)}\left[\hat{\nabla}\times\left(\mathcal{I}^{n_{\mathrm{c,1D}}}\left(\Phi_l(\hat{\boldsymbol{x}})\hat{\nabla}\Phi_m(\hat{\boldsymbol{x}})\right)\right)\right]_i,$$

with $i = 1, \ldots, d$, and a cyclic permutation for the indices (j, m, l). Here, $\mathcal{I}^{n_{\mathrm{c,1D}}}$ denotes the interpolation of the argument onto the $n_{\mathrm{c,1D}}$ points of a quadrature formula used to approximate the integrals. In this text, we assume the inverse form $\left(\boldsymbol{J}^{(e)}\right)^{-T}$ to hold some suitable approximation, e.g., $\boldsymbol{J}^{\mathrm{cov},(e)}$.

The polynomial mapping (1.4) can also be used to construct other geometrical quantities, such as the outer normal vector $\boldsymbol{n}$ on the surfaces of the elements needed for the boundary integrals in the discontinuous Galerkin scheme derived below. It is given by transforming the two reference tangential vectors, $\hat{\boldsymbol{t}}_1$ and $\hat{\boldsymbol{t}}_2$, on the respective surface of the element to the real space and take the normalized vector orthogonal to that vector,

$$\boldsymbol{n} = \frac{\boldsymbol{J}_e\hat{\boldsymbol{t}}_1 \times \boldsymbol{J}_e\hat{\boldsymbol{t}}_2}{|\boldsymbol{J}_e\hat{\boldsymbol{t}}_1 \times \boldsymbol{J}_e\hat{\boldsymbol{t}}_2|}. \tag{1.7}$$

The scaling when transforming integrals over a face of the element Ω_e to reference coordinates is exactly given by the quantity in the denominator of (1.7). For more details on the differential geometry, see, e.g., Kopriva (2009).

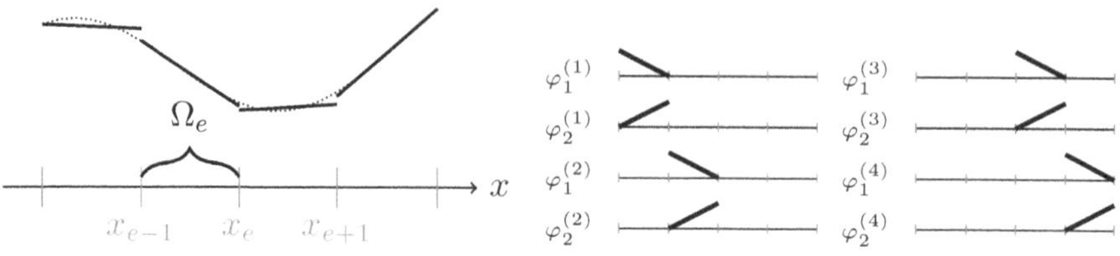

Fig. 1.3 Discontinuous representation of the solution on four elements (left) and the eight basis functions represented by nodal polynomials

Polynomial Approximation

Like for finite element methods, the solution on each element Ω_e is represented by a polynomial expansion

$$u_h^{(e)}(\boldsymbol{x},t) = \sum_{j=1}^{n_\mathrm{p}} \varphi_j^{(e)}(\boldsymbol{x}) u_j^{(e)}(t), \tag{1.8}$$

where φ_j denotes the jth basis function and $u_j^{(e)}$ is the coefficient that is to be found such that the weak form (1.3) is fulfilled. In this formula, a discrete representation in space with n_p polynomials is implied via the basis functions φ_j, whereas the time dependency is contained in the coefficient values $u_j^{(e)}$. Figure 1.3 shows a one-dimensional setup with 4 elements of polynomial degree $k = 1$. A continuous finite element method where the interpolation is concatenated from one interval to the next, only $n + 1 = 5$ different basis functions are involved on n elements. Conversely, the DG function space is completely discontinuous between the elements, involving separate basis functions for each element, giving a dimension of $2n$ unknowns in total. The discontinuous function space increases the flexibility of the method to cope with barely resolved scenarios as indicated by the figure: the absence of continuity between the elements allows each element to produce a local best fit to the solution. However, the additional functions appear unnecessary at least from the point of view of interpolating a smooth function once the resolution increases, because the value on both sides of the interface will converge to the same value.

The absence of continuity requirements between neighboring elements make discontinuous Galerkin methods very flexible in the choice of the polynomial spaces. The two most important classes of polynomials are nodal basis functions based on Lagrange polynomials in interpolating nodal points as well as modal polynomials.

Nodal polynomials of degree k are constructed by defining a set of points $\{\hat{x}_0, \ldots, \hat{x}_k\}$, called nodes, and setting up the Lagrangian interpolants,

$$\ell_i(\hat{x}) = \prod_{j=0, j\neq i}^{k} \frac{\hat{x} - \hat{x}_j}{\hat{x}_i - \hat{x}_j}.$$

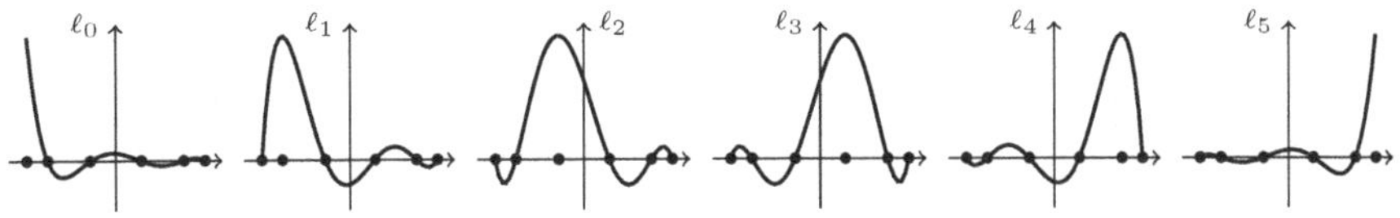

Fig. 1.4 Nodal basis functions ℓ_i defined on Gauss–Lobatto points of polynomial degree $k = 5$

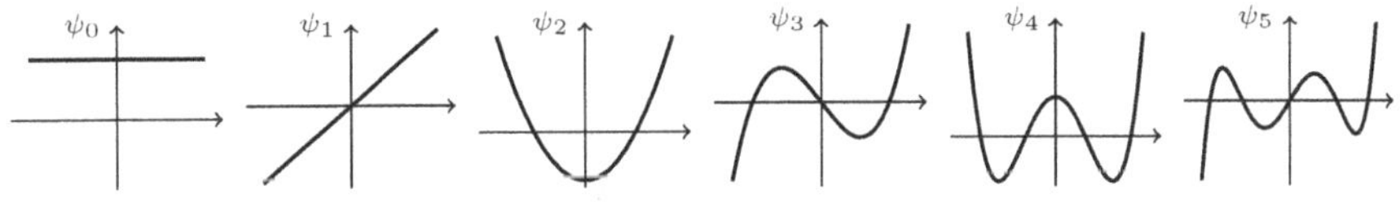

Fig. 1.5 Modal basis functions ψ_i up to polynomial degree $k = 5$

Since the naive choice of equidistant points gives rise to an exponential rise in the conditioning of interpolation as the degree k increases, node distributions that cluster points more closely to the element boundary are widespread. The two most common setups are the points underlying the Gauss–Lobatto quadrature formula, which include the interval end points, and the points underlying the Gauss–Legendre quadrature formula, which are based on the zeros of the Legendre polynomials. The former simplify the computation of face integrals because a single shape functions evaluates to non-zero on each of the two element boundaries, ℓ_0 and ℓ_k, respectively. Figure 1.4 shows the Lagrange basis of degree $k = 5$ on Gauss–Lobatto points. Polynomials defined in the Gauss–Legendre points are orthogonal in the L_2 inner product on the reference interval $[-1, 1]$ and thus yield a diagonal mass matrix with $(\ell_i, \ell_j)_{\hat{\Omega}} = w_i \delta_{ij}$ with w_i the ith weight of the Gaussian quadrature formula and δ_{ij} the Kronecker delta. The stable evaluation of Lagrange polynomials in terms of roundoff errors is implemented by the product form above, a variant of the barycentric formula (Berrut and Trefethen 2004).

Modal polynomials are based on a decomposition of the polynomial space into contributions of various degrees. Instead of the naive monomial basis $\{1, \hat{x}, \hat{x}^2, \dots\}$ with its exponential increase in the condition number, the most common choice are Legendre polynomials or Jacobi polynomials. Legendre polynomials, denoted ψ_j, fulfill an orthogonality condition $(\psi_i, \psi_j)_{\hat{\Omega}} = \alpha_i \delta_{ij}$ for some α_i positive or, by appropriate scaling, $(\psi_i, \psi_j)_{\hat{\Omega}} = \delta_{ij}$. Figure 1.5 exemplarily shows the Legendre polynomials up to degree $k = 5$.

Nodal bases have the advantage of a simple interpolation of functions to coefficient values and identification of the entries in solution vectors, including the point-wise evaluation of nonlinear flux functions $\boldsymbol{f}(u_i)$ (Hesthaven and Warburton 2008). In modal representations, integrals and inversions of the mass matrix are necessary for this kind of operations. On the other hand, modal representations give a natural interpretation of the slowly changing low-degree content of a solution on the element and the higher degree oscillatory content, as e.g., used for filters (Hesthaven and Warburton 2008, Sect. 5.6) or as indicator for limiters (Persson and Peraire 2006).

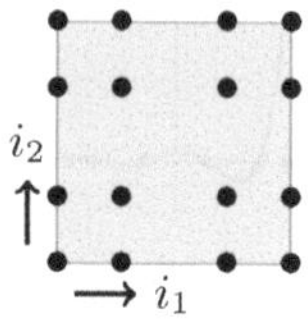

Fig. 1.6 Definition of node points of tensor product polynomials of degree $k = 3$ in two dimensions

Furthermore, modal representations of lower degree can easily be embedded into a higher order representation by adding zero coefficients in the orthogonal expansion.

It is straight-forward to change between different polynomial representations. Starting from a modal representation of a function,

$$u_h(\hat{x}) = \sum_{j=0}^{k} \psi_j(\hat{x})\tilde{u}_j,$$

with given coefficients $\tilde{\boldsymbol{u}} = [\tilde{u}_j]_{j=0,\ldots,k}$, the nodal representation

$$u_h(\hat{x}) = \sum_{j=0}^{k} \ell_j(\hat{x})u_j,$$

with unknown coefficients $\boldsymbol{u} = [u_j]_{j=0,\ldots,k}$ can be found by the matrix-vector product

$$\boldsymbol{u} = \mathcal{V}\tilde{\boldsymbol{u}},$$

where the Vandermonde matrix $\mathcal{V}$ is defined by the entries $\mathcal{V}_{ij} = \psi_j(\hat{x}_i)$ for the node points $\hat{x}_i$. The matrix-vector product simply evaluates the nodal polynomial representation in each point by the underlying representation of the sum.

When going to higher dimensions, the standard approach on quadrilateral and hexahedra is a tensor product of one-dimensional functions,

$$\varphi_i(\hat{x}_1, \ldots, \hat{x}_d) = \varphi_{i_1}^{1\mathrm{D}}(\hat{x}_1) \ldots \varphi_{i_d}^{1\mathrm{D}}(\hat{x}_d), \tag{1.9}$$

where the indices $i_1, \ldots, i_d$ each run between 0 and k. Figure 1.6 shows an two-dimensional example with indices i_1 and i_2 that combine in a lexicographic way as $i = i_2(k+1) + i_1$, running from 0 to the total number of polynomials, $n_{\mathrm{p}} = (k+1)^d$ (exclusive). For modal polynomials, an alternative to a tensor product is to define polynomials of complete degree k by

$$\varphi_i(\hat{x}_1, \ldots, \hat{x}_d) = \prod_{\substack{e=1 \\ i_1+\ldots+i_d \le k}}^{d} \varphi_{i_e}^{1\mathrm{D}}(\hat{x}_e),$$

where the sum of polynomial degrees in all d dimensions is at most k. Polynomials of complete degree are most common on triangles and tetrahedra, but they are also used on quadrilaterals and hexahedra. Due to the non-trivial effort to define Lagrange polynomials in a non-tensor product way, nodal polynomials on triangles are typically found by a two-step procedure, going via orthogonal (Jacobi) polynomials as an auxiliary space and using a Vandermonde matrix to switch to a nodal representation (Hesthaven and Warburton 2008). The node points on the triangles are either based on analytical formulas giving rise to the Fekete points (Taylor et al. 2000), or use approximate techniques discussed, e.g., in Hesthaven and Warburton (2008).

Polynomial spaces are often defined in terms of the reference coordinates $\hat{\boldsymbol{x}}$ introduced in the previous subsection, even though there also exist DG realizations defining polynomials in real space (Noventa et al. 2016). In the setup of coordinate transformations, the definition (1.8) turns into

$$u_h^{(e)}(\boldsymbol{x}, t) = \sum_{j=1}^{n_p} \varphi_j\left(\left(\boldsymbol{\Phi}^{(e)}\right)^{-1}(\boldsymbol{x})\right) u_j^{(e)}(t), \tag{1.10}$$

where φ_j is given positions in reference coordinates $\hat{\boldsymbol{x}}$. Using the integral transformation (1.5), an integral over, e.g., the product $\varphi_i \boldsymbol{a} \cdot \nabla \varphi_j$ for some vector $\boldsymbol{a}$ becomes

$$\begin{aligned} &\int_{\Omega_e} \varphi_i^{(e)}(\boldsymbol{x})\boldsymbol{a}(\boldsymbol{x}) \cdot \nabla \varphi_j^{(e)}(\boldsymbol{x})\,\mathrm{d}\boldsymbol{x} \\ &\quad = \int_{\hat{\Omega}} \varphi_i(\hat{\boldsymbol{x}})\boldsymbol{a}\left(\boldsymbol{\Phi}_e(\hat{\boldsymbol{x}})\right) \cdot \boldsymbol{J}_e^{-T}\hat{\nabla}\varphi_j(\hat{\boldsymbol{x}}) \det\left(\boldsymbol{J}_e(\hat{\boldsymbol{x}})\right)\,\mathrm{d}\hat{\boldsymbol{x}}, \end{aligned} \tag{1.11}$$

where all polynomials are evaluated in the reference coordinates and the geometry is added by some metric terms and, in case of space-dependent coefficients, the mapped location of the integrand $\boldsymbol{x} = \boldsymbol{\Phi}_e(\hat{\boldsymbol{x}})$. In the remainder of this chapter, the notation on the left-hand side or the associated bilinear form $(\cdot, \cdot)_{\Omega_e}$ is used as an abstraction for the terms on the right-hand side, using a specified quadrature formula and mapping $\boldsymbol{\Phi}_e$, as common in the finite element literature.

Using the solution representation (1.10) on each element Ω_e, we define the space of admissible solutions in a DG discretization as

$$\mathcal{U}_h = \left\{u_h \in L_2(\Omega_h) \,:\, u_h|_{\Omega_e} \in \mathcal{P}_k \circ \boldsymbol{\Phi}_e^{-1}\right\}, \tag{1.12}$$

where $\mathcal{P}_k$ denotes a polynomial space of degree k (e.g., polynomials of complete degree k or of tensor degree k), which is composed with $\boldsymbol{\Phi}_e^{-1}$ to express the definition of polynomials in the reference coordinates. The function space (1.12) is denoted a broken finite element space, highlighting that the approximation is only piecewise continuous with jumps allowed between the elements.

Derivation of the DG Weak Form

For a discrete representation with DG, the weak form (1.3) is restricted to the admissible solution space (1.12) and tested by only those functions in $\mathcal{U}_h$. As opposed to continuous finite element methods where the weak form can be posed globally, the discontinuous nature of the ansatz spaces introduced in the previous subsection with derivatives not defined across the element boundary makes the weak form only valid on a single element Ω_e. The problem statement is hence to find $u_h^{(e)}$ of the form (1.10), such that for all test functions $v_h^{(e)}$ it holds

$$\left(v_h^{(e)}, \frac{\partial u_h^{(e)}}{\partial t}\right)_{\Omega_e} - \left(\nabla v_h^{(e)}, \boldsymbol{f}\left(u_h^{(e)}\right)\right)_{\Omega_e} + \left\langle v_h^{(e)} \boldsymbol{n}, \boldsymbol{f}^*\right\rangle_{\partial\Omega_e} = \left(v_h^{(e)}, b\right)_{\Omega_e}.$$

Here, the boundary term $\boldsymbol{f}\left(u_h^{(e)}\right)$ is replaced by an abstract quantity $\boldsymbol{f}^*$, the numerical flux, to be specified in section "Connecting Neighboring Elements: Numerical Fluxes" below. The boundary $\partial\Omega_e$ is defined by a set of *faces*, with a unique neighboring element behind each face (for conforming meshes). When including the interface terms over $\partial\Omega_e$, the integrals on all elements can be added and a global description of the DG problem on Ω_h can be stated: We find the numerical solution $u_h \in \mathcal{U}_h$ such that

$$\left(v_h, \frac{\partial u_h}{\partial t}\right)_{\Omega_h} - \left(\nabla v_h, \boldsymbol{f}\left(u_h\right)\right)_{\Omega_h} + \left\langle v_h \boldsymbol{n}, \boldsymbol{f}^*\right\rangle_{\partial\Omega_h} = \left(v_h, b\right)_{\Omega_h} \tag{1.13}$$

holds for all test functions $v_h \in \mathcal{U}_h$. In this equation, $(a, b)_{\Omega_h} = \sum_{e=1}^{n}(a, b)_{\Omega_e}$ is defined as the sum of all element integrals, and $\langle a, b\rangle_{\partial\Omega_h} = \sum_{e=1}^{n} \langle a, b\rangle_{\partial\Omega_e}$ as the sum all integrals over element boundaries, both those located on the physical boundary of Ω and those at the interface between two elements. In $\partial\Omega_h$, each inner face is visited twice from the two adjacent elements with different direction of the normal vector $\boldsymbol{n}$.

Connecting Neighboring Elements: Numerical Fluxes

In Eq. (1.13), the quantity $\boldsymbol{f}^*$ at the element interfaces $\partial\Omega_h$ takes the role of coupling the initially separate elemental problems on individual Ω_e. At each point $\boldsymbol{x}$ on the surface $\partial\Omega_e$, the numerical flux combines the solution value inside the element, denoted u_h^-, with an outside value u_h^+. The outside value is either given by the numerical solution interpolated from the neighbor to the same position $\boldsymbol{x}$, or the boundary condition as specified below. The result of the numerical flux also depends on the normal vector $\boldsymbol{n}^-$, pointing from the interior value u_h^- to the outside value u_h^+. In this text, the dependence of the numerical flux on the three quantities is denoted by $\boldsymbol{f}^*(u_h^-, u_h^+)$, with the direction of the normal vector $\boldsymbol{n}^-$ implied by the order of the two arguments. According to this convention, the evaluation of $\boldsymbol{f}^*(u_h^+, u_h^-)$ implies the normal vector $\boldsymbol{n}^+ = -\boldsymbol{n}^-$. In this work, we write $\boldsymbol{n} = \boldsymbol{n}^-$ when the orientation is implied from the underlying integrals.

In the definition of the numerical flux $\boldsymbol{f}^*$, the physical properties of the equation at hand, i.e., the physical flux $\boldsymbol{f}$, are taken into account. The task to find a connection between two discontinuous neighboring states u_h^- and u_h^+ with an orientation $\boldsymbol{n}^-$ consistent with a PDE is a Riemann problem. The solvers of these problems, called Riemann solvers, form the basis of finite volume schemes. There exists a large body of numerical fluxes for many important equations, for which we refer to the finite volume literature (LeVeque 2002; Ketcheson et al. 2020). The main difference between a discontinuous Galerkin method and a finite volume method is the fact that the DG scheme evaluates the numerical flux at every point $\boldsymbol{x}$ along the boundary $\partial\Omega_e$ and subsequently performs an integration, whereas finite volumes impose the balance between volume averages in two neighbors. This definition means that the DG method can retain higher formal orders of accuracy implied by higher degree polynomials, as the balance is made between two polynomial states evaluated at the same site, only using local information. Higher order finite volume methods, on the other hand, rely on certain reconstructions that span to more than the immediate neighbors. As a result, the numerical flux in DG methods is not as crucial as for finite volume methods, allowing for a more approximate character in fluxes without affecting the consistency.

In order to get a well-defined value for the numerical flux, the following two conditions are assumed:

$$
\begin{aligned}
\boldsymbol{f}^*(u_h, u_h) &= \boldsymbol{f}(u_h) \qquad &\text{(consistency)},\\
\boldsymbol{f}^*\left(u_h^-, u_h^+\right) &= \boldsymbol{f}^*\left(u_h^+, u_h^-\right) \qquad &\text{(conservation)}.
\end{aligned}
$$

The first condition expresses the fact that in case the solution u_h is continuous across the element boundary with $u_h^- = u_h^+$, the numerical flux $\boldsymbol{f}^*$ needs to reproduce the physical flux for consistency with integration by parts in Eq. (1.13). The second condition ensures mass conservation of the DG scheme. To see this, first look at the mass balance of the continuous problem (1.1) without source terms, $b = 0$, i.e., the time derivative of the integral over the solution $u(\boldsymbol{x}, t)$. By Gauss' divergence theorem, we find that

$$
\frac{\mathrm{d}}{\mathrm{d}t}\int_{\Omega} u(\boldsymbol{x}, t)\,\mathrm{d}\boldsymbol{x} = -\int_{\Omega} \nabla\cdot \boldsymbol{f}(u)\,\mathrm{d}\boldsymbol{x} = -\int_{\partial\Omega} \boldsymbol{n}\cdot \boldsymbol{f}(u)\,\mathrm{d}\boldsymbol{x}, \tag{1.14}
$$

i.e., the mass is changed by what leaves the computational domain along the part with $\boldsymbol{n}\cdot\frac{\partial f}{\partial u} > 0$ on $\partial\Omega \setminus \partial\Omega_{\text{in}}$ and by what enters the computational domain at the inflow part of the boundary $\partial\Omega_{\text{in}}$. This balance can also be seen by evaluating the weak form (1.3) on the domain Ω with the function $v \equiv 1$.

In the discrete setting (1.13) with $b = 0$, insert the test functions $v_h \equiv 1$ on the whole domain to get the balance

$$\frac{\mathrm{d}}{\mathrm{d}t}\int_{\Omega_h} u_h(\boldsymbol{x},t)\,\mathrm{d}\boldsymbol{x} = \left(1, \frac{\partial u_h}{\partial t}\right)_{\Omega_h} = -\left\langle \boldsymbol{n}, \boldsymbol{f}^*\right\rangle_{\partial\Omega_h},$$

where the spatial derivative term vanishes due to $\nabla v_h = 0$ for v_h constant. For the interface terms on $\partial\Omega_h$, we split the faces into interior faces $\gamma \in \Gamma_\mathrm{i}$ where two adjacent elements Ω_{e^-} and Ω_{e^+} meet, and boundary faces $\gamma \in \Gamma_\mathrm{b}$ on the domain boundary. This gives

$$\frac{\mathrm{d}}{\mathrm{d}t}\int_{\Omega_h} u_h(\boldsymbol{x},t)\,\mathrm{d}\boldsymbol{x} = -\left\langle \boldsymbol{n}^-, \boldsymbol{f}^*\left(u_h^-, u_h^+\right) - \boldsymbol{f}^*\left(u_h^+, u_h^-\right)\right\rangle_{\Gamma_\mathrm{i}} - \left\langle \boldsymbol{n}, \boldsymbol{f}^*\right\rangle_{\Gamma_\mathrm{b}}.$$

The terms on both sides of interior faces in Γ_i have been collected into one expression by using $\boldsymbol{n}^- = -\boldsymbol{n}^+$. In this equation, the equation for the conservativity $\boldsymbol{f}^*\left(u_h^-, u_h^+\right) = \boldsymbol{f}^*\left(u_h^+, u_h^-\right)$ makes the interior contribution drop and result in similar boundary contributions as for the continuous problem (1.14), with the physical flux $\boldsymbol{f}$ replaced by the numerical flux $\boldsymbol{f}^*$ whose value depends on the imposed boundary conditions.

For the definition of numerical fluxes, some notation is introduced. We write the average of the interior value u_h^- and the exterior (neighboring) value u_h^+ as

$$\{\{u_h\}\} = \frac{u_h^- + u_h^+}{2}.$$

Furthermore, the jump between the two states is written as

$$[\![u_h]\!] = \boldsymbol{n}^- u_h^- + \boldsymbol{n}^+ u_h^+.$$

Due to $\boldsymbol{n}^+ = -\boldsymbol{n}^-$, this is equivalent to

$$[\![u_h]\!] = \boldsymbol{n}^-(u_h^- - u_h^+),$$

highlighting the fact we indeed look at the difference in the two solution values, equipped with a direction. This definition extends to vector-valued variables $\boldsymbol{u}_h$, where the multiplication by $\boldsymbol{n}^-$ increases the tensor rank by one, i.e., $\boldsymbol{u}_h\boldsymbol{n}^-$ is the outer product of the two constituents and is a tensor of rank two.

Finally, the directed jump operator represents merely the difference between the two states,

$$[u_h] = u_h^- - u_h^+.$$

Central and Upwind Fluxes for Linear Transport

For the case of linear transport with $\boldsymbol{f}(u) = \boldsymbol{a}u$ with a possibly space- and time-dependent transport direction $\boldsymbol{a}$, the straight-forward choice is the *central flux* as the average between the two states,

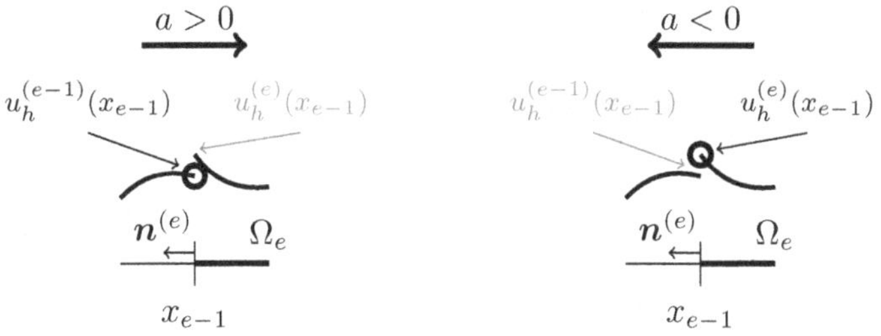

Fig. 1.7 Visualization of upwind flux in 1D for positive and negative transport speed a and normal $n = -1$ pointing to the left. The selected value is indicated by a black circle

$$(au)^* = \{\{au_h\}\}\,.$$

While being simple, the central flux is not optimal for the underlying physics: The linear transport equation moves the solution u along the characteristics, i.e., along the path described by the speed $\boldsymbol{a}$. For constant $\boldsymbol{a}$, the analytical solution is $u(\boldsymbol{x}, t) = u_0(\boldsymbol{x} - \boldsymbol{a}t)$. By the symmetric choice $\{\{\boldsymbol{a}u_h\}\}$, the directionality in the analytical solution is ignored entirely.

In the *upwind flux*, the choice is based on the characteristics, selecting always the value that is *upstream* in terms of the flow direction. Given a flow direction $\boldsymbol{a}$ and the orientation of the normal, the selection is

$$(\boldsymbol{a}u)^* = \begin{cases} \boldsymbol{a}u_h^-, & \boldsymbol{n}^- \cdot \boldsymbol{a} \geq 0,\\ \boldsymbol{a}u_h^+, & \boldsymbol{n}^- \cdot \boldsymbol{a} < 0. \end{cases} \tag{1.15}$$

In case the transport points in the same direction as the normal $\boldsymbol{n}^-$ in the sense $\boldsymbol{n}^- \cdot \boldsymbol{a} > 0$, the interior value u_h^- is upstream and selected for the numerical flux. If the flow points into the element with $\boldsymbol{n}^- \cdot \boldsymbol{a} < 0$, the outside value is chosen. This selection aligns naturally with the boundary conditions in the hyperbolic equation (1.1), using the boundary condition g on inflow boundaries and choosing the interior solution on the outflow boundaries. For the case of flow tangential to the face, $\boldsymbol{n}^- \cdot \boldsymbol{a} = 0$, the definition appears to be ambiguous. However, this is uncritical because the contribution cancels in the final weak form (1.13) upon multiplication by $\boldsymbol{n}$. Figure 1.7 shows the selection procedure for the upwind flux.

It is easy to verify that the definition (1.15) is conservative. As the order of arguments u_h^-, u_h^+ and thus the direction of the normal is changed, the selection whether the interior or exterior value is chosen is switched as well, making the definition unique from both sides of an interior face.

The upwind flux (1.15) can be written equivalently as

$$(\boldsymbol{a}u)^* = \{\{\boldsymbol{a}u_h\}\} + \frac{|\boldsymbol{n} \cdot \boldsymbol{a}|}{2} [\![u_h]\!]\,. \tag{1.16}$$

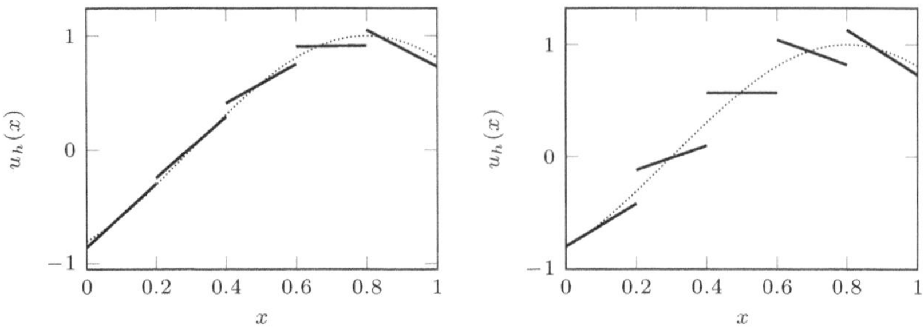

Fig. 1.8 Comparison of solution with upwind flux (left) and central flux (right) for linear transport with $a = 1$ in 1D with 5 elements and linear polynomials, with the analytical solution indicated by the dotted line

Upon multiplication by $\boldsymbol{n}$ in (1.13), we find that

$$\begin{aligned}\boldsymbol{n} \cdot (\boldsymbol{a}u)^* &= \boldsymbol{n} \cdot \{\{\boldsymbol{a}u_h\}\} + \frac{|\boldsymbol{n} \cdot \boldsymbol{a}|}{2}[u_h] \\ &= \left(\frac{\boldsymbol{n} \cdot \boldsymbol{a}}{2} + \frac{|\boldsymbol{n} \cdot \boldsymbol{a}|}{2}\right) u_h^- + \left(\frac{\boldsymbol{n} \cdot \boldsymbol{a}}{2} - \frac{|\boldsymbol{n} \cdot \boldsymbol{a}|}{2}\right) u_h^+.\end{aligned}$$

For outflow with $\boldsymbol{n} \cdot \boldsymbol{a} > 0$, the modulus returns the same sign as its argument, and only the interior contribution $(\boldsymbol{n} \cdot \boldsymbol{a})u_h^-$ is retained, whereas for inflow with $\boldsymbol{n} \cdot \boldsymbol{a} < 0$ the first term becomes zero and the outside information $(\boldsymbol{n} \cdot \boldsymbol{a})u_h^+$ gets used.

In Fig. 1.8, the numerical solution for the upwind flux is compared to the result with the central flux on a coarse grid of $n = 5$ elements with linear polynomials for the analytical solution $u(x, t) = \sin(\pi(x - t))$ with transport speed $a = 1$ in 1D, evaluated at time $t = 0.3$. While the upwind flux follows the analytical solution relatively well, the central flux develops big jumps in the solution between the elements. These jumps can be seen as insufficient control on the jumps between the elements, which is part of the numerical flux according to Eq. (1.16).

It is possible to blend the central flux and the upwind flux by a parameter α as

$$(\boldsymbol{a}u)^* = \{\{\boldsymbol{a}u_h\}\} + \frac{\alpha|\boldsymbol{n} \cdot \boldsymbol{a}|}{2} [\![u_h]\!] , \tag{1.17}$$

with $\alpha = 0$ denoting the central flux and $\alpha = 1$ the upwind flux.

The Lax–Friedrichs Flux as an Example for Nonlinear Equations

A popular general-purpose flux for nonlinear fluxes f is the Lax–Friedrichs flux,

$$f^*(u_h^-, u_h^+) = \{\{f(u_h)\}\} + \frac{C}{2}[\![u_h]\!] = \frac{f(u_h^-) + f(u_h^+)}{2} + \frac{C}{2}\boldsymbol{n}^-(u_h^- - u_h^+). \tag{1.18}$$

The constant is set to

$$C \geq \max_{\inf u_h(\boldsymbol{x}) \leq s \leq \sup u_h(\boldsymbol{x})} \left| \boldsymbol{n} \cdot \frac{\partial \boldsymbol{f}}{\partial u}(s) \right|$$

for a so-called global Lax–Friedrichs flux, i.e., the largest possible value for the normal derivative of $\boldsymbol{f}$ with the expected range of the solution u_h. Finding this maximum involves extra effort, so one often resorts to the *local Lax–Friedrichs flux*

$$C \geq \max_{\min(u^-, u^+) \leq s \leq \max(u^-, u^+)} \left| \boldsymbol{n} \cdot \frac{\partial \boldsymbol{f}}{\partial u}(s) \right|.$$

The local Lax–Friedrichs flux (also called Rusanov flux) is defined from the two adjacent states u_h^- and u_h^+, making it a simple choice to implement. For linear transport $\boldsymbol{f}(u) = \boldsymbol{a}u$, the Lax–Friedrichs flux is equivalent to the upwind flux because $\partial \boldsymbol{f}/\partial u = \boldsymbol{a}$. However, the flux does not always deliver optimal accuracy. The Lax–Friedrichs flux is conservative with $\boldsymbol{f}^*(u^-, u^+) = \boldsymbol{f}^*(u^+, u^-)$.

Riemann Solvers for Burgers' Equation

Consider the inviscid Burgers equation with $f(u) = \frac{1}{2}u^2$ in one dimension. The local Lax–Friedrichs flux is given by

$$f^{*,\mathrm{LF}}(u_h^-, u_h^+) = \left\{\!\!\left\{ \frac{u_h^2}{2} \right\}\!\!\right\} + \frac{\max\left(|u_h^-|, |u_h^+|\right)}{2} [\![u_h]\!]. \tag{1.19}$$

For the Burgers equation, an exact Riemann solver can be constructed as detailed, e.g., in Ketcheson et al. (2020, Sects. 4 & 11). Denote by u_{left} the solution on the left (i.e., u_h^+ if $\boldsymbol{n} = -1$ or u_h^- if $\boldsymbol{n} = +1$) and by u_{right} the solution on the right. The 1D Burgers equation produces a shock if $u_{\mathrm{left}} > u_{\mathrm{right}}$ or a rarefaction wave if $u_{\mathrm{left}} < u_{\mathrm{right}}$. In the former case, the speed $s = \frac{u_{\mathrm{left}} + u_{\mathrm{right}}}{2} > 0$ determines whether the state u_{left} or u_{right} is chosen. For the rarefaction wave, the selection is based on the value of the exact solution along the ray $x/t = 0$, choosing either u_{left} or u_{right}, or zero in case the two states go into opposite directions. This gives the flux

$$f^{*,\mathrm{exact}}(u_h^-, u_h^+) = \begin{cases} \text{if } u_L > u_R \begin{cases} \frac{1}{2}u_{\mathrm{left}}^2 & \text{if } u_{\mathrm{left}} + u_{\mathrm{right}} > 0, \\ \frac{1}{2}u_{\mathrm{right}}^2 & \text{if } u_{\mathrm{left}} + u_{\mathrm{right}} \leq 0, \end{cases} \\ \text{if } u_{\mathrm{left}} \leq u_{\mathrm{right}} \begin{cases} \frac{1}{2}u_{\mathrm{left}}^2 & \text{if } u_{\mathrm{left}} > 0, \\ \frac{1}{2}u_{\mathrm{right}}^2 & \text{if } u_{\mathrm{right}} < 0, \\ 0 & \text{if } u_{\mathrm{left}}u_{\mathrm{right}} \leq 0. \end{cases} \end{cases} \tag{1.20}$$

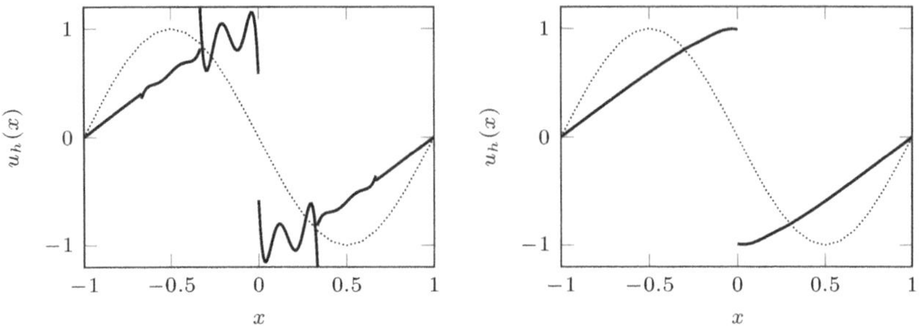

Fig. 1.9 Comparison of solution of Burgers' equation on $n = 6$ elements with degree $k = 5$ for the local Lax–Friedrichs flux (1.19) (left) and the exact Riemann solver (1.20) (right), with the initial condition indicated by the dotted line

In Fig. 1.9, the numerical solution of Burgers' equation, started with initial condition $u_0(x) = -\sin(\pi x)$ on the domain $\Omega = (-1, 1)$, is considered for the local Lax–Friedrichs flux (1.19) and the exact Riemann solver (1.20), solved on a domain with $n = 6$ elements and polynomial degree $k = 5$, and evaluated at time $t = 0.5$. While the local Lax–Friedrichs flux exhibits strong oscillations near the discontinuity that develops at $x = 0$, the exact Riemann solver nicely captures the discontinuity. In terms of the wave speed applied by the Lax–Friedrichs method, the reason is the choice of the maximal speed $\max\left(|u_h^-|, |u_h^+|\right)$, leading to too much upwinding. If it were chosen to $\left|\frac{u_h^- + u_h^+}{2}\right|$, the resulting approximate Riemann solver

$$f^{*,\mathrm{Roe}}(u_h^-, u_h^+) = \begin{cases} \frac{1}{2}u_{\mathrm{left}}^2 & \text{if } u_{\mathrm{left}} + u_{\mathrm{right}} > 0, \\ \frac{1}{2}u_{\mathrm{right}}^2 & \text{if } u_{\mathrm{left}} + u_{\mathrm{right}} \leq 0, \end{cases}$$

becomes the Roe solver (Ketcheson et al. 2020), which produces essentially the same result as the exact Riemann solver in this (simple) case.

Note, however, that exact Riemann solvers are not generally free from oscillations like in the present example with a discontinuity always located at the interface between elements. As soon as the discontinuity moves into the element where a polynomial expansion is applied, also the exact Riemann solver leads to oscillatory solutions.

Weak Form and Fluxes for Systems of Equations

The weak form (1.13) extends to systems of e equations in $\mathbb{R}^d$. In the polynomial representation (1.10), each component gets a separate coefficient value in the solution expansion and similarly for the test function. The Lax–Friedrichs flux can be defined as in the scalar case for a vector-valued solution $\boldsymbol{u}_h$ given a suitable definition of the constant C by the maximum eigenvalue of the flux Jacobian,

$$C = \max_{u} |\lambda (\boldsymbol{n} \cdot \nabla_{u} \boldsymbol{f})| .$$

In this expression, the flux $\boldsymbol{f}$ is an $e \times d$ matrix, which becomes an $e \times e \times d$ tensor upon taking all partial derivatives with respect to the components of $\boldsymbol{u}$. The multiplication by the normal vector, $\boldsymbol{n}\cdot$, then contracts along the d components representing the directions of the flux. The Lax–Friedrichs flux imposes upwinding by the maximal wave speed in the hyperbolic system along all components. This can result in a too dissipative behavior for systems where only one or a few components pertain to the fast waves. A more balanced approximate Riemann solver is the Roe flux, where upwinding is imposed in terms of the linearization $\mathcal{A}_f$ of $\boldsymbol{f}$ about an intermediate state. The linearization is also used for defining the Harten–Lax–van Leer (HLL) or the Harten–Lax–van Leer–contact (HLLC) flux for the Euler equations (LeVeque 2002), for example.

A Strong DG Form

Applying integration by parts on the spatial integral of Equation (1.13) leads to

$$\left(v_h, \frac{\partial u_h}{\partial t}\right)_{\Omega_h} + (v_h, \nabla \cdot \boldsymbol{f}(u_h))_{\Omega_h} + \left\langle v_h \boldsymbol{n}, \boldsymbol{f}^* - \boldsymbol{f}(u_h)\right\rangle_{\partial\Omega_h} = (v_h, b)_{\Omega_h} . \quad (1.21)$$

This form of the equation is called the strong form of the DG method because the derivative in front of $\boldsymbol{f}(u)$ assumes differentiability of the solution inside the element. For linear functions $\boldsymbol{f}$ and smooth solutions, the two forms are equivalent. The strong form (1.21) illustrates the way the discontinuous Galerkin method extends continuous finite elements. Assuming a continuous approximation, the boundary term $\boldsymbol{f}^*(u_h^-, u_h^+) - \boldsymbol{f}(u_h)$ disappears because of the consistency of the numerical flux. As the DG solutions u_h^- and u_h^+ start to deviate from each other, the difference in the flux acts as a penalty to the gap in the solution. This is a weak enforcement of continuity over the elements as compared to strong enforcement by specific construction of function spaces in continuous elements.

Boundary Conditions

An essential ingredient in the discontinuous Galerkin method is its treatment of boundary conditions, which follows the mechanisms as the transport of information between neighboring elements. As indicated by the weak form (1.13), the same numerical flux $\boldsymbol{f}^*$ with two slots for the interior solution u_h^- and an exterior value u_h^+ is used. The crucial part for boundary conditions is hence the definition of the outside value u_h^+ at the boundary. In DG schemes, we distinguish between the inflow

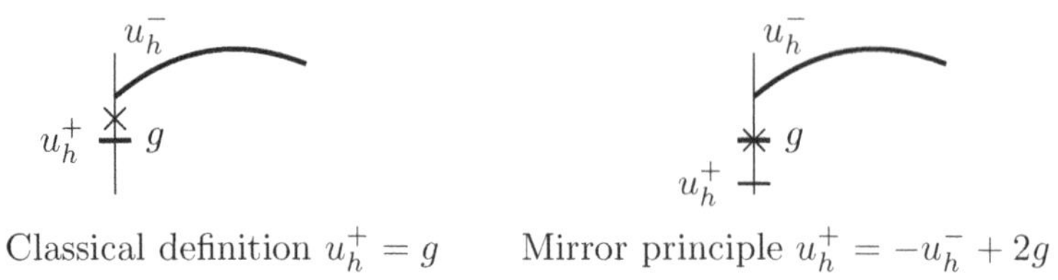

Fig. 1.10 Illustration of two ways to impose outside values u_h^+ at a Dirichlet boundary in a DG scheme, with the The cross defining the value of $\{\{u_h\}\}$

portion of the boundary Ω_{in} where a function g is given, and the outflow part. On the outflow boundary where no information is given, the natural choice is

$$u_h^+ = u_h^-.$$

This ensures that only the locally available information is used. On the inflow boundary, the two options are to define either

$$u_h^+ = g,$$

or the so-called mirror principle,

$$u_h^+ = -u_h + 2\,g,$$

see Fig. 1.10 for an illustration. The mirror principle imposes symmetry in the sense $\{\{u_h\}\} = g$, which is beneficial to the rate of convergence in certain schemes especially in the elliptic context. However, both variants are in use for transport problems.

Discrete-in-Space/Continuous-in-Time System

The weak form (1.13) on n elements with polynomial description (1.10) using n_{p} unknowns per element gives rise to a discrete-in-space, continuous-in-time system in $n_{\text{p}}n$ unknowns. As usual for finite elements, $n_{\text{p}}n$ equations are derived from the weak form by going through the individual basis functions $\varphi_i^{(e)}$ on all elements Ω_e. Assuming a linear flux $\boldsymbol{f}(u) = \boldsymbol{a}u$, the discrete system can be expressed as a matrix system,

$$\mathcal{M}\frac{\mathrm{d}\boldsymbol{U}_h}{\mathrm{d}t} - \mathcal{C}\boldsymbol{U}_h = \boldsymbol{B}_h. \tag{1.22}$$

Here, the vector $\boldsymbol{U}_h \in \mathbb{R}^{n_{\text{p}}n}$ collects the coefficients in the polynomial expansion (1.10), which are the node values in the case of nodal polynomials. The vector $\boldsymbol{B}_h$ is the result of the right-hand side integral $(v_h, b)_{\Omega_h}$ as well as the inhomogeneous part of the boundary integrals moved to the right-hand side. For the upwind flux (1.16) and imposition of boundary conditions by the mirror principle, the contribution is

$$\langle v_h, (\boldsymbol{n} \cdot \boldsymbol{a} + |\boldsymbol{n} \cdot \boldsymbol{a}|)g\rangle_{\Gamma_b}.$$

The mass matrix $\mathcal{M}$ is the result of the integrals $\left(\varphi_i^{(e)}, \varphi_j^{(e)}\right)_{\Omega_h}$ for all elements e and test functions $\varphi_i^{(e)}, \varphi_j^{(e)}$. Since the basis functions are independent for each element, there is no coupling between the elements and the mass matrix is block-diagonal,

$$\mathcal{M} = \begin{bmatrix} \mathcal{M}^{(1)} & & & \\ & \mathcal{M}^{(2)} & & \\ & & \ddots & \\ & & & \mathcal{M}^{(n)} \end{bmatrix}.$$

with $\mathcal{M}^{(e)}$ the mass matrix on element Ω_e. Using a transformation formula similar to (1.11), an entry (i, j) of the $\mathcal{M}^{(e)}$ is given by

$$\mathcal{M}_{ij}^{(e)} = \int_{\hat{\Omega}} \varphi_i(\hat{\boldsymbol{x}})\varphi_j(\hat{\boldsymbol{x}}) \det\left(\boldsymbol{J}_e(\hat{\boldsymbol{x}})\right) \,\mathrm{d}\hat{\boldsymbol{x}}. \tag{1.23}$$

The matrix $\mathcal{C}$ contains the transport terms from the space integral and the homogeneous part of the face integrals. For a test function with a global index $\underline{i} = n_\mathrm{p}e + i$ on element Ω_e with local index i and tentative solution component $\underline{j} = n_\mathrm{p}\bar{e} + j$, it is given by

$$\mathcal{C}_{\underline{i}\,\underline{j}} = \delta_{e\bar{e}} \left(\nabla\varphi_i^{(e)}, \boldsymbol{a}\varphi_j^{(e)}\right)_{\Omega_h} - \left\langle \varphi_i^{(e)}, \left(\frac{\boldsymbol{n}\cdot\boldsymbol{a}}{2} \pm \frac{|\boldsymbol{n}\cdot\boldsymbol{a}|}{2}\right)\varphi_j^{(\bar{e})}\right\rangle_{\partial\Omega_e\cap\partial\Omega_{\bar{e}}}.$$

The cell integral is only non-zero for global indices $\underline{i}$ and $\underline{j}$ belonging to the same element, whereas the face integral gives contributions from the element e (using a positive sign with $\bar{e} = e$) and all its neighbors (with a negative sign when $\bar{e} \neq e$).

Time Integration

System (1.22) is a system of ordinary differential equations in time for the unknown coefficients $\boldsymbol{U}_h$ in the method-of-lines setting, which can be solved by classical time integration schemes (Hairer et al. 1993; Hairer and Wanner 1991). There also exist space-time DG methods where both the spatial and temporal coordinates using a DG solution expansion, but the larger dimension of the solution space has restricted their use to special cases, such as moving domains discussed in Sudirham et al. (2006).

An attractive feature of DG discretizations is the fact that the block-diagonal mass matrix $\mathcal{M}$ renders explicit time stepping of (1.22) very efficient, which lead to a large body of literature of schemes started by the work by Cockburn and Shu (1991), Cockburn and Shu (1998b), see also the history presented by Cockburn et al. (2000). By contrast, continuous finite element methods involve basis functions spanning several elements, which leads to a globally coupled linear system also for explicit time stepping. In explicit time stepping, the system is written as

$$\frac{\mathrm{d}\boldsymbol{U}_h}{\mathrm{d}t} = \mathcal{M}^{-1}\left(\mathcal{C}\boldsymbol{U}_h + \boldsymbol{B}_h(t)\right), \tag{1.24}$$

and the work in each time step boils down to the evaluation of the source term $\boldsymbol{B}_h$ and the matrix-vector multiplication $\mathcal{C}\boldsymbol{U}_h$. The most common explicit time integrator are Runge–Kutta schemes, such as the classical Runge–Kutta scheme of order four, which progresses from t_m to $t_{m+1} = t_m + \Delta t$ by the update procedure

$$\begin{aligned}
\boldsymbol{K}^{(1)} &= \mathcal{M}^{-1}\left(\mathcal{C}\boldsymbol{U}_h^m + \boldsymbol{B}_h(t_m)\right),\\
\boldsymbol{K}^{(2)} &= \mathcal{M}^{-1}\left(\mathcal{C}\left(\boldsymbol{U}_h^m + \frac{1}{2}\Delta t\boldsymbol{K}^{(1)}\right) + \boldsymbol{B}_h\left(t_m + \frac{1}{2}\Delta t\right)\right),\\
\boldsymbol{K}^{(3)} &= \mathcal{M}^{-1}\left(\mathcal{C}\left(\boldsymbol{U}_h^m + \frac{1}{2}\Delta t\boldsymbol{K}^{(2)}\right) + \boldsymbol{B}_h\left(t_m + \frac{1}{2}\Delta t\right)\right),\\
\boldsymbol{K}^{(4)} &= \mathcal{M}^{-1}\left(\mathcal{C}\left(\boldsymbol{U}_h^m + \Delta t\boldsymbol{K}^{(3)}\right) + \boldsymbol{B}_h\left(t_m + \Delta t\right)\right),\\
\boldsymbol{U}_h^{m+1} &= \boldsymbol{U}_h^m + \frac{1}{6}\Delta t\left(\boldsymbol{K}^{(1)} + 2\boldsymbol{K}^{(2)} + 2\boldsymbol{K}^{(3)} + \boldsymbol{K}^{(4)}\right).
\end{aligned}$$

A general explicit Runge–Kutta method of s stages follows the update procedure

$$\begin{aligned}
\boldsymbol{K}^{(i)} &= \mathcal{M}^{-1}\left(\mathcal{C}\left(\boldsymbol{U}_h^m + \Delta t\sum_{j=1}^{i} a_{ij}\boldsymbol{K}^{(j)}\right) + \boldsymbol{B}_h\left(t_m + c_i\Delta t\right)\right),\\
\boldsymbol{U}_h^{m+1} &= \boldsymbol{U}_h^m + \sum_{i=1}^{s} b_i\boldsymbol{K}^{(i)},
\end{aligned}$$

with some coefficients b_i, a_{ij}, c_i. For the example of a $s = 5$-stage explicit Runge–Kutta scheme, the coefficients are summarized in the form of a Butcher tableau

$$\begin{array}{c|ccccc}
0 & & & & & \\
c_2 & a_{21} & & & & \\
c_3 & a_{31} & a_{32} & & & \\
c_4 & a_{41} & a_{42} & a_{43} & & \\
c_5 & a_{51} & a_{52} & a_{53} & a_{54} & \\
\hline
 & b_1 & b_2 & b_3 & b_4 & b_5
\end{array}$$

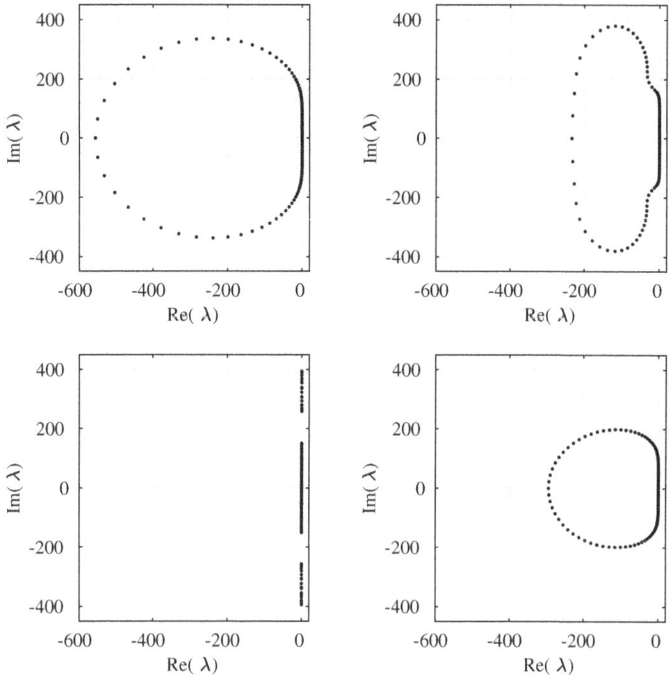

Fig. 1.11 Distribution of eigenvalues for the DG discretization of linear transport in 1D with $a = 1$, $n = 20$ elements, degree $k = 4$, for the upwind flux (upper left), the blending of upwind and central flux with $\alpha = 0.5$ in (1.17) (upper right), the upwind flux (lower left) and the central flux with the inexact Gauss–Lobatto integration (lower right)

Explicit time integration comes at the price of a limit on the maximum stable time step size, the so-called Courant–Friedrichs–Lewy (CFL) condition. The critical time step size can be expressed in terms of the mesh size h, the polynomial degree k, the maximal transport speed $|\partial f/\partial u|$, as well as the dimensionless Courant number Cr as

$$\Delta t_{\text{crit}} = \text{Cr}_{\text{crit}} \frac{h}{k^{1.5}|\partial f/\partial u|}. \tag{1.25}$$

In this formula, the power 1.5 in terms of the polynomial degree is a heuristic value to make the number Cr almost independent of k. It was proposed for spectral element methods in Karniadakis and Sherwin (2005) and verified to be accurate for discontinuous Galerkin schemes in Schoeder et al. (2018a), Fehn et al. (2018, 2019b), see also the discussion in Hesthaven and Warburton (2008). The time step constraint for higher order DG schemes is more restrictive than for other discretization methods, see also the discussion in (Hesthaven and Warburton 2008, Sect. 4.8).

The time step limit also depends on the type of boundary conditions, such as the mirror principle or the classical definition, and on the choice of the numerical flux. Figure 1.11 shows the eigenvalue spectrum of $\mathcal{M}^{-1}\mathcal{C}$ for a DG discretization of the

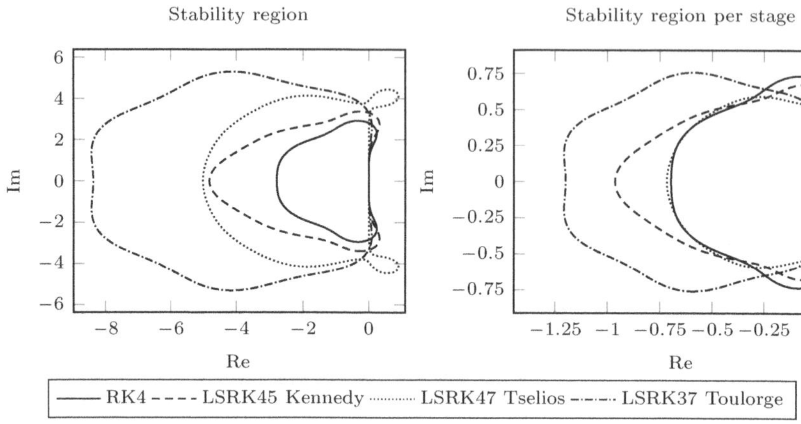

Fig. 1.12 Stability regions of the classical Runge–Kutta scheme of order 4 (RK4), the low-storage variant of order 4 with 5 stages of Kennedy et al. (2000), the 4th order 7-stage scheme of Tselios and Simos (2007), and the 3rd order 7-stage scheme of Toulorge and Desmet (2012). The left panel shows the stability region and the right panel the stability region scaled by the number of stages, which is representative of work

one-dimensional version of Eq. (1.1) with $f(u) = au$ and transport speed $a = 1$ on a mesh of 20 elements of polynomial degree 4 for the upwind flux, the central flux, and a blending of the upwind and central flux with $\alpha = 0.5$ in Eq. (1.17). Periodic boundary conditions are set to eliminate effects of the boundary linking the left value in the leftmost element to the right value of the last element. While all eigenvalues of the central flux have a zero real part due to the skew-symmetry of $\mathcal{C}$ for a constant coefficient $\boldsymbol{a}$, there are strong negative real parts especially for the upwind flux. The selection of the time step size Δt aims to fit the spectrum of $\Delta t \mathcal{M}^{-1}\mathcal{C}$ into the stability region of the respective time integrator, see Fig. 1.12 for the stability region of four explicit Runge–Kutta schemes. In the present example, the upwind flux will force the classical Runge–Kutta scheme to take a smaller time step than the central flux, with a limit of $\mathrm{Cr}_{\mathrm{crit}} \approx 1.15$ for the central flux, $\mathrm{Cr}_{\mathrm{crit}} \approx 1.09$ for the blended flux, and $\mathrm{Cr}_{\mathrm{crit}} \approx 0.80$ for the upwind flux.

Among the general-purpose Runge–Kutta schemes, so-called low-storage variants have emerged as one of the most attractive representatives in the context of wave-like phenomena in DG. They aim to avoid the storage cost of having all vectors $\boldsymbol{K}^{(i)}$ in memory by restricting the range of coefficients in the Runge–Kutta scheme. This saves both memory for auxiliary vectors and can improve performance on modern hardware as discussed in Schoeder et al. (2018a). As an example, the schemes presented in Kennedy et al. (2000) set $a_{ij} = b_j$ for $j < i - 1$. This allows to share the memory between updates to $\boldsymbol{U}_h^m + \sum_{i=1}^r b_i K^{(i)}$ and the right-hand side for stage $r+2$ as $\boldsymbol{U}_h^m + \Delta t \sum_{i=1}^r b_i \boldsymbol{K}^{(i)} + \Delta t a_{(r+2)(r+1)} \boldsymbol{K}^{(r+1)}$, requiring only two or three vectors to be kept in memory at the same time. Low-storage Runge–Kutta schemes optimized to wave propagation have been presented in (Tselios and Simos 2007).

Recent work has focused on optimizing the stability region of Runge–Kutta schemes especially for the eigenvalue spectrum of upwind-discretized discontinuous Galerkin schemes according to the upper left panel in Fig. 1.11, such as the work by Toulorge and Desmet (2012) who proposed a seven-stage scheme of order 3 with a large stability region per Runge–Kutta stage. Figure 1.12 show the stability regions of three low-storage variants and compares them to the classical Runge–Kutta scheme of order four, including a scaled version listing the limit per stage of the scheme, which is representative of the work when operating close to the stability limit.

For applications of the DG method to problems with shocks like the Euler or compressible Navier–Stokes equations at high Mach numbers, strong stability preserving explicit Runge–Kutta methods are often used, which aim to preserve monotonicity properties in individual Runge–Kutta stages as proposed by Shu (1988), see also Kubatko et al. (2014) for variants optimized for DG discretizations.

Computation of Integrals

Up to this point, it has been assumed that the integrals in the weak forms are computed exactly. Given the difficulty of finding closed-form expressions for the integrals involving, e.g., rational terms for a general curvilinear geometry representation discussed in section "Mapping to Curved Elements", it is common to compute the integrals exemplified by Eq. (1.11) by quadrature in the form

$$\begin{aligned}\int_{\Omega_e} &\varphi_i^{(e)}(\boldsymbol{x})\boldsymbol{a}(\boldsymbol{x})\cdot\nabla\varphi_j^{(e)}(\boldsymbol{x})\,\mathrm{d}\boldsymbol{x}\\ &\approx\sum_{q=1}^{n_\mathrm{c}}\varphi_i(\hat{\boldsymbol{x}}_q)\boldsymbol{a}\left(\boldsymbol{\Phi}_e(\hat{\boldsymbol{x}}_q)\right)\cdot\boldsymbol{J}_e^{-T}\hat{\nabla}\varphi_j(\hat{\boldsymbol{x}}_q)\det\left(\boldsymbol{J}_e(\hat{\boldsymbol{x}}_q)\right)w_q,\end{aligned}\tag{1.26}$$

with all terms evaluated at a set of quadrature points $\hat{\boldsymbol{x}}_q$ on the reference domain $\hat{\Omega}$ with associated quadrature weights w_q, $q = 1, \ldots, n_\mathrm{c}$. A similar evaluation is used for all integrals in the DG weak forms and enables an abstraction: Here and in the following, we state all ingredients according to the left-hand side, which is eventually transformed to reference coordinates via a mapping $\boldsymbol{\Phi}^{-1}$, whereas the implementation relies on the sum on the right-hand side.

As to the choice of the quadrature points and weights, the most common integration formula in 1D is the Gauss–Legendre quadrature. By using n_c points, integrands up to polynomial degree $2n_\mathrm{c} - 1$ can be integrated exactly. For the case that $\boldsymbol{a}$ is constant, φ_i is a polynomial of degree k, $\nabla\varphi_j = \frac{\partial\varphi_j}{\partial x}$ (in 1D) is of polynomial degree $k - 1$, and the metric term $\boldsymbol{J}_e$ is constant within the element (i.e., $\boldsymbol{\Phi}$ is an affine mapping), the integrand is a polynomial of degree $2k - 1$ and $n_\mathrm{c} = k$ points suffice to produce exact integrals. For the mass matrix (1.23), the maximal polynomial degree is $2k$, requiring $n_\mathrm{c} = k + 1$ points for exact integration.

On quadrilateral and hexahedral elements, the tensor product of the one-dimensional Gaussian quadrature formulas is the de-facto standard, which fits well with the polynomial shape functions constructed as tensor products. The formulas derived in 1D regarding the number of points can be straightforwardly applied to the multi-dimensional case. Note that due to the tensor product shape functions, the maximal polynomial degree of the product $\varphi_i \boldsymbol{a} \cdot \hat{\nabla} \varphi_j$ with constant $\boldsymbol{a}$ is $2k$ because the partial derivatives only reduce the polynomial degree in one direction. For triangles and tetrahedra, there exist either specialized formulas or tensor products on with a suitable transformation such as the Duffy transformation, see, e.g., Karniadakis and Sherwin (2005).

Collocation of Node Points and Quadrature Points

An important special case are quadrature formulas whose points coincide with the node points of the Lagrange polynomials defined in section "Polynomial Approximation". In that case, the evaluation of polynomials fulfills a Kronecker delta property, $\varphi_i(\hat{\boldsymbol{x}}_q) = \delta_{iq}$. This simplifies the computations of the integrals such as (1.26). Furthermore, the mass matrix $\mathcal{M}^{(e)}$ becomes diagonal because the integral (1.23) is only sampled in the node points.

The spectral element method (Deville et al. 2002; Kopriva 2009) with a continuous function space places the node and quadrature points in the same positions, a collocation, by using Lagrange polynomials on the points of the Gauss–Lobatto formula and using the same Gauss–Lobatto quadrature formula for the approximation of integrals. Gauss–Lobatto integration on $k + 1$ points per dimension only integrates polynomials up to degree $2k - 1$ exactly. In other words, the mass matrix becomes diagonal at the price of an integration error in the highest degree. At the same time, Gauss–Lobatto quadrature reduces the size of the largest eigenvalues in the discrete operator as shown by the lower right panel of Fig. 1.11, which in turn allows for larger time steps in explicit time stepping. Note that diagonal mass matrices have also been used in the spectral element context with hanging nodes on adaptively refined meshes (Kormann 2016).

Without inter-element continuity, the DG method allows for both the Gauss–Lobatto quadrature as well as the more accurate Gauss–Legendre formula in a collocation setting. In the latter case, Lagrange polynomials are defined in the points of the Gauss quadrature formula and the mass matrix is consistent, ensuring more accurate results. Conversely, the reduced order of the Gauss–Lobatto integration has been shown to affect convergence orders especially on curved geometries, see Durufle et al. (2009) for an evaluation with spectral elements. This topic is evaluated numerically in section "Stability and Convergence" below.

The spectral element setup simplifies a series of operations because no separate interpolation step between quadrature points and shape function values is necessary. While the reduction in operation counts has been considered essential at the time when spectral elements were introduced in 1980s, the situation is much less pressing today because the additional arithmetic operations of a non-collocated setup can often be hidden behind the time the data arrives from memory to the CPU core, see Schoeder et al. (2018a), Kronbichler and Kormann (2019) for a recent evaluation in the HPC context.

Over-Integration

For deformed geometries or for nonlinear fluxes $\boldsymbol{f}(u)$, the minimal number of quadrature points $k+1$ is often not enough to ensure accurate integration. For the example of the one-dimensional Burgers equations with $f(u) = u^2/2$ on affine elements, the required number of quadrature points to ensure exact integration is $\lceil \frac{3k}{2} \rceil$, because the integrand ${}^{\partial\varphi_i}/_{\partial x} u_h^2$ is of polynomial degree $3k-1$. Similarly, the integration of a $\boldsymbol{u} \otimes \boldsymbol{u}$ term in the multi-dimensional vector-valued case necessitates $\lfloor \frac{3k}{2} \rfloor + 1$ points for consistent integration. For other equations like the Euler equations discussed in the second chapter of this book, rational expressions in terms of the basis functions appear, for which an integration error with standard quadrature formula is unavoidable, irrespective of the number of points n_{c} in the Gaussian quadrature.

While the error due to inaccurate integration can often be neglected for elliptic operators as a variational crime (Brenner and Scott 2002), the situation with transport phenomena can be more severe, especially for barely resolved simulations where the contributions in higher polynomial degrees are not negligible. Using too few quadrature points correspond to an insufficient sampling of the function, which can give rise to instability of the simulation, see, e.g., Hesthaven and Warburton (2008), an effect called aliasing. There has been a large body of literature to address aliasing problems, such as filtering of high frequencies or specially designed schemes that are energy stable also in the presence of integration errors. We refer to Mengaldo et al. (2015); Winters et al. (2018) and references therein for the state of the art and recent developments. Quadrature with more points than polynomials, $n_{\text{c}} > (k+1)^d$, is also a form of de-aliasing. The number of points is often chosen based on the dominating nonlinearity, for example, the quadratic nonlinearity for convective term in compressible flows, for robust results (Fehn et al. 2019b).

Stability and Convergence

This section summarizes the most important results on the stability and convergence of DG schemes for smooth solutions. For a more detailed overview of theoretical results, the reader is referred to as Hesthaven and Warburton (2008, Chap. 4) and references therein. Furthermore, the application of high-order DG schemes to problems involving shocks necessitates additional ingredients, such as limiters to avoid unphysical over- and undershoots, for which we refer to the respective literature (Hesthaven and Warburton 2008; Shu 2018) and references therein.

Computational results are based on a C++ implementation following the procedure outline in the previous section, based on the open-source finite element library deal.II (Alzetta et al. 2018; Arndt et al. 2020), using variations of the code https://github.com/kronbichler/advection_miniapp.[1]

Stability by the Energy Method

The first ingredient for a convergent numerical scheme is its stability, i.e., the boundedness of the discrete operator. The energy method is a classical tool for verifying stability, established initially in the finite difference context as discussed in the textbook by Gustafsson et al. (2013).

In this section, we consider the stability for the linear transport equation, Eq. (1.1) with $\boldsymbol{f}(u) = \boldsymbol{a}u$ for a divergence-free transport speed, $\nabla \cdot \boldsymbol{a} = 0$, and without sources, i.e., $b(\boldsymbol{x}, t) = 0$. We consider the energy

$$E_u(t) = \int_\Omega u(\boldsymbol{x}, t)^2 \, \mathrm{d}\boldsymbol{x} = \|u(\cdot, t)\|_\Omega^2,$$

whose time derivative is given by

$$\frac{\mathrm{d}E_u}{\mathrm{d}t} = \frac{\mathrm{d}}{\mathrm{d}t}\|u\|_\Omega^2 = \left(u, \frac{\partial u}{\partial t}\right)_\Omega + \left(\frac{\partial u}{\partial t}, u\right)_\Omega = 2\left(u, \frac{\partial u}{\partial t}\right)_\Omega. \tag{1.27}$$

Inserting the partial differential equation $\frac{\partial u}{\partial t} + \nabla \cdot (\boldsymbol{a}u) = 0$ allows to substitute the term on the right-hand side. Consider now the spatial term and apply integration by parts,

$$-\left(u, \nabla \cdot (\boldsymbol{a}u)\right)_\Omega = (\nabla u, \boldsymbol{a}u)_\Omega - \langle u\boldsymbol{n}, \boldsymbol{a}u\rangle_{\partial\Omega}, \tag{1.28}$$

Due to $\nabla \cdot \boldsymbol{a} = 0$ and hence $\nabla \cdot (\boldsymbol{a}u) = \boldsymbol{a} \cdot \nabla u$ (product rule), the term on the left-hand side and the first term on the right-hand side are the same and can be added. Thus,

$$2\left(u, \nabla \cdot (\boldsymbol{a}u)\right)_\Omega = 2\left(\nabla u, \boldsymbol{a}u\right)_\Omega = \langle u, (\boldsymbol{n} \cdot \boldsymbol{a})u\rangle_{\partial\Omega}.$$

Inserting this relation into the right-hand side of Eq. (1.27) gives the final evolution of the energy

$$\frac{\mathrm{d}}{\mathrm{d}t}\|u\|_\Omega^2 = -\langle u, (\boldsymbol{n} \cdot \boldsymbol{a})u\rangle_{\partial\Omega}. \tag{1.29}$$

This equation states that the energy is only changed by the flux over the domain boundary. The contribution that flows into the domain on the inflow boundary $\partial\Omega_{\mathrm{in}}$ with $\boldsymbol{n} \cdot \boldsymbol{a} < 0$ increases the energy according to the square term u^2, whereas the contribution that leaves the domain via the outflow boundary decreases the energy by a non-positive term $\int_{\partial\Omega\setminus\partial\Omega_{\mathrm{in}}} (\boldsymbol{n} \cdot \boldsymbol{a})u^2 \, \mathrm{d}\boldsymbol{x}$. In this form, the energy method also

[1] Commit 385c588, retrieved on August 21, 2020.

reveals that boundary conditions need to be set precisely on the inflow boundary, as only a given data can provide a physically valid influx of energy, whereas no information is needed on the outflow boundary where the energy is non-increasing already.

For the discrete DG version (1.13), the same manipulations are applied element by element. The generic flux (1.17) is assumed, together with the boundary condition $u_h^+ = g$ on $\partial\Omega_{\text{in}}$ and $u_h^+ = u_h^-$ on $\partial\Omega \setminus \partial\Omega_{\text{in}}$. On element Ω_e, we obtain in analogy to (1.27) that

$$\frac{\mathrm{d}}{\mathrm{d}t}\|u_h\|^2_{\Omega_e} = 2\left(\nabla u_h, \boldsymbol{a} u_h\right)_{\Omega_e} - 2\left\langle u_h \boldsymbol{n}, \{\{\boldsymbol{a} u_h\}\} + \frac{\alpha|\boldsymbol{n}\cdot\boldsymbol{a}|}{2}[\![u_h]\!]\right\rangle_{\partial\Omega_e}.$$

Using integration by parts on the element similar to (1.28) transforms this equation into

$$\frac{\mathrm{d}}{\mathrm{d}t}\|u_h\|^2_{\Omega_e} = \langle u_h \boldsymbol{n}, \boldsymbol{a} u_h\rangle_{\partial\Omega_e} - 2\left\langle u_h \boldsymbol{n}, \{\{\boldsymbol{a} u_h\}\} + \frac{\alpha|\boldsymbol{n}\cdot\boldsymbol{a}|}{2}[\![u_h]\!]\right\rangle_{\partial\Omega_e}.$$

Summing up over all elements and splitting integrals into those over interior faces Γ_{i} and those over boundary faces Γ_{b} yields

$$\begin{aligned}\frac{\mathrm{d}}{\mathrm{d}t}\|u_h\|^2_{\Omega_h} = &\\ &+\left\langle \boldsymbol{n}^- \cdot \boldsymbol{a}, (u_h^-)^2 - (u_h^+)^2\right\rangle_{\Gamma_{\text{i}}} - 2\left\langle [\![u_h]\!], \{\{\boldsymbol{a} u_h\}\} + \frac{\alpha|\boldsymbol{n}\cdot\boldsymbol{a}|}{2}[\![u_h]\!]\right\rangle_{\Gamma_{\text{i}}}\\ &+\left\langle \boldsymbol{n}\cdot\boldsymbol{a}, u_h^2\right\rangle_{\Gamma_{\text{b}}\setminus\partial\Omega_{\text{in}}} - 2\left\langle u_h\boldsymbol{n}, \boldsymbol{a} u_h\right\rangle_{\Gamma_{\text{b}}\setminus\partial\Omega_{\text{in}}}\\ &+\left\langle \boldsymbol{n}\cdot\boldsymbol{a}, u_h^2\right\rangle_{\Gamma_{\text{b}}\cap\partial\Omega_{\text{in}}} - \left\langle u_h\boldsymbol{n}, \boldsymbol{a}(u_h+g) + \alpha|\boldsymbol{n}\cdot\boldsymbol{a}|(u_h-g)\boldsymbol{n}\right\rangle_{\Gamma_{\text{b}}\cap\partial\Omega_{\text{in}}}.\end{aligned}$$

For the interior faces, the summation from both sides has resulted in the term $u_h^-\boldsymbol{n}^- + u_h^+\boldsymbol{n}^+ = [\![u_h]\!]$. On outflow boundaries $\Gamma_{\text{b}} \setminus \partial\Omega_{\text{in}}$, the jump term $[\![u_h]\!]$ drops out because of $u_h^- = u_h^+$.

For the integrals over interior faces, we note that $2\left\langle [\![u_h]\!], \{\{\boldsymbol{a} u_h\}\}\right\rangle_{\Gamma_{\text{i}}} = \left\langle u_h^- - u_h^+, \boldsymbol{n}\cdot\boldsymbol{a}(u_h^- + u_h^+)\right\rangle_{\Gamma_{\text{i}}}$, which makes this average term cancel with the first term on the right-hand side involving $(u_h^-)^2 - (u_h^+)^2$. Similarly, part of the boundary terms involving the average cancel. The resulting discrete energy is thus

$$\begin{aligned}E_{u_h}(t) = \frac{\mathrm{d}}{\mathrm{d}t}\|u_h\|^2_{\Omega_h} &= -\alpha\left\langle [\![u_h]\!], |\boldsymbol{n}\cdot\boldsymbol{a}|\,[\![u_h]\!]\right\rangle_{\Gamma_{\text{i}}} - \left\langle u_h, \boldsymbol{n}\cdot\boldsymbol{a} u_h\right\rangle_{\Gamma_{\text{b}}\setminus\partial\Omega_{\text{in}}}\\ &\quad - \alpha\left\langle u_h, |\boldsymbol{n}\cdot\boldsymbol{a}| u_h\right\rangle_{\Gamma_{\text{b}}\cap\partial\Omega_{\text{in}}} + \left\langle u_h, |\boldsymbol{n}\cdot\boldsymbol{a}|(1+\alpha) g\right\rangle_{\Gamma_{\text{b}}\cap\partial\Omega_{\text{in}}}.\end{aligned} \tag{1.30}$$

The first and third terms involve a square of $[\![u_h]\!]$ and u_h, respectively, and thus give a non-increasing contribution to the discrete energy for $\alpha \geq 0$. The second term reflects the respective contribution on the outflow boundary in the continuous energy estimate (1.29), whereas the fourth term describes the increase in energy due to the inhomogeneous boundary condition. Neglecting the contribution from the boundary, we have shown that the discrete energy is non-increasing.

Non-positive terms on the right-hand side of the discrete energy estimate (1.30) imply the *stability* of the spatial semi-discretization. We note from this equation that, for $\alpha > 0$, there is some negative contribution proportional to α and the magnitude of the jumps in the DG solution. This negative contribution acts as a numerical dissipation and illustrates the stabilization mechanism of the method. In the eigenvalue plots shown in Fig. 1.11, the negative real parts in the eigenvalues can be associated with these jump contributions.

Numerical Verification

We study the linear transport problem on the domain $\Omega = (0, 1)^2$ with the transport speed

$$\boldsymbol{a}(\boldsymbol{x}, t) = 2\cos\left(\frac{\pi t}{T_f}\right)\begin{pmatrix}\sin(2\pi x_2)\sin^2(\pi x_1)\\ -\sin(2\pi x_1)\sin^2(\pi x_2)\end{pmatrix} \tag{1.31}$$

and the initial condition

$$u_0(\boldsymbol{x}) = \mathrm{e}^{-400\left[(x_1-0.5)^2+(x_2-0.75)^2\right]}.$$

The transport field is solenoidal, $\nabla \cdot \mathbf{a} = 0$, and represents motion that is initially in counterclockwise direction around the center (0.5, 0.5) of the domain. The time-dependent factor is $+2$ for $t = 0$ and reverses to -2 for the final time $t = T_f$. The switch of the transport direction implies that $u(x, y, T_f) = u(x, y, 0)$, which allows the assessment of the accuracy of the numerical scheme. Figure 1.13 shows the initial condition and the solution at time $t = 4$ for the case with $T_f = 8$. The rotating flow gives rise to very steep gradients, despite the smooth initial condition and velocity field. The transport speed $\boldsymbol{a}$ is zero on the boundary, so no boundary condition is set.

Figure 1.14 shows the numerical solution obtained on a 32^2 mesh with polynomial degree $k = 4$, a discretization with 25,600 unknowns. This resolution is insufficient to adequately represent the structures of the reference result shown in Fig. 1.13. The upwind flux leads to a solution that still resembles the reference result, whereas the central flux is polluted by oscillations throughout the domain, an effect of the insufficient resolution and accumulated dispersion errors. When comparing the solution at the final time, $u_h(\boldsymbol{x}, T_f)$, to the initial condition, the upwind flux shows a deviation in the maximum norm of 0.76 because of a loss in energy, going from $E_{u_h}(0) = 0.00393$ for the initial condition to $E_{u_h}(8) = 0.00123$. This loss is explained by the energy estimate (1.30) with $\partial\Omega_{\text{in}} = \emptyset$ and $\boldsymbol{n} \cdot \boldsymbol{a} = 0$ at the boundary, showing that the jumps $[\![u_h]\!]$ over the interior element boundaries decrease the energy.

For the central flux, the energy estimate (1.30) predicts exact conservation of energy. Indeed, the measured energy coincides with the one from the initial condition up to six leading digits. Furthermore, the error at the final time $\|u_h(\boldsymbol{x}, T_f) - u_h(\boldsymbol{x}, 0)\|_\infty$ is $1.02 \cdot 10^{-4}$, despite the completely unphysical results at

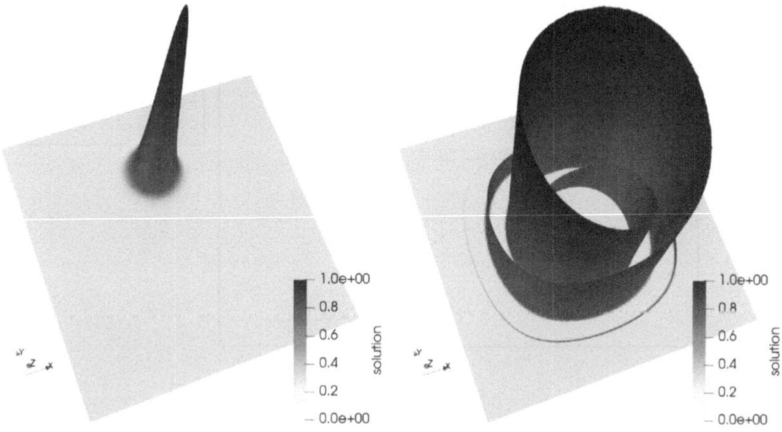

Fig. 1.13 Test case for the assessment of stability with the initial condition (left panel) and the solution at the time of the maximal deviation, $t = 4$ (right panel)

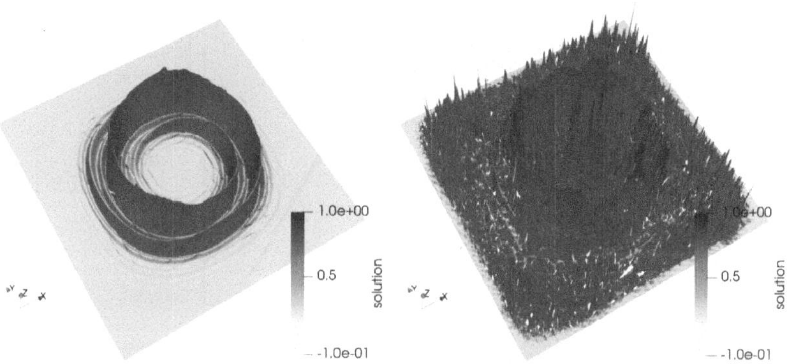

Fig. 1.14 Deformation by a vortex at time $t = 4$ solved on a 32^2 mesh with polynomial degree $k = 4$ for the upwind flux (left panel) and the central flux (right panel)

the intermediate time $t = 4$ shown in Fig. 1.14. This behavior can be attributed to the exactly reversed flow pattern in the time span $(4, 8]$. Since exact reversibility rarely occurs on real physical models such as fluid flow governed by the Navier–Stokes equations, the energy-conserving central flux is primarily of theoretical value in this example.

Figure 1.15 shows the evolution of the discrete energy as a function of time for the upwind flux compared to the central flux for three different combinations of the polynomial degree and the mesh size. For polynomial degree $k = 2$, a 512^2 mesh is chosen, for $k = 4$, the mesh is 256^2, whereas the mesh is 128^2 for the highest degree $k = 8$. Assuming the interpolation of a continuous field, each degree has the same number of node points in this setup. Note that in DG it is also common to consider

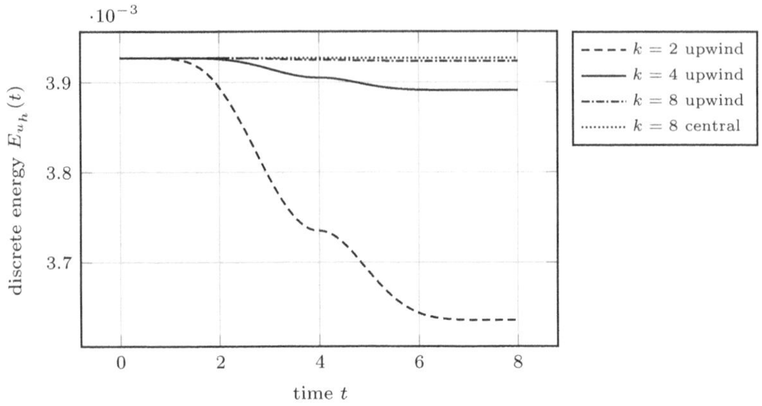

Fig. 1.15 Development of discrete energy over time for a 512^2 mesh with $k = 2$, a 256^2 mesh with $k = 4$, and a 128^2 mesh with $k = 8$, all with the upwind flux, and for the central flux with 128^2 elements and $k = 8$

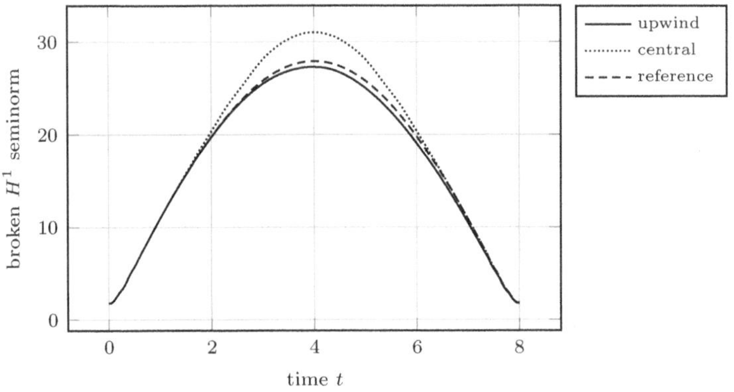

Fig. 1.16 Evolution of discrete H^1 seminorm, $\sqrt{\int_{\Omega_h} |\nabla u_h|^2 \, d\boldsymbol{x}}$, of the numerical solution for $k = 4$ on the 256^2 mesh with central and upwind flux

the number of unknowns as a metric (Fehn et al. 2018), where $k = 1$, $k = 3$, and $k = 7$ would yield the same number of unknowns. Despite the higher number of unknowns in the lower order methods, the numerical loss in energy and thus the numerical dissipation is nonetheless smaller for the higher polynomial degrees with the higher resolution capability. This is also reflected in the distance to the initial condition, which is 0.15 for $k = 2$, 0.045 for $k = 4$, and 0.013 for $k = 8$.

Figure 1.16 displays the evolution of the discrete H^1 seminorm of the solution, computed as the square root of the integral of the square of the gradient ∇u_h. This metric picks up the unnatural oscillations seen for the central flux in Fig. 1.14. As expected, the central flux performs considerably worse than the upwind flux. While this metric is somewhat artificial, it gives an indication that exact energy conservation, a variational (integral) quantity balancing the solution content over the computational domain, does not quantify the quality of the solution apart from stability. For many conservation laws, point-wise behavior such as maximum-principle preserving behavior or local entropy balances are more important.

Exact Energy Conservation

A closer look at the behavior of the central flux reveals that the energy still changes in the sixth to seventh digit over time, despite a predicted exact energy conservation. This is because the energy estimate (1.30) assumes *exact integration*, which is not fulfilled for the chosen Gaussian quadrature with $k+1$ points for the trigonometric functions in the velocity field (1.31). More precisely, the integration by parts underlying Eq. (1.28) is violated for a discrete summation like (1.26). The situation can be fixed by a so-called *skew-symmetric* form, a convex combination of the weak DG form (1.13) and the strong DG form (1.21) as

$$\begin{aligned}\left(v_h, \frac{\partial u_h}{\partial t}\right)_{\Omega_h} &- \frac{1}{2}\left(\nabla v_h, \boldsymbol{f}(u_h)\right)_{\Omega_h} + \frac{1}{2}\left(v_h, \nabla\cdot\boldsymbol{f}(u_h)\right)_{\Omega_h} \\ &+ \left\langle v_h\boldsymbol{n}, \boldsymbol{f}^* - \frac{1}{2}\boldsymbol{f}(u_h)\right\rangle_{\partial\Omega_h} = (v_h, b)_{\Omega_h}.\end{aligned} \tag{1.32}$$

The skew-symmetric form is equivalent to the conservative weak DG form (1.13) for exact integration with a solenoidal field $\nabla\cdot\boldsymbol{a}=0$, but differs in the discrete case with inexact integration. In the skew-symmetric form, the second and third term on the left-hand side cancel when considering the energy with test function $v_h = u_h$ at each quadrature point. Thus, only the boundary terms are left which can be bounded pointwise, irrespective the quadrature. By using the skew-symmetric version (1.32) with a central flux, the discrete energy remains constant up to errors of the time stepping scheme or, when Δt is small enough, to roundoff precision. For upwind fluxes, the energy is guaranteed to be non-increasing also in the presence of integration errors. Skew-symmetric forms also exist for other conservation laws and for vector-valued systems. However, the form (1.32) is only mass conserving up to integration errors, because manipulations as in section "Connecting Neighboring Elements: Numerical Fluxes" necessitate integration by parts.

A systematic approach to construct energy stable schemes (or entropy-stable schemes in the nonlinear context) are so-called split-form DG schemes. The basic variant utilizes the analogy between DG methods using the Gauss–Lobatto integration formula with finite difference methods, see Gassner (2013), Kopriva and Gassner (2016) for the general methodology including curved meshes, Gassner (2014), Gassner et al. (2016) for applications to the Euler equations, and Chan (2018) for methods with more general quadrature rules, as well as the chapter by Winters, Kopriva, Gassner and Hindenlang later in this book.

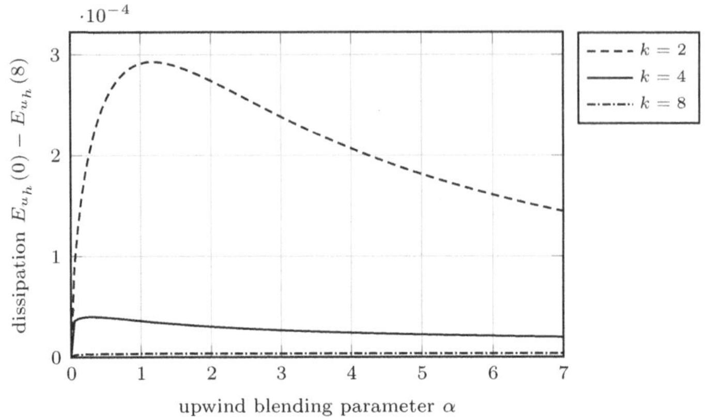

Fig. 1.17 Dissipation the numerical solution measured as $E_{u_h}(0) - E_{u_h}(8)$ for the upwind flux with a 512^2 mesh for $k = 2$, a 256^2 mesh for $k = 4$, and a 128^2 mesh for $k = 8$

Dissipation Versus Upwind Parameter

From the energy estimate (1.30), one might be tempted to expect a stronger damping of energy as the upwind blending parameter α in (1.17) increases. However, the dissipation also depends on the magnitude of the jump in the solution, which decreases as α is growing. Figure 1.17 shows that the dissipation takes a maximum for a finite value of α before it decreases again. For the 256^2 mesh with $k = 4$, the maximum dissipation is observed for $\alpha = 0.3$, whereas for the 512^2 mesh with $k = 2$ the maximum dissipation is for $\alpha = 1.1$. In the limit $\alpha \to \infty$, the DG method approaches a continuous finite element method without any numerical dissipation in the absence of boundary conditions. The observations made in this simple setting are also relevant for turbulent flow, as experiments varying penalty parameters in Fehn et al. (2019a) have shown that, beyond a certain threshold, more penalization translates to less dissipation rather than more.

Theoretical Convergence Results

Given the stability of the DG approximation shown in the previous subsection, the convergence of the numerical approximation is closely related to the consistency, which in turn is governed by the interpolation properties of the polynomials. The main interpolation result (Hesthaven and Warburton 2008, Theorem 4.8) is that the interpolation $u_{h,\mathrm{I}}$ is close to the analytical solution u in terms of the mesh size h and the polynomial degree k as

$$\|u - u_{h,\mathrm{I}}\|_{L_2} \leq C \frac{h^\sigma}{k^{p-1/2}} |u|_{\Omega,\sigma}, \tag{1.33}$$

where p denotes the regularity of the analytic solution u in terms of the number of derivatives of u that are square-integrable, $\sigma = \min(k+1, p)$, the norm $|u|_{\Omega,\sigma}$ denotes the L_2 norm of all partial derivatives of order σ of u, and C is a constant that does not depend on the mesh size.

For discontinuous Galerkin methods, there is usually a gap between these consistency results and the actually observed convergence speed. A simple analysis would suggest rates of $\mathcal{O}(h^k)$, i.e., one order less compared to consistency result. This estimate is however not optimal. General results strongly depend on the choice of the numerical flux. For example, rates of order $\mathcal{O}(h^{k+1/2})$ are the best result for all kinds of meshes (Peterson 1991). On the other hand, for advection with a pure upwind fluxes it has been shown that optimal convergence with errors $\mathcal{O}(h^{k+1})$ is possible (Lesaint and Raviart 1974; Johnson and Pitkäranta 1986).

Convergence on an Analytical Test Case

In this subsection, we study numerical solutions to the multi-dimensional transport equation (1.1) with $f(u) = \boldsymbol{a}u$ on the domain $\Omega = [0, 1]^d$ in $d = 1, 2, 3$ dimensions with the transport speed given by the first d components of $\boldsymbol{a} = [1.1, 0.15, -0.05]^T$ and zero source $b \equiv 0$. The analytical solution at time t is given by

$$u(\boldsymbol{x}, t) = \sin(4\pi(x_1 - a_1 t)) \prod_{i=2}^{d} \cos(4\pi(x_i - a_i t)), \tag{1.34}$$

which is evaluated at $t = 0$ for the initial condition. We consider both periodic boundary conditions and Dirichlet conditions at the respective inflow boundary with $x_1 = 0$, $x_2 = 0$, or $x_3 = 1$, the latter set from the analytical solution.

For all experiments in this section, the relative L_2 error

$$\|u - u_h\|_{L_2} = \frac{\sqrt{\int_{\Omega_h} (u - u_h)^2 \, \mathrm{d}\boldsymbol{x}}}{\sqrt{\int_{\Omega_h} u^2 \, \mathrm{d}\boldsymbol{x}}},$$

is reported at the final time $t = 2$. The spatial discretization is based on a discontinuous Galerkin scheme with nodal polynomials in the Gauss–Lobatto points as described in the previous section. As a numerical flux, the upwind scheme (1.16) is used. For integration, either the consistent Gauss–Legendre quadrature on $k+1$ points or the Gauss–Lobatto quadrature formula on $k+1$ is used as described in section "Computation of Integrals". The initial condition is applied by projection using the underlying quadrature formula. Time integration is done with an explicit fourth-order Runge–Kutta scheme of five stages with low-storage according to the two-register variant of Kennedy et al. (2000). The time step size calculated according

Table 1.1 Relative L_2 error measured at time $t = 2$ for linear transport on a uniform mesh for $k = 2, 4, 6$ in 1D with Cr = 0.5

	$k = 2$		$k = 4$		$k = 6$	
# elements	L_2 error	Rate	L_2 error	Rate	L_2 error	Rate
4	$5.55 \cdot 10^{-1}$	—	$4.53 \cdot 10^{-3}$	—	$6.14 \cdot 10^{-5}$	—
8	$3.68 \cdot 10^{-2}$	3.92	$1.40 \cdot 10^{-4}$	5.01	$9.33 \cdot 10^{-7}$	6.04
16	$2.60 \cdot 10^{-3}$	3.82	$4.45 \cdot 10^{-6}$	4.98	$4.91 \cdot 10^{-8}$	4.24
32	$2.98 \cdot 10^{-4}$	3.13	$1.43 \cdot 10^{-7}$	4.96	$3.06 \cdot 10^{-9}$	4.00
64	$3.70 \cdot 10^{-5}$	3.01	$4.85 \cdot 10^{-9}$	4.88	$1.93 \cdot 10^{-10}$	3.99
128	$4.62 \cdot 10^{-6}$	3.00	$1.91 \cdot 10^{-10}$	4.67	$1.31 \cdot 10^{-11}$	3.87
256	$5.77 \cdot 10^{-7}$	3.00	$1.39 \cdot 10^{-11}$	3.78	$8.20 \cdot 10^{-12}$	0.68
512	$7.21 \cdot 10^{-8}$	3.00	$1.62 \cdot 10^{-12}$	3.10	$1.26 \cdot 10^{-11}$	−0.63

to the formula (1.25) for a fixed Courant number Cr as specified in the text, leading to refinement in space and time simultaneously as is common in practical applications.

Table 1.1 reports the errors of a one-dimensional experiment run with polynomial degrees $k = 2$, $k = 4$, $k = 6$ for mesh sizes between $h = \frac{1}{4}$ and $h = \frac{1}{512}$, using periodic boundary conditions and a Courant number Cr = 0.5. For both $k = 2$ and $k = 4$, optimal convergence rates $\mathcal{O}(h^{k+1})$ are recorded, with a slight deviation for coarse grids with $k = 2$ due to pre-asymptotic behavior and for very fine grids with $k = 4$ as the error level approaches the roundoff accuracy. For $k = 6$, however, the observed convergence rate quickly reduces to fourth order, despite the expected seventh order of accuracy. Indeed, for $n = 32$ and more elements, the error is almost entirely governed by the error of the time integration. As the mesh size is halved, the time step size is halved according to the relation (1.25) with fixed Courant number, which results in a 16 times smaller error according to the error relation $\mathcal{O}(\Delta t^4)$. Optimal observed convergence rates governed by the error in space can be restored by choosing smaller time step sizes. An alternative would be higher order time integrators.

Figure 1.18 reports the L_2 error as a function of the mesh size in 1D for polynomial degrees $k = 1, \ldots, 7$ for a Courant number Cr = 0.1 in a double-logarithmic plot. In the left panel, the situation with periodic boundary conditions, i.e., treating all faces as interior ones, is depicted, whereas Dirichlet conditions on the inflow boundary are considered for the data on the right panel. The former is slightly more challenging in terms of dissipation behavior of the schemes, as the information of the analytical solution is only injected by the initial condition. The latter shows that the choice of the exterior value $u_h^+ = g$ with the classical definition yields accurate results. In all cases, optimal rates of convergence $\mathcal{O}(h^{k+1})$ are observed, which appear as straight lines of slope h^{k+1}.

The figure reveals that higher order schemes give more accurate results for the same mesh size. This is not surprising, as the weak form (1.13) searches the solution in a strictly larger space for higher polynomial degrees and the fluxes are optimal for the chosen scenario. As opposed to the data of Table 1.1, the Courant number is

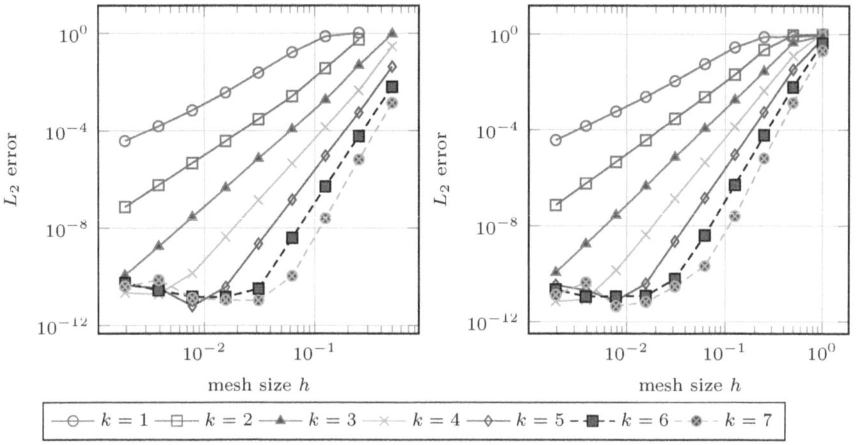

Fig. 1.18 Convergence with mesh refinement for degrees $k = 1$ to $k = 7$ in one dimension for periodic boundary conditions (left panel) and Dirichlet boundary conditions (right panel)

Fig. 1.19 Convergence for polynomial refinement on three different meshes in 1D with Cr = 0.1

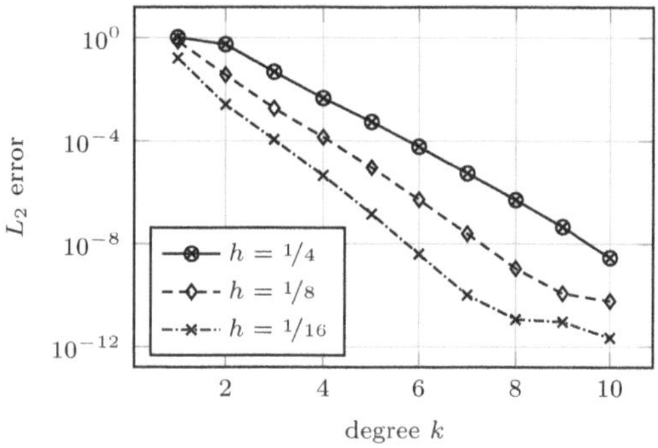

small enough to make the spatial error dominate for all data points except one for $k = 7$. For higher polynomial degrees and fine meshes, the error is dominated by roundoff errors due to the 16 digits of precision in the machine numbers.

Figure 1.19 presents the results of a p-refinement study. Here, the mesh size is kept constant and the polynomial degree k is increased. In the chosen semi-logarithmic scaling of axes, the error appears to follow a linear trend, which indicates an exponential convergence in the degree, see, e.g., Hesthaven et al. (2006). Comparing Fig. 1.19 with Fig. 1.18 reveals that increasing the polynomial degree is more efficient for reducing the error than refining the mesh. Even though higher degrees come with a disproportionally reduced time step size due to the power 1.5 in the time step formula (1.25), simulations with smooth solutions and moderate to high accuracy requirements are solved more efficiently with higher polynomial degrees as compared to finer meshes. Section "Refining the Mesh or Increasing the Polynomial Degree?" below gives a different, more practically relevant, viewpoint on this topic.

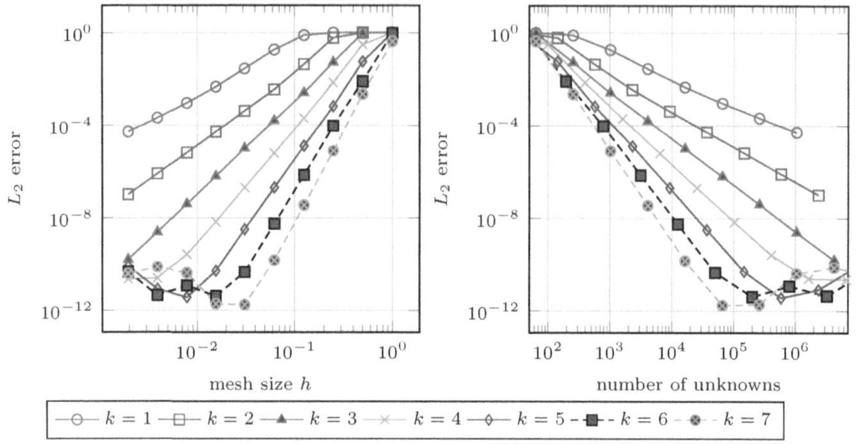

Fig. 1.20 Convergence with mesh refinement for degrees $k = 1$ to $k = 7$ in two dimensions as a function of mesh size (left) and as a function of the number of unknowns (right)

Figure 1.20 reports the results of an h-convergence experiment in two space dimensions. The convergence behavior is almost identical to the one-dimensional case, which is also confirmed by the errors reported in Table 1.2. In order to account for the fact that higher order methods involve more unknowns on the same mesh as lower order methods, the right panel of Fig. 1.20 plots the error against the number of unknowns. Even though the topic of the actual execution on hardware performance is postponed to later chapters of this book, it can be anticipated that the work will at least be proportional to the number of unknowns as the polynomial degree is increased, possibly more due to the denser coupling between the unknowns (Chang et al. 2018; Kronbichler and Wall 2018). In this metric, higher order methods are still considerably more efficient than lower order schemes. To reach an error of 10^{-4}, the solver with $k = 2$ and convergence order needs around 25,000 unknowns, whereas only about 2,000 unknowns would be needed for $k = 4$, or 800 unknowns for $k = 6$. A coarser representation allows to run with a larger time step and consumes less memory overall.

Table 1.2 reports the error and measured convergence rates between the results of mesh size h and $h/2$ for the consistent Gauss–Legendre quadrature as well as the (spectral) Gauss–Lobatto case, using a polynomial degree $k = 4$. While both deliver optimal convergence rates, the errors are approximately two times smaller for the consistent integration. The table reports the errors for two different time step sizes in the case of Gauss–Legendre quadrature, one obtained with $\mathrm{Cr} = 0.5$ and one with $\mathrm{Cr} = 0.1$. While the former switches from being dominated by the spatial discretization on coarser meshes to being dominated by the time error with convergence rates dropping from 5 to 4, the latter is completely dominated by the spatial discretization error.

Table 1.2 Relative L_2 error measured at time $t = 2$ for linear transport on a uniform mesh of quadrilaterals with $k = 4$ in 2D

Quadrature	Gauss–Legendre		Gauss–Legendre		Gauss–Lobatto	
Time step	Cr = 0.1		Cr = 0.5		Cr = 0.5	
Mesh	L_2 error	Rate	L_2 error	Rate	L_2 error	Rate
4^2	$7.24 \cdot 10^{-3}$	—	$7.24 \cdot 10^{-3}$	—	$1.89 \cdot 10^{-2}$	—
8^2	$2.01 \cdot 10^{-4}$	5.16	$2.01 \cdot 10^{-4}$	5.16	$4.27 \cdot 10^{-4}$	5.46
16^2	$6.27 \cdot 10^{-6}$	5.01	$6.31 \cdot 10^{-6}$	5.00	$1.33 \cdot 10^{-5}$	5.00
32^2	$1.98 \cdot 10^{-7}$	4.99	$2.03 \cdot 10^{-7}$	4.96	$4.17 \cdot 10^{-7}$	5.00
64^2	$6.13 \cdot 10^{-9}$	5.01	$6.78 \cdot 10^{-9}$	4.91	$1.31 \cdot 10^{-8}$	5.00
128^2	$1.92 \cdot 10^{-10}$	5.00	$2.66 \cdot 10^{-10}$	4.67	$4.40 \cdot 10^{-10}$	4.89
256^2	$1.58 \cdot 10^{-11}$	3.60	$2.45 \cdot 10^{-11}$	3.44	$2.31 \cdot 10^{-11}$	4.25
512^2	$5.03 \cdot 10^{-11}$	−1.67	$2.26 \cdot 10^{-11}$	0.12	$1.76 \cdot 10^{-11}$	0.39

Table 1.3 Relative L_2 error measured at time $t = 2$ for linear transport on a uniform mesh of hexahedra with $k = 4$ in 3D

Quadrature	Gauss–Legendre		Gauss–Legendre		Gauss–Lobatto	
Time step	Cr = 0.2		Cr = 0.5		Cr = 0.5	
Mesh	L_2 error	Rate	L_2 error	Rate	L_2 error	Rate
4^3	$8.56 \cdot 10^{-3}$	—	$8.56 \cdot 10^{-3}$	—	$2.26 \cdot 10^{-2}$	—
8^3	$2.47 \cdot 10^{-4}$	5.11	$4.28 \cdot 10^{-4}$	5.11	$5.31 \cdot 10^{-4}$	5.41
16^3	$7.79 \cdot 10^{-6}$	4.99	$7.83 \cdot 10^{-6}$	4.98	$1.71 \cdot 10^{-5}$	4.95
32^3	$2.41 \cdot 10^{-7}$	5.01	$2.46 \cdot 10^{-7}$	4.99	$4.98 \cdot 10^{-7}$	5.10
64^3	$7.55 \cdot 10^{-9}$	5.00	$8.13 \cdot 10^{-9}$	4.92	$1.60 \cdot 10^{-8}$	4.96
128^3	$2.35 \cdot 10^{-10}$	5.01	$3.04 \cdot 10^{-10}$	4.74	$5.26 \cdot 10^{-10}$	4.92
256^3	$2.09 \cdot 10^{-11}$	3.49	$1.43 \cdot 10^{-11}$	4.41	$1.76 \cdot 10^{-11}$	4.90

Table 1.3 shows the three-dimensional version of this experiment on meshes with 64 to 16.8 million elements of degree $k = 4$, corresponding to 8,000 to 2.1 billion unknowns. (Despite the considerable size of the largest problem, the implementation concepts from Kronbichler and Kormann (2019) make a run of this size on a moderate machine with 40 cores and 192 GB of memory complete in around a day, or much less time when run on a parallel cluster.) Optimal convergence with errors proportional to h^5 is observed for the Gauss–Legendre quadrature with a small Courant number Cr = 0.2. The spectral Gauss–Lobatto quadrature formula also leads to optimal spatial convergence, albeit with an error approximately twice as large as with consistent integration.

On the Accuracy of Time Stepping

Given the observed decay in convergence rates due to the error made by the time stepping, it would seem natural to go to time stepping of order higher than four.

However, such schemes in time are less widespread in most application areas. This is because the spatial discretization is often the deciding factor and chosen as coarse as possible. With explicit time integration, the time error is small as soon as a CFL condition like (1.25) is met when aiming for engineering accuracies in the range of an error level of a few percent. For long-time integration, certain conservation properties of the time steppers are more important than the order alone, leading to so-called symplectic integrators (Hairer et al. 2006). For explicit Runge–Kutta methods, the stability region per stage tends to become smaller compared to the variants considered in Fig. 1.12, decreasing efficiency. Furthermore, the so-called Butcher barriers necessitate the number of stages to increase more quickly as compared to the convergence order.

An alternative approach to time stepping in the DG context are arbitrary-derivative (ADER) schemes with arbitrarily high order in time integration using a predictor–corrector setup (Dumbser and Käser 2006). The predictor stage propagates the solution on each DG element over one time step without taking the neighboring contributions into account, e.g., using a Taylor expansion in time or a Runge–Kutta scheme of appropriate accuracy. In the corrector stage, the numerical fluxes to the neighbors are taken into account. The stability region per time step of these methods is typically smaller than for optimized Runge–Kutta methods presented in the left panel of Fig. 1.12. However, only a single flux computation plus some element-local operations are needed as compared to multiple stages of Runge–Kutta schemes. The higher arithmetic intensity in the predictor stage runs well on modern hardware where computations are cheap compared to memory access, which can often compensate for the smaller time step, a topic analyzed in Schoeder et al. (2018a), Dumbser et al. (2018), Reinarz et al. (2020). Moreover, the predictor/corrector setup also facilitates the implementation of local time stepping (Dumbser et al. 2007; Schoeder et al. 2018b).

Convergence on Deformed Mesh

As a next test, we assess the problem with analytical solution (1.34) on two meshes with non-affine element shapes. In the first setup, the domain $\Omega = [0, 1]^d$ is subjected to the deformation

$$\boldsymbol{x} + \beta \prod_{e=1}^{d} \sin(2\pi x_e),$$

which is exemplarily shown for the 8^2 mesh in the left panel of Fig. 1.21. The parameter β is set to $\beta = 0.12$ in 2D and to $\beta = 0.1$ in 3D. The second example considers a disk of radius 0.2 and center $(0.5, 0.5)^T$ that is inserted in the square $(0, 1)^2$. The circle is described by polar coordinates, and transfinite interpolation is used to smoothly transition to the straight-side edges in the center and on the outer domain boundaries. In both cases, the curved geometry is described by polynomials of degree

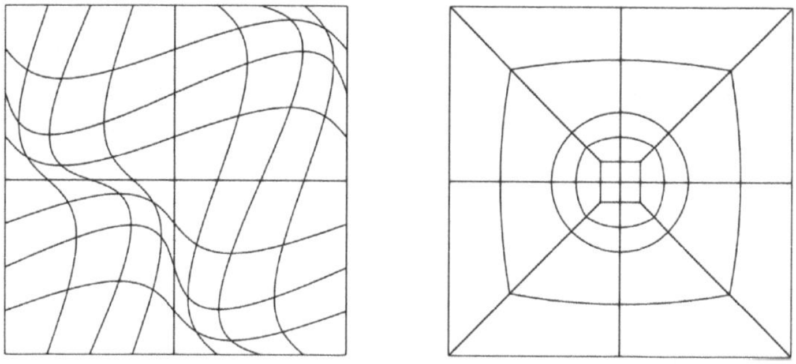

Fig. 1.21 Exemplary deformed mesh with 8×8 elements using a sine perturbation (left) and with 9×2^2 elements of a circle inscribed within a square (right)

Table 1.4 Relative L_2 error measured at time $t = 2$ for linear transport on a deformed domain via a sine function with $k = 4$ in 2D

Quadrature	Gauss–Legendre		Gauss–Lobatto		Gauss–Lobatto	
Time step	Conservative		Conservative		Skew-symmetric	
Mesh	L_2 error	Rate	L_2 error	Rate	L_2 error	Rate
4^2	$2.03 \cdot 10^{-1}$	—	$3.95 \cdot 10^{-1}$	—	$3.72 \cdot 10^{-1}$	—
8^2	$6.79 \cdot 10^{-3}$	4.90	$2.42 \cdot 10^{-2}$	4.03	$1.59 \cdot 10^{-2}$	4.55
16^2	$2.16 \cdot 10^{-4}$	4.97	$1.77 \cdot 10^{-3}$	3.77	$5.95 \cdot 10^{-4}$	4.74
32^2	$7.82 \cdot 10^{-6}$	4.79	$1.08 \cdot 10^{-4}$	4.03	$3.79 \cdot 10^{-5}$	3.97
64^2	$2.02 \cdot 10^{-7}$	5.27	$3.89 \cdot 10^{-6}$	4.79	$1.14 \cdot 10^{-6}$	5.06
128^2	$5.81 \cdot 10^{-9}$	5.12	$1.12 \cdot 10^{-7}$	5.12	$3.80 \cdot 10^{-8}$	4.90
256^2	$1.77 \cdot 10^{-10}$	5.04	$3.33 \cdot 10^{-9}$	5.07	$1.38 \cdot 10^{-9}$	4.78
512^2	$1.64 \cdot 10^{-11}$	3.43	$1.15 \cdot 10^{-10}$	4.85	$5.29 \cdot 10^{-11}$	4.71

4, the same as the interpolation of the numerical solution, and all points are placed on the respective deformed shape using a manifold description (Heltai et al. 2019).

Tables 1.4 and 1.5 report the errors for a sequence of refined meshes similar to the experiment from Table 1.2 for the sine deformation and inscribed circle, respectively. Due to the deformed geometries, the Jacobian determinant is a polynomial of degree $4^d - 1$, and the metric term times the Jacobian is a polynomial of degree four in 2D and eight in 3D, respectively. As a consequence, the Gaussian quadrature is not exact. Apart from a somewhat higher error level due to the change in element sizes, the convergence behavior is nonetheless optimal at a rate of h^5 for the Gauss–Legendre integration. For the Gauss–Lobatto integration, however, the convergence rate drops to slightly more than four in coarser meshes, before accelerating towards 5 as the mesh becomes more refined. Overall, the error level of the conservative form (1.13) with Gauss–Lobatto quadrature is more than an order of magnitude larger compared to the one with the Gaussian quadrature. The skew-symmetric variant with Gauss–Lobatto quadrature (1.32) is more accurate, with errors only around five times larger than

Table 1.5 Relative L_2 error measured at time $t = 2$ for linear transport on a square with inscribed circle with $k = 4$ in 2D

Quadrature	Gauss–Legendre		Gauss–Lobatto		Gauss–Lobatto	
Time step	Conservative		Conservative		Skew-symmetric	
Mesh	L_2 error	Rate	L_2 error	Rate	L_2 error	Rate
9×2^2	$7.91 \cdot 10^{-2}$	—	$4.33 \cdot 10^{-1}$	—	$2.00 \cdot 10^{-1}$	—
9×4^2	$3.31 \cdot 10^{-3}$	4.57	$1.38 \cdot 10^{-2}$	4.97	$7.07 \cdot 10^{-3}$	4.82
9×8^2	$1.08 \cdot 10^{-4}$	4.94	$1.23 \cdot 10^{-3}$	3.49	$3.42 \cdot 10^{-4}$	4.38
9×16^2	$2.71 \cdot 10^{-6}$	5.31	$6.07 \cdot 10^{-5}$	4.34	$1.50 \cdot 10^{-5}$	4.51
9×32^2	$7.35 \cdot 10^{-8}$	5.21	$2.49 \cdot 10^{-6}$	4.61	$6.82 \cdot 10^{-7}$	4.46
9×64^2	$2.61 \cdot 10^{-9}$	4.82	$9.95 \cdot 10^{-8}$	4.65	$2.68 \cdot 10^{-8}$	4.66
9×128^2	$6.17 \cdot 10^{-10}$	2.08	$4.89 \cdot 10^{-9}$	4.35	$1.16 \cdot 10^{-9}$	4.54
9×256^2	$8.48 \cdot 10^{-10}$	−0.46	$7.45 \cdot 10^{-10}$	2.72	$3.64 \cdot 10^{-10}$	1.66

Table 1.6 Relative L_2 error measured at time $t = 2$ for linear transport on a deformed domain via a sine function with $k = 4$ in 3D

Quadrature	Gauss–Legendre		Gauss–Lobatto		Gauss–Lobatto	
Time step	Conservative		Conservative		Skew-symmetric	
Mesh	L_2 error	Rate	L_2 error	Rate	L_2 error	Rate
4^3	$1.57 \cdot 10^{-1}$	—	$2.88 \cdot 10^{-1}$	—	$2.55 \cdot 10^{-1}$	—
8^3	$6.66 \cdot 10^{-3}$	4.55	$2.32 \cdot 10^{-2}$	3.63	$1.47 \cdot 10^{-2}$	4.12
16^3	$2.33 \cdot 10^{-4}$	4.84	$1.21 \cdot 10^{-3}$	4.26	$5.59 \cdot 10^{-4}$	4.72
32^3	$7.16 \cdot 10^{-6}$	5.02	$6.85 \cdot 10^{-5}$	4.14	$2.93 \cdot 10^{-5}$	4.25
64^3	$2.01 \cdot 10^{-7}$	5.15	$3.31 \cdot 10^{-6}$	4.37	$1.26 \cdot 10^{-6}$	4.54
128^3	$5.70 \cdot 10^{-9}$	5.14	$1.30 \cdot 10^{-7}$	4.67	$4.74 \cdot 10^{-8}$	4.73
256^3	$1.68 \cdot 10^{-10}$	5.08	$4.39 \cdot 10^{-9}$	4.89	$1.68 \cdot 10^{-9}$	4.82

for the consistent integration. Overall, under-integration clearly affects the solution quality also with an energy-conserving variant, even though the loss in convergence order is not as pronounced as for the spectral element results from Durufle et al. (2009). A loss in accuracy of Gauss–Lobatto versus consistent integration has been also been observed in the practical application of a turbulent flow in Klose et al. (2020).

Table 1.6 reports the results of the sine mesh deformation in three dimensions, confirming the results from the 2D case. The consistent Gauss–Legendre integration on $k + 1$ points delivers optimal convergence rates of order five, whereas Gauss–Lobatto quadrature achieves a rate closer to four. As in 2D, the skew-symmetric form is able to restore some of the loss of accuracy, but it is nonetheless a factor of five less accurate than the consistent integration when considering the same mesh. For Gauss–Legendre integration, the skew-symmetric form does not significantly alter accuracy, showing that the integration errors due to the deformed geometry are already very small in this example. Given these results, the observed larger permissible time step of Gauss–Lobatto integration observed in the lower right panel of Fig. 1.11 must be set in relation: The consistent integration enables the same accuracy on a coarser mesh, which also comes with a larger admissible time step size.

Refining the Mesh or Increasing the Polynomial Degree?

In practical applications of the DG method, especially for the simulation of fluid flow at high Reynolds numbers, the topic of convergence rate is more delicate than the academic convergence tests on the smooth solution above suggest. This is because the underlying solution fields are typically not smooth enough when compared to the chosen resolution. The methods are often applied with the numerical approximation barely good enough, meaning that some suitable metrics of the solution are represented to a few percent of accuracy. In the context of simulation of fluid flow, this scenario is referred to as *under-resolved* computations in the pre-asymptotic regime. The primary factors to judge the quality of a scheme is its stability to avoid a blow-up of the numerical approximation, and the accuracy in terms of dispersion and dissipation. A high-quality scheme is able to transport features with moderate to high-frequency over long-time intervals without significantly altering the shape or losing amplitude. Discontinuous Galerkin method is very competitive in this regard, as shown by results on dispersion accuracy by Hu et al. (1999), Ainsworth (2004), Ainsworth and Wajid (2009), Gassner and Kopriva (2011), Moura et al. (2015), as well as the impact on turbulence given in Wang (2007), Gassner and Beck (2012), Fehn et al. (2018).

In order to assess the capabilities of higher order methods, consider a slight modification of the setup (1.34) in two dimensions with the solution

$$\begin{aligned} u(\boldsymbol{x}, t) &= \sin(4\pi(x_1 - a_1 t))\cos(4\pi(x_e - a_1 t)) \\ &+ \frac{1}{10}\prod_{i=1}^{2}\sin(16\pi(x_i - a_i t)) + \frac{1}{80}\prod_{i=1}^{2}\sin(64\pi(x_i - a_i t)). \end{aligned}$$

The two oscillatory terms have a smaller magnitude than the low-frequency content, mimicking the energy decay at smaller scales in turbulent flows. Figure 1.22 shows the observed convergence for polynomial degrees between $k = 1$ and $k = 9$. The theoretical convergence rate is only observed once all frequencies are resolved. For intermediate resolutions, dispersion and dissipation effects on the high frequencies play a substantial role. For the experiment, it appears as if higher order methods are only marginally better in the regime with relative errors between 10^{-2} and 10^{-3}, when compared in the metric of the number of unknowns. This reasoning, considered, e.g., in Brown (2010) and evaluated for incompressible turbulent flow in Fehn et al. (2018), shows that the advantage of higher order methods over linear or quadratic schemes with the same number of unknowns is limited to around an order of magnitude, see also Moura et al. (2015). As a consequence, the implementation of higher order

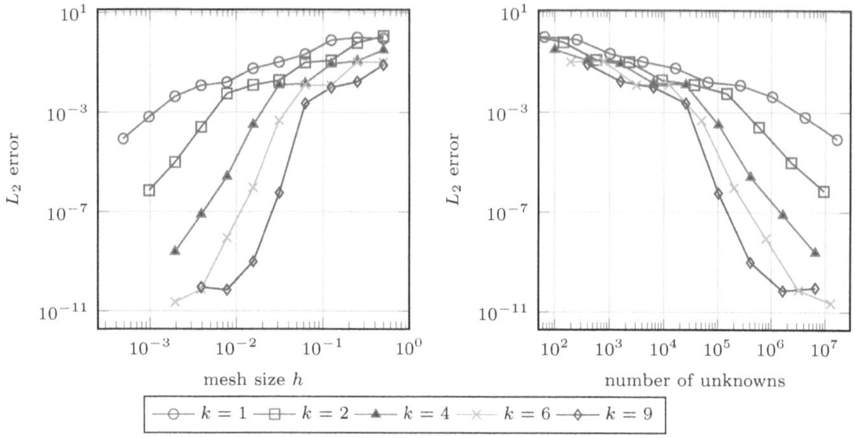

Fig. 1.22 Convergence with mesh refinement for a solution with oscillatory components in two dimensions as a function of mesh size (left) and as a function of the number of unknowns (right)

methods needs to deliver a similar throughput in unknowns computed per second as the best low-order methods, a topic discussed in more details in later chapters of this book. Similar behavior is observed for non-smooth solutions where the regularity p in the consistency result (1.33) reduces the convergence rates; high-order schemes often still deliver several times more accurate results for the same number of unknowns.

DG Discretizations for Second Derivatives

For problems with second spatial derivatives of elliptic or parabolic type, continuous finite elements are optimal methods due to the best-approximation property (Strang and Fix 1988). Nonetheless, discontinuous Galerkin methods have emerged as viable alternatives in a range of applications. Problems of mixed first and second derivatives with a strong contribution from the hyperbolic part are ideal candidates for DG methods, such as the compressible or incompressible Navier–Stokes equations at moderate and high Reynolds numbers. Furthermore, elliptic equations with strongly varying diffusivity such the equations of subsurface flow also profit from conservation properties of DG schemes (Bastian 2014).

As opposed to continuous finite elements where the H^1 regularity of the ansatz space makes the representation of second derivative operators natural by the weak form with integration by parts, DG methods with discontinuities in the basis functions over element boundaries and their associated L_2 regularity need a different approach to create stable and efficient methods. This section considers Poisson's equation as the prototype elliptic problem,

$$
\begin{aligned}
-\nabla \cdot (\kappa \nabla u) &= b, && \text{in } \Omega, \\
u &= g_{\mathrm{D}}, && \text{on } \partial\Omega_{\mathrm{D}}, \\
\boldsymbol{n} \cdot \nabla u &= g_{\mathrm{N}}, && \text{on } \partial\Omega_{\mathrm{N}},
\end{aligned}
\tag{1.35}
$$

where $\kappa > 0$ is the diffusivity bounded away from zero. The boundary $\partial\Omega$ is assumed to be split into a Dirichlet part $\partial\Omega_{\mathrm{D}}$ and a Neumann part $\partial\Omega_{\mathrm{N}}$. Problems with mixed first and second spatial derivatives rely on one of the methods developed in this section for the second derivative, and on one of the methods discussed in section "The Main Concepts" for the first derivative.

The canonical approach to derive a DG discretization is to rewrite the equation into an equivalent first-order system of equations with an auxiliary variable $\boldsymbol{q} = [q_1, \ldots, q_d]$ in d space dimensions,

$$
\begin{aligned}
-\nabla \cdot \boldsymbol{q} &= b, \\
\boldsymbol{q} &= \kappa \nabla u.
\end{aligned}
$$

Note that different formulations are in use for the variable κ, such as placing $\sqrt{\kappa}$ in both the first and second equation to ensure symmetry in the discrete representation, see, e.g., Hesthaven and Warburton (2008), or to include κ only in the equation for u_h. For u and each component of $\boldsymbol{q}$, a polynomial approximation according to (1.10) is used. The equations are multiplied by test functions v_h and $\boldsymbol{w}_h$, integrated over the element Ω_e and integrated by parts. The resulting weak form seeks the fields $u_h \in \mathcal{U}_h$ and $\boldsymbol{q}_h \in \mathcal{U}_h^d$ such that it holds for all test functions $(v_h, \boldsymbol{w}_h) \in \mathcal{U}_h^{d+1}$ that

$$
\left(\nabla v_h, \boldsymbol{q}_h\right)_{\Omega_h} - \left\langle v_h \cdot \boldsymbol{n}, \boldsymbol{q}^*\right\rangle_{\partial\Omega_h} = \left(v_h, b\right)_{\Omega_h}, \tag{1.36}
$$

$$
\left(\boldsymbol{w}_h, \boldsymbol{q}_h\right)_{\Omega_h} = -\left(\nabla \cdot (\kappa \boldsymbol{w}_h), u_h\right)_{\Omega_h} + \left\langle \boldsymbol{w}_h \cdot \boldsymbol{n}, \kappa u^*\right\rangle_{\partial\Omega_h}. \tag{1.37}
$$

In these equations, numerical fluxes need to be provided for u and $\boldsymbol{q}$. Given the coupled nature of the problem, both fluxes u^* and $\boldsymbol{q}^*$ can depend on the four quantities $u_h^-, u_h^+, \boldsymbol{q}_h^-, \boldsymbol{q}_h^+$. The three most common fluxes are

- the central flux, often referred to as Bassi–Rebay flux (Bassi and Rebay 1997b),

$$
\boldsymbol{q}^* = \{\{\boldsymbol{q}_h\}\} - \tau \left[\!\left[u_h\right]\!\right], \qquad u_h^* = \{\{u_h\}\}, \tag{1.38}
$$

  which simply takes the average of the two states to derive a symmetric approximation,
- the local discontinuous Galerkin (LDG) method (Cockburn and Shu 1998a), which adds upwinding in opposite directions for the two components along the interface,

$$
\boldsymbol{q}^* = \{\{\boldsymbol{q}_h\}\} + \hat{\boldsymbol{\beta}} \cdot \left[\!\left[\boldsymbol{q}_h\right]\!\right] - \tau \left[\!\left[u_h\right]\!\right], \qquad u_h^* = \{\{u_h\}\} - \hat{\boldsymbol{\beta}} \cdot \left[\!\left[\boldsymbol{q}_h\right]\!\right], \tag{1.39}
$$

where $\hat{\beta} = \pm \boldsymbol{n}/2$ is chosen as either the normal or negative normal for a fixed orientation of interior faces; the judicious choice of the direction makes the final scheme symmetric in the interior of the computational domain,

- and the symmetric interior penalty method (Arnold 1982), which is based on ideas proposed earlier by Nitsche (1971),

$$\boldsymbol{q}^* = \{\!\{\nabla u_h\}\!\} - \tau \left[\!\left[u_h \right]\!\right], \qquad u_h^* = \{\!\{u_h\}\!\}. \tag{1.40}$$

In all three fluxes, the term $\tau \left[\!\left[u_h \right]\!\right]$ is a stabilization to prevent spurious zero eigenmodes in the discrete operator with some $\tau > 0$. The first two methods have originally been used for parabolic problems without stabilization, where they are stable but with possibly non-optimal convergence orders. For more details about the derivation of various methods and their similarities, see Arnold et al. (2002).

The "external" values $(u_h^+, \boldsymbol{q}_h^+)$ for imposing boundary conditions are found using the mirror principle,

- on Dirichlet boundaries

$$u_h^+ = 2g_{\mathrm{D}} - u_h^-, \; \boldsymbol{q}_h^+ = \boldsymbol{q}_h^-,$$

- and on Neumann boundaries

$$u_h^+ = u_h^-, \; \boldsymbol{q}_h^+ = 2g_{\mathrm{N}}\boldsymbol{n} - \boldsymbol{q}_h^-.$$

Boundary data is imposed only on one of the two variables at a time for the Dirichlet and Neumann case, respectively. The other variable is defined by the interior value only.

The choice of the fluxes (1.38)–(1.40) is explained by the computational properties. Exemplarily, the weak form with the central flux (1.38) gives the following contributions to a linear system:

$$\underbrace{\left\langle v_h \cdot \boldsymbol{n}, \tau \left[\!\left[u_h \right]\!\right] \right\rangle_{\partial\Omega_h}}_{\rightsquigarrow \mathcal{E}} + \underbrace{\left(\nabla v_h, \boldsymbol{q}_h\right)_{\Omega_h} - \left\langle v_h \cdot \boldsymbol{n}, \{\!\{\boldsymbol{q}_h\}\!\}\right\rangle_{\partial\Omega_h}}_{\rightsquigarrow \mathcal{B}} = \underbrace{(v_h, b)_{\Omega_h}}_{\rightsquigarrow \boldsymbol{B}_h},$$

$$\underbrace{\left(\nabla \cdot (\kappa \boldsymbol{w}_h), u_h\right)_{\Omega_h} - \left\langle \boldsymbol{w}_h \cdot \boldsymbol{n}, \kappa \{\!\{u_h\}\!\}\right\rangle_{\partial\Omega_h}}_{\rightsquigarrow \mathcal{C}} + \underbrace{\left(\boldsymbol{w}_h, \boldsymbol{q}_h\right)_{\Omega_h}}_{\rightsquigarrow \mathcal{M}} = 0.$$

Collecting the matrices and vectors, a block linear system

$$\begin{bmatrix} \mathcal{E} & \mathcal{B} \\ \mathcal{C} & \mathcal{M} \end{bmatrix} \begin{bmatrix} \boldsymbol{U}_h \\ \boldsymbol{Q}_h \end{bmatrix} = \begin{bmatrix} \boldsymbol{B}_h \\ \boldsymbol{0} \end{bmatrix} \tag{1.41}$$

is obtained. Due to the absence of a dependence of the flux u^* on $\boldsymbol{q}_h$, the mass matrix in the lower right matrix block does not couple between neighboring elements, which allows to substitute the variable $\boldsymbol{Q}_h$ from the final linear system by

$$\boldsymbol{Q}_h = -\mathcal{M}^{-1}\mathcal{C}\boldsymbol{U}_h,$$

such that the linear system only involves the vector of primal coefficients,

$$\left(\mathcal{E} - \mathcal{B}\mathcal{M}^{-1}\mathcal{C}\right)\boldsymbol{U}_h = \boldsymbol{B}_h.$$

This system is considerably smaller and, for constant κ, symmetric positive definite because $\mathcal{C} = -\mathcal{B}^{-1}$, as can be seen by integration by parts. As a result, common direct or iterative solvers can be applied for the solution of this system. The condition number of the matrix depends on the penalty parameter τ, with $\tau \sim 1$ representing a good balance between the various terms, whereas the conditioning deteriorates for very small or very large τ. This affects the achievable accuracy in floating point arithmetic and the performance of iterative solvers. For the LDG flux, not only the conditioning but also the accuracy suffers as τ approaches zero (Hesthaven and Warburton 2008).

The special construction of the numerical flux for the interior penalty method enables a substitution of $\boldsymbol{q}_h$ already on the level of the differential equation. By integrating Equation (1.37) by parts again to retrieve the strong form, we obtain

$$\left(\boldsymbol{w}_h, \boldsymbol{q}_h\right)_{\Omega_e} = (\boldsymbol{w}_h, \kappa\nabla u_h)_{\Omega_e} + \left\langle \boldsymbol{w}_h \cdot \boldsymbol{n}, \kappa(u^* - u_h)\right\rangle_{\partial\Omega_e}. \tag{1.42}$$

The first term on the left-hand side of (1.36) can be replaced by the integrals on the right-hand side of (1.42), because all functions ∇v_h appearing as test functions in (1.36) can be represented by a linear combination of test functions $\boldsymbol{w}_h$ due to the construction of the polynomial spaces. Inserting the expression of the numerical flux yields

$$\begin{aligned}(\nabla v_h, \kappa\nabla u_h)_{\Omega_e} &+ \langle \nabla v_h \boldsymbol{n}, \kappa(\{\{u_h\}\} - u_h)\rangle_{\partial\Omega_e} \\ &- \left\langle v_h \boldsymbol{n}, \kappa(\{\{\nabla u_h\}\} - \tau\, [\![u_h]\!])\right\rangle_{\partial\Omega_e} = (v_h, b)_{\Omega_e}.\end{aligned}$$

Substituting the expression $\boldsymbol{n}(\{\{u_h\}\} - u_h) = -\frac{1}{2}\,[\![u_h]\!]$ for the second term on the left-hand side gives the final form

$$\begin{aligned}(\nabla v_h, \kappa\nabla u_h)_{\Omega_e} - \langle v_h \boldsymbol{n}, \kappa\, \{\{\nabla u_h\}\}\rangle_{\partial\Omega_e} - \left\langle \nabla v_h, \frac{\kappa}{2}\, [\![u_h]\!]\right\rangle_{\partial\Omega_e} \\ + \left\langle v_h \boldsymbol{n}, \kappa\tau\, [\![u_h]\!]\right\rangle_{\partial\Omega_e} = (v_h, b)_{\Omega_e}.\end{aligned} \tag{1.43}$$

The form (1.43) is an equation only in the primal variable u_h and what is typically implemented for the interior penalty method. The auxiliary variable $\boldsymbol{q}_h$ has been eliminated completely and needs not be set up at all. The three face integrals in the equation are typically referred to as the primal consistency term, the adjoint consistency term, and the penalty term (Arnold et al. 2002). The price to pay is the larger value of the penalty parameter, which depends on the mesh size h and polynomial degree k as

$$\tau \gtrsim \frac{(k+1)^2}{h}. \tag{1.44}$$

This restriction on the penalty parameter is needed to make the weak form coercive and relies on inverse estimates that bound the derivative with scaling $\frac{1}{h}$ by the value or, more precisely, the ratio between the surface area and the volume (Warburton and Hesthaven 2003; Shahbazi 2005; Epshteyn and Rivière 2007). The size of τ also affects the conditioning of the linear system and performance of iterative solvers. A common alternative is the non-symmetric interior penalty method which switches the sign of the adjoint consistency term (Rivière et al. 1999), necessitating only $\tau > 0$ but with the loss of optimal convergence orders.

Convergence Test with Manufactured Solution

In order to assess the behavior of an elliptic solver, we consider the 3D Poisson equation with analytical solution

$$u(\boldsymbol{x}) = \left(\frac{3}{2\pi}\right)^3 \mathrm{e}^{-|\boldsymbol{x}-\boldsymbol{x}_\mathrm{c}|^2/3^2}$$

with $\boldsymbol{x}_\mathrm{c} = (-0.2, 0.1, 0.3)^T$ on the domain $\Omega = [-1, 1]^3$. Dirichlet boundary conditions are set according to the analytical solution and the right-hand side $b(\boldsymbol{x})$ in Poisson's equation (1.35) is set to the negative Laplacian of the analytical solution in the spirit of the method of manufactured solutions. The numerical solution is computed with the symmetric interior penalty method (1.43) on a mesh of uniform hexahedral elements of mesh size h with polynomial degree k and the penalty parameter $\tau = (k+1)^2/h$.

Figure 1.23 displays the L_2 error for polynomial degrees k between 1 and 7 on meshes constructed by alternately refining a 2^3 and 3^3 base mesh. All degrees show optimal $\mathcal{O}(h^{k+1})$ order convergence. Higher order methods appear more efficient in terms of the number of unknowns employed to reach a certain level of accuracy. Experiments considering the accuracy against the computing time on a modern implementation and comparisons to continuous finite elements are found in Kronbichler and Wall (2018).

Concluding Remarks

In the last decades, discontinuous Galerkin methods have been successfully applied to a wide range of problems. High-order discontinuous Galerkin inherit ingredients from high-order finite and spectral element methods in terms of the polynomial approximation and representation of curved computational domains by mappings, as well as the concept of Riemann solvers from finite volume methods. This flexibility allows to tailor the schemes to the problem at hand. Particular practical interest in DG schemes concerns their dispersion and dissipation behavior, given that applications often drive numerical approximations at the coarsest possible resolution. By a solution composed of low- and high-frequency content, high-order methods have been demonstrated to be more accurate per unknown than low-order schemes, but the advantage is at most one to two orders of magnitudes, and much less dramatic than in the regime of asymptotic convergence. A lot of research activities have been on optimizing the computational properties of DG methods with its intensive computations on locally cacheable data and compact nearest-neighbor communication. Considering both accuracy, implementation, and practical aspects, the following guidelines can be given, compare also with Wang et al. (2013):

- For the same number of unknowns, discontinuous Galerkin schemes of moderate orders $k = 3, 4, 5$ are much more accurate in terms of their dispersion and dissipation behavior than linear methods with $k = 1$ or low-order finite volumes, representing DG with $k = 0$.
- High-order schemes add additional structure by more unknowns within (fewer) elements. Depending on the target geometry and the number of elements needed for meshing, very high degrees might be excluded due to their higher numbers of unknowns. Using high-order elements also necessitates the use of high orders for the geometry representation to not destroy the possible convergence orders.
- Higher order schemes have a denser coupling between the unknowns. To achieve a similar throughput as the best low-order methods, specific algorithms like sum factorization described in Kronbichler and Kormann (2012) as well as Kronbichler and Kormann (2019) with a linear increase in arithmetic work per unknown are

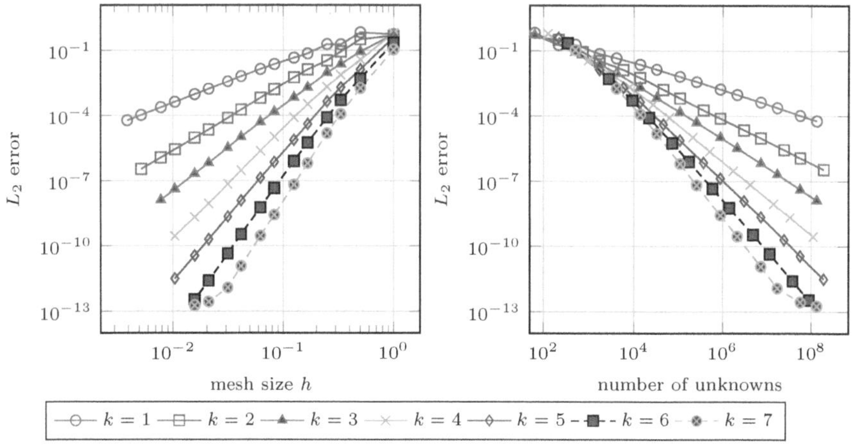

Fig. 1.23 Convergence of the symmetric interior penalty discretization of Poisson's equation with mesh refinement for degrees $k = 1$ to $k = 7$ in three dimensions as a function of mesh size (left) and as a function of the number of unknowns (right)

necessary. In the intermediate range of the degree for $2 \leq k \leq 10$, the polynomial degree is mostly a parameter to be weighted against the flexibility of meshing.

- Very high-order discontinuous Galerkin schemes with $k > 8$ are often less efficient. On the one hand, the CFL constraint of explicit time integration becomes over-proportionally restrictive, which increases the computational cost. Furthermore, apart from specialized implementations currently restricted to constant coefficients on Cartesian meshes (Huismann et al. 2019) or approaches using concepts like the fast Fourier transform for very high degrees, the work per unknown increases at least as the degree k, leading to over-proportional cost.
- The design order of the most popular explicit time integrators is between two and four because stability requirements force small time steps where the error in space dominates, especially for barely resolved simulations and with engineering accuracy.

Finally, despite more than three decades of intensive research, the application of high-order discontinuous Galerkin method to problems involving shocks has remained a challenge. Like any linear discretization method, DG schemes are susceptible to oscillatory behavior due to the Gibbs phenomenon when discontinuities pass through the elements. Given that the built-in dissipation cannot address these oscillations, the identification of the most appropriate limiter for a given problem set has remained a hot research topic, equally relevant as when the work by Cockburn and Shu (1991) was written or highlighted by the large body of references given in Hesthaven and Warburton (2008) or Shu (2018).

References

Ainsworth, M. (2004). Dispersive and dissipative behaviour of high order discontinuous Galerkin finite element methods. *Journal of Computational Physics*, *198*(1), 106–130. https://doi.org/10.1016/j.jcp.2004.01.004.

Ainsworth, M., & Wajid, H. A. (2009). Dispersive and dissipative behavior of the spectral element method. *SIAM Journal on Numerical Analysis*, *47*(5), 3910–3937. https://doi.org/10.1137/080724976.

Alzetta, G., Arndt, D., Bangerth, W., Boddu, V., Brands, B., Davydov, D., Gassmoeller, R., Heister, T., Heltai, L., Kormann, K., Kronbichler, M., Maier, M., Pelteret, J.-P., Turcksin, B., & Wells, D. (2018). The `deal.II` library, version 9.0. Journal of Numerical Mathematics, *26*(4), 173–184. https://doi.org/10.1515/jnma-2018-0054.

Arndt, D., Bangerth, W., Davydov, D., Heister, T., Heltai, L., Kronbichler, M., Maier, M., Pelteret, J.-P., Turcksin, B., & Wells, D. (2020). The deal.II finite element library: design, features, and insights. *Computers & Mathematics with Applications*. In press. https://doi.org/10.1016/j.camwa.2020.02.022.

Arnold, D. N. (1982). An interior penalty finite element method with discontinuous elements. *SIAM Journal on Numerical Analysis*, *19*(4), 742–760. https://doi.org/10.1137/0719052.

Arnold, D. N., Brezzi, F., Cockburn, B., & Marini, L. D. (2002). Unified analysis of discontinuous Galerkin methods for elliptic problems. *SIAM Journal on Numerical Analysis*, *39*(5), 1749–1779. https://doi.org/10.1137/s0036142901384162.

Bassi, F., & Rebay, S. (1997a). High-order accurate discontinuous finite element solution of the 2D Euler equations. *Journal of Computational Physics*, *138*(2), 251–285. https://doi.org/10.1006/jcph.1997.5454.

Bassi, F., & Rebay, S. (1997b). A high-order accurate discontinuous finite element method for the numerical solution of the compressible Navier-Stokes equations. *Journal of Computational Physics*, *131*(2), 267–279. https://doi.org/10.1006/jcph.1996.5572.

Bastian, P. (2014). A fully-coupled discontinuous Galerkin method for two-phase flow in porous media with discontinuous capillary pressure. *Computational Geosciences*, *18*, 779–796. https://doi.org/10.1007/s10596-014-9426-y.

Berrut, J.-P., & Trefethen, L. N. (2004). Barycentric Lagrange interpolation. *SIAM Review*, *46*(3), 501–517. https://doi.org/10.1137/s0036144502417715.

Brenner, S. C., & Scott, R. L. (2002). *The mathematical theory of finite elements* (2nd ed.). Berlin: Springer.

Brown, J. (2010). Efficient nonlinear solvers for nodal high-order finite elements in 3D. *Journal of Scientific Computing*, *45*(1–3), 48–63.

Chan, J. (2018). On discretely entropy conservative and entropy stable discontinuous Galerkin methods. *Journal of Computational Physics*, *362*, 346–374. https://doi.org/10.1016/j.jcp.2018.02.033.

Chang, J., Fabien, M. S., Knepley, M. G., & Mills, R. T. (2018). Comparative study of finite element methods using the time-accuracy-size (TAS) spectrum analysis. *SIAM Journal on Scientific Computing*, *40*(6), C779–C802. https://doi.org/10.1137/18m1172260.

Cockburn, B., & Shu, C.-W. (1991). The Runge-Kutta local projection p^1-discontinuous-Galerkin finite element method for scalar conservation laws. *Mathematical Modelling and Numerical Analysis*, *25*(3), 337–361.

Cockburn, B., & Shu, C.-W. (1998a). The local discontinuous Galerkin method for time-dependent convection-diffusion systems. *SIAM Journal on Numerical Analysis*, *35*(6), 2440–2463. https://doi.org/10.1137/s0036142997316712.

Cockburn, B., & Shu, C.-W. (1998b). The Runge-Kutta discontinuous Galerkin method for conservation laws V: Multidimensional systems. *Journal of Computational Physics*, *141*(2), 199–224. https://doi.org/10.1006/jcph.1998.5892.

Cockburn, B., Karniadakis, G. E., & Shu., C.-W. (2000). The development of discontinuous Galerkin methods. Lecture notes in computational science and engineering (pp. 3–50). Berlin: Springer. https://doi.org/10.1007/978-3-642-59721-3_1.

Coons, S. A. (1967). *Surfaces for computer-aided design of space forms*. Technical report MAC-TR-41, MIT.

Deville, M. O., Fischer, P. F., & Mund, E. H. (2002). *High-order methods for incompressible fluid flow* (Vol. 9). Cambridge: Cambridge University Press.

Dumbser, M., & Käser, M. (2006). An arbitrary high-order discontinuous Galerkin method for elastic waves on unstructured meshes - II. The three-dimensional isotropic case. *Geophysical Journal International*, *167*(1), 319–336. https://doi.org/10.1111/j.1365-246x.2006.03120.x.

Dumbser, M., Käser, M., & Toro, E. F. (2007). An arbitrary high-order discontinuous Galerkin method for elastic waves on unstructured meshes - V. Local time stepping and p-adaptivity. *Geophysical Journal International*, *171*(2), 695–717. https://doi.org/10.1111/j.1365-246x.2007.03427.x.

Dumbser, M., Fambri, F., Tavelli, M., Bader, M., & Weinzierl, T. (2018). Efficient implementation of ADER discontinuous Galerkin schemes for a scalable hyperbolic PDE engine. *Axioms*, *7*(3), 63. https://doi.org/10.3390/axioms7030063.

Durufle, M., Grob, P., & Joly, P. (2009). Influence of Gauss and Gauss-Lobatto quadrature rules on the accuracy of a quadrilateral finite element method in the time domain. *Numerical Methods for Partial Differential Equations*, *25*(3), 526–551. https://doi.org/10.1002/num.20353.

Epshteyn, Y., & Rivière, B. (2007). Estimation of penalty parameters for symmetric interior penalty Galerkin methods. *Journal of Computational and Applied Mathematics*, *206*, 843–872. https://doi.org/10.1016/j.cam.2006.08.029.

Fehn, N., Wall, W. A., & Kronbichler, M. (2018). Efficiency of high-performance discontinuous Galerkin spectral element methods for under-resolved turbulent incompressible flows. *International Journal for Numerical Methods in Fluids*, *88*(1), 32–54. https://doi.org/10.1002/fld.4511.

Fehn, N., Kronbichler, M., Lehrenfeld, C., Lube, G., & Schroeder, P. W. (2019a). High-order DG solvers for under-resolved turbulent incompressible flows: A comparison of L^2 and $H(\mathrm{div})$ methods. *International Journal for Numerical Methods in Fluids*, *91*(11), 533–556. https://doi.org/10.1002/fld.4763.

Fehn, N., Wall, W. A., & Kronbichler, M. (2019b). A matrix-free high-order discontinuous Galerkin compressible Navier-Stokes solver: A performance comparison of compressible and incompressible formulations for turbulent incompressible flows. *International Journal for Numerical Methods in Fluids*, *89*(3), 71–102. https://doi.org/10.1002/fld.4683.

Fischer, P., Min, M., Rathnayake, T., Dutta, S., Kolev, T., Dobrev, V., Camier, J.-S., Kronbichler, M., Warburton, T., Świrydowicz, K., & Brown, J. (2020). Scalability of high-performance PDE solvers. *International Journal of High Performance Computing Applications*, *34*(5), 562–586. https://doi.org/10.1177/1094342020915762.

Gassner, G., & Kopriva, D. A. (2011). A comparison of the dispersion and dissipation errors of Gauss and Gauss-Lobatto discontinuous Galerkin spectral element methods. *SIAM Journal on Scientific Computing*, *33*(5), 2560–2579. https://doi.org/10.1137/100807211.

Gassner, G. J. (2013). A skew-symmetric discontinuous Galerkin spectral element discretization and its relation to SBP-SAT finite difference methods. *SIAM Journal on Scientific Computing*, *35*(3), A1233–A1253. https://doi.org/10.1137/120890144.

Gassner, G. J. (2014). A kinetic energy preserving nodal discontinuous Galerkin spectral element method. *International Journal for Numerical Methods in Fluids*, *76*(1), 28–50. https://doi.org/10.1002/fld.3923.

Gassner, G. J., & Beck, A. D. (2012). On the accuracy of high-order discretizations for underresolved turbulence simulations. *Theoretical and Computational Fluid Dynamics*, *27*(3–4), 221–237. https://doi.org/10.1007/s00162-011-0253-7.

Gassner, G. J., Winters, A. R., & Kopriva, D. A. (2016). Split form nodal discontinuous Galerkin schemes with summation-by-parts property for the compressible Euler equations. *Journal of Computational Physics*, *327*, 39–66. https://doi.org/10.1016/j.jcp.2016.09.013.

Gordon, W. J., & Thiel, L. C. (1982). Transfinite mappings and their application to grid generation. *Applied Mathematics and Computation*, *10*, 171–233. https://doi.org/10.1016/0096-3003(82)90191-6.

Gustafsson, B. (2008). *High order difference methods for time dependent PDE*. Berlin: Springer. https://doi.org/10.1007/978-3-540-74993-6.

Gustafsson, B., Kreiss, H.-O., & Oliger, J. (2013). *Time dependent problems and difference methods* (2nd ed.). New York: Wiley.

Hairer, E., & Wanner, G. (1991). *Solving ordinary differential equations II. Stiff and differential-algebraic problems*. Berlin: Springer.

Hairer, E., Nørsett, S. P., & Wanner, G. (1993). *Solving ordinary differential equations I. Nonstiff problems* (2nd ed.). Berlin: Springer.

Hairer, E., Lubich, C., & Wanner, G. (2006). *Geometric numerical integration: Structure-preserving algorithms for ordinary differential equations* (2nd ed.). Berlin: Springer.

Heltai, L., Bangerth, W., Kronbichler, M., & Mola, A. (2019). *Using exact geometry information in finite element computations*. Technical report. arXiv:1910.09824.

Hesthaven, J. S., & Warburton, T. (2008). *Nodal discontinuous Galerkin methods: algorithms, analysis, and applications*. Berlin: Springer. https://doi.org/10.1007/978-0-387-72067-8.

Hesthaven, J. S., Gottlieb, S., & Gottlieb, D. (2006). *Spectral methods for time-dependent problems*. Cambridge: Cambridge University Press.

Hu, F. Q., Hussaini, M. Y., & Rasetarinera, P. (1999). An analysis of the discontinuous Galerkin method for wave propagation problems. *Journal of Computational Physics*, *151*(2), 921–946. https://doi.org/10.1006/jcph.1999.6227.

Hughes, T. J. R., Cottrell, J. A., & Bazilevs, Y. (2005). Isogeometric analysis: CAD, finite elements, NURBS, exact geometry and mesh refinement. *Computer Methods in Applied Mechanics and Engineering*, *194*(39–41), 4135–4195. https://doi.org/10.1016/j.cma.2004.10.008.

Huismann, I., Stiller, J., & Fröhlich, J. (2019). Scaling to the stars - a linearly scaling elliptic solver for p-multigrid. *Journal of Computational Physics*, *398*, 108868. https://doi.org/10.1016/j.jcp.2019.108868.

Huynh, H. T., Wang, Z. J., & Vincent, P. E. (2014). High-order methods for computational fluid dynamics: A brief review of compact differential formulations on unstructured grids. *Computers & Fluids*, *98*, 209–220. https://doi.org/10.1016/j.compfluid.2013.12.007.

Johnson, C., & Pitkäranta, J. (1986). An analysis of the discontinuous Galerkin method for a scalar hyperbolic equation. *Mathematics of Computation*, *46*(173), 1. https://doi.org/10.1090/s0025-5718-1986-0815828-4.

Karniadakis, G., & Sherwin, S. J. (2005). *Spectral/hp element methods for computational fluid dynamics* (2nd ed.). Oxford: Oxford University Press. https://doi.org/10.1093/acprof:oso/9780198528692.001.0001.

Kennedy, C. A., Carpenter, M. H., & Lewis, R. M. (2000). Low-storage, explicit Runge-Kutta schemes for the compressible Navier-Stokes equations. *Applied Numerical Mathematics*, *35*(3), 177–219. https://doi.org/10.1016/s0168-9274(99)00141-5.

Ketcheson, D. I., LeVeque, R. J., & del Razo, M. J. (2020). *Riemann problems and Jupyter solutions*. Philadelphia: Society for Industrial and Applied Mathematics. *10*(1137/1), 9781611976212.

Klose, B. F., Jacobs, G. B., & Kopriva, D. A. (2020). Assessing standard and kinetic energy conserving volume fluxes in discontinuous Galerkin formulations for marginally resolved Navier-Stokes flows. *Computers & Fluids*, *205*, 104557. https://doi.org/10.1016/j.compfluid.2020.104557.

Kopriva, D. A. (2006). Metric identities and the discontinuous spectral element method on curvilinear meshes. *Journal of Scientific Computing*, *26*(3), 301–327. https://doi.org/10.1007/s10915-005-9070-8.

Kopriva, D. A. (2009). *Implementing spectral methods for partial differential equations*. Berlin: Springer.

Kopriva, D. A., & Gassner, G. J. (2016). Geometry effects in nodal discontinuous Galerkin methods on curved elements that are provably stable. *Applied Mathematics and Computation*, *272*, 274–290. https://doi.org/10.1016/j.amc.2015.08.047.

Kormann, K. (2016). A time-space adaptive method for the Schrödinger equation. *Communications in Computational Physics*, *20*(1), 60–85.

Kronbichler, M., & Kormann, K. (2012). A generic interface for parallel cell-based finite element operator application. *Computers & Fluids*, *63*, 135–147.

Kronbichler, M., & Kormann, K. (2019). Fast matrix-free evaluation of discontinuous Galerkin finite element operators. *ACM Transactions on Mathematical Software*, *45*(3), 29:1–29:40. https://doi.org/10.1145/3325864.

Kronbichler, M., & Wall, W. A. (2018). A performance comparison of continuous and discontinuous Galerkin methods with fast multigrid solvers. *SIAM Journal on Scientific Computing*, *40*(5), A3423–A3448. https://doi.org/10.1137/16M110455X.

Kubatko, E. J., Yeager, B. A., & Ketcheson, D. I. (2014). Optimal strong-stability-preserving Runge-Kutta time discretizations for discontinuous Galerkin methods. *Journal of Scientific Computing*, *60*(2), 313–344. https://doi.org/10.1007/s10915-013-9796-7.

Lesaint, P., & Raviart, P. A. (1974). On a finite element method for solving the neutron transport equation. In *Mathematical aspects of finite elements in partial differential equations* (pp. 89–123). Amsterdam: Elsevier. https://doi.org/10.1016/b978-0-12-208350-1.50008-x.

LeVeque, R. J. (2002). *Finite volume methods for hyberbolic problems*. Cambridge Texts in Applied Mathematics. Cambridge.

Löhner, R. (2011). Error and work estimates for high-order elements. *International Journal for Numerical Methods in Fluids*, *67*(12), 2184–2188. https://doi.org/10.1002/fld.2488.

Löhner, R. (2013). Improved error and work estimates for high-order elements. *International Journal for Numerical Methods in Fluids*, *72*(11), 1207–1218. https://doi.org/10.1002/fld.3783.

Mengaldo, G., De Grazia, D., Moxey, D., Vincent, P. E., & Sherwin, S. J. (2015). Dealiasing techniques for high-order spectral element methods on regular and irregular grids. *Journal of Computational Physics, 299*, 56–81. https://doi.org/10.1016/j.jcp.2015.06.032.

Moura, R. C., Sherwin, S. J., & Peiró, J. (2015). Linear dispersion-diffusion analysis and its application to under-resolved turbulence simulations using discontinuous Galerkin spectral/*hp* methods. *Journal of Computational Physics, 298*, 695–710. https://doi.org/10.1016/j.jcp.2015.06.020.

Nitsche, J. (1971). Über ein Variationsprinzip zur Lösung von Dirichlet-Problemen bei Verwendung von Teilräumen, die keinen Randbedingungen unterworfen sind. *Abhandlungen aus dem Mathematischen Seminar der Universität Hamburg, 36*(1), 9–15.

Noventa, G., Massa, F., Bassi, F., Colombo, A., Franchina, N., & Ghidoni, A. (2016). A high-order discontinuous Galerkin solver for unsteady incompressible turbulent flows. *Computers & Fluids, 139*, 248–260. https://doi.org/10.1016/j.compfluid.2016.03.007.

Persson, P.-O., & Peraire, J. (2006). Sub-cell shock capturing for discontinuous Galerkin methods. In *44th AIAA Aerospace Sciences Meeting and Exhibit*. American Institute of Aeronautics and Astronautics. https://doi.org/10.2514/6.2006-112.

Peterson, T. E. (1991). A note on the convergence of the discontinuous Galerkin method for a scalar hyperbolic equation. *SIAM Journal on Numerical Analysis, 28*(1), 133–140. https://doi.org/10.1137/0728006.

Reed, W.H., & Hill, T. R. (1973). Triangluar mesh methods for the neutron transport equation. Technical report, Los Alamos Scientific Lab, New Mexico (USA). http://www.osti.gov/scitech/servlets/purl/4491151.

Reinarz, A., Charrier, D. E., Bader, M., Bovard, L., Dumbser, M., Duru, K., Fambri, F., Gabriel, A.-A., Gallard, J.-M., Köppel, S., Krenz, L., Rannabauer, L., Rezzolla, L., Samfass, P., Tavelli, M., & Weinzierl, T. (2020). ExaHyPE: An engine for parallel dynamically adaptive simulations of wave problems. *Computer Physics Communications, 254*, 107251. https://doi.org/10.1016/j.cpc.2020.107251.

Rivière, B., Wheeler, M. F., & Girault, V. (1999). Improved energy estimates for interior penalty, constrained and discontinuous Galerkin methods for elliptic problems. Part I. *Computational Geosciences, 3*(3/4), 337–360. https://doi.org/10.1023/a:1011591328604.

Schoeder, S., Kormann, K., Wall, W. A., & Kronbichler, M. (2018a). Efficient explicit time stepping of high order discontinuous Galerkin schemes for waves. *SIAM Journal on Scientific Computing, 40*(6), C803–C826. https://doi.org/10.1137/18M1185399.

Schoeder, S., Kronbichler, M., & Wall, W. A. (2018b). Arbitrary high-order explicit hybridizable discontinuous Galerkin methods for the acoustic wave equation. *Journal of Scientific Computing, 76*, 969–1006. https://doi.org/10.1007/s10915-018-0649-2.

Sevilla, R., Fernández-Méndez, S., & Huerta, A. (2008). NURBS-enhanced finite element method (NEFEM). *International Journal for Numerical Methods in Engineering, 76*(1), 56–83. https://doi.org/10.1002/nme.2311.

Shahbazi, K. (2005). An explicit expression for the penalty parameter of the interior penalty method. *Journal of Computational Physics, 205*, 401–407. https://doi.org/10.1016/j.jcp.2004.11.017.

Shu, C.-W. (1988). Total-variation-diminishing time discretizations. *SIAM Journal on Scientific and Statistical Computing, 9*(6), 1073–1084. https://doi.org/10.1137/0909073.

Shu, C.-W. (2018). Bound-preserving high-order schemes for hyperbolic equations: Survey and recent developments. In *Theory, numerics and applications of hyperbolic problems II* (pp. 591–603). Berlin: Springer International Publishing. https://doi.org/10.1007/978-3-319-91548-7_44.

Šolín, P., Segeth, K., & Doležel, I. (2004). *High-order finite element methods*. Boca Raton, FL, USA: Chaptman & Hall/CRC.

Strang, G., & Fix, G. F. (1988). *An analysis of the finite element method*. Wellesley, MA, USA: Wellesley-Cambridge Press.

Sudirham, J. J., van der Vegt, J. J. W., & van Damme, R. M. J. (2006). Space-time discontinuous Galerkin method for advection-diffusion problems on time-dependent domains. *Applied Numerical Mathematics, 56*(12), 1491–1518. https://doi.org/10.1016/j.apnum.2005.11.003.

Taylor, M. A., Wingate, B. A., & Vincent, R. E. (2000). An algorithm for computing Fekete points in the triangle. *SIAM Journal on Numerical Analysis*, *38*(5), 1707–1720. https://doi.org/10.1137/s0036142998337247.

Toulorge, T., & Desmet, W. (2012). Optimal Runge-Kutta schemes for discontinuous Galerkin space discretizations applied to wave propagation problems. *Journal of Computational Physics*, *231*(4), 2067–2091. https://doi.org/10.1016/j.jcp.2011.11.024.

Tselios, K., & Simos, T. E. (2007). Optimized Runge-Kutta methods with minimal dispersion and dissipation for problems arising from computational acoustics. *Physics Letters A*, *363*(1–2), 38–47. https://doi.org/10.1016/j.physleta.2006.10.072.

Wang, Z. J. (2007). High-order methods for the Euler and Navier-Stokes equations on unstructured grids. *Progress in Aerospace Sciences*, *43*(1–3), 1–41. https://doi.org/10.1016/j.paerosci.2007.05.001.

Wang, Z. J., Fidkowski, K., Abgrall, R., Bassi, F., Caraeni, D., Cary, A., Deconinck, H., Hartmann, R., Hillewaert, K., Huynh, H. T., Kroll, N., May, G., Persson, P.-O., van Leer, B., & Visbal, M. (2013). High-order CFD methods: current status and perspective. *International Journal for Numerical Methods in Fluids*, *72*(8), 811–845. https://doi.org/10.1002/fld.3767.

Warburton, T., & Hesthaven, J. S. (2003). On the constants in hp-finite element trace inverse inequalities. *Computer Methods in Applied Mechanics and Engineering*, *192*, 2765–2773. https://doi.org/10.1016/S0045-7825(03)00294.

Winters, A. R., Moura, R. C., Mengaldo, G., Gassner, G. J., Walch, S., Peiró, J., & Sherwin, S. J. (2018). A comparative study on polynomial dealiasing and split form discontinuous Galerkin schemes for under-resolved turbulence computations. *Journal of Computational Physics*, *372*, 1–21. https://doi.org/10.1016/j.jcp.2018.06.016.

Chapter 2
High-Performance Implementation of Discontinuous Galerkin Methods with Application in Fluid Flow

Martin Kronbichler

Abstract In this book chapter, the high-performance implementation of discontinuous Galerkin methods is reviewed, with the main focus on sum factorization algorithms. The main computational properties of the algorithms are compared to capabilities of modern computer hardware, highlighting the opportunities and limitations of discontinuous Galerkin discretizations. The chapter closes with a presentation of how to apply these algorithms to the compressible Euler equations, the acoustic wave equation, and the incompressible Navier–Stokes equations.

Introduction

Discontinuous Galerkin (DG) schemes are a very powerful class of discretization schemes for the mathematical models describing fluid flow. Hence, the computational fluid dynamics (CFD) community has been increasingly relying on DG methods especially in the medium and high accuracy regime for challenging simulations that exhibit a large range of relevant scales, such as turbulent flows or wave propagation, followed over long time intervals on complex geometries.

Besides the attractive mathematical behavior of discontinuous Galerkin schemes, they also match well with the properties of modern computer hardware. As a result, DG-based simulations are being performed routinely on some of the largest supercomputers. This book chapter presents a selection of DG algorithms, specialized for quadrilateral and hexahedral element shapes with tensor product shape functions. Besides describing a high performance implementation, the resulting algorithms

The author acknowledges joint algorithm and code development with Niklas Fehn, Katharina Kormann, Benjamin Krank, Karl Ljungkvist, Peter Munch, and Svenja Schoeder, as well as collaboration with the deal.II community.

M. Kronbichler (✉)
Institute for Computational Mechanics, Technical University of Munich, Boltzmannstr. 15, 85748 Garching bei München, Germany
e-mail: kronbichler@lnm.mw.tum.de

M. Kronbichler and P.-O. Persson (eds.), *Efficient High-Order Discretizations for Computational Fluid Dynamics*, CISM International Centre for Mechanical Sciences 602, https://doi.org/10.1007/978-3-030-60610-7_2

are exemplified in the context of a few applications of fluid flow, bringing together method development, implementation, and engineering demands.

The structure of this chapter is as follows. Section "Discontinuous Galerkin Algorithms" gives a background on selected high-order DG schemes as well as their main computational properties and data access. Section "Background on Computer Architecture" presents the main concepts of computer architecture that guide the development of efficient implementations and parallelization. Section "Fast Computation of Integrals with Sum Factorization" explains the realization of sum factorization schemes, which are a particularly efficient algorithm for higher order methods, and presents the main performance characteristics. In section "The Euler Equations", the application to the Euler equations and the acoustic wave equation is presented, before section "Solving Linear Systems" discusses how to approach implicit systems using sum factorization as a matrix-free evaluation scheme in iterative solvers. The efficient discretization of the incompressible Navier–Stokes equations, including a detailed presentation of time stepping with a dual-splitting scheme, is topic of section "The Incompressible Navier–Stokes Equations". The last four sections give an outlook to active research directions and possible trends for the next decade.

Discontinuous Galerkin Algorithms

Discontinuous Galerkin schemes rely on a subdivision of a d-dimensional computational domain into a mesh of possibly curvilinear elements,

$$\Omega \approx \Omega_h = \bigcup_{e=1}^{n} \Omega_e,$$

see the first chapter of this book for details. The present chapter assumes the elements to be intervals in 1D, quadrilaterals in 2D, or hexahedra in 3D, but most algorithms also apply to other shapes of the elements. On each element Ω_e, polynomial solutions of the form

$$u_h^{(e)}(\boldsymbol{x}, t) = \sum_{j=1}^{n_\mathrm{p}} \varphi_j\left(\boldsymbol{\Phi}_e^{-1}(\boldsymbol{x})\right) u_j^{(e)}(t) \tag{1}$$

are assumed, where φ_j are polynomial basis functions depending on the spatial variables $\boldsymbol{x}$ and $u_j^{(e)}$ are the unknown coefficient values to be determined by the Galerkin procedure. For time-dependent problems, these coefficients depend on time t and are handled by separate time integrators in a methods-of-line approach, whereas they are constants for stationary problems. The basis functions are expressed in terms of reference coordinates $\hat{\boldsymbol{x}}$, which map to the real coordinates $\boldsymbol{x}$ by a function $\boldsymbol{\Phi}_e$, and are constructed by the tensor product of one-dimensional polynomials

$$\varphi_j(\hat{x}_1, \ldots, \hat{x}_d) = \varphi_{j_1}^{\mathrm{1D}}(\hat{x}_1) \ldots \varphi_{j_d}^{\mathrm{1D}}(\hat{x}_d), \tag{2}$$

where the multi-indices $(j_i, \ldots, j_d)$ of the one-dimensional polynomials are related to index j in a bijective way. The polynomials represent a space $\mathcal{P}_k$ of degree k with $n_\mathrm{p} = |\mathcal{P}_k|$ basis functions. The one-dimensional polynomials are defined by Lagrange polynomials with nodes specified by the $(k+1)$-point Gauss–Lobatto quadrature formula.

Up to section "Fast Computation of Integrals with Sum Factorization", two sample operators are considered, the advection of the numerical solution u_h along a given solenoidal velocity field $\boldsymbol{a}(\boldsymbol{x}, t)$ with $\nabla \cdot \boldsymbol{a} = 0$ in skew-symmetric form and using an upwind-like flux ($\alpha \geq 0$, classical upwinding with $\alpha = 1$),

$$\begin{aligned}\left(v_h, \frac{\partial u_h}{\partial t}\right)_{\Omega_h} &= \frac{1}{2}\left(\nabla v_h, \boldsymbol{a} u_h\right)_{\Omega_h} - \frac{1}{2}\left(v_h, \boldsymbol{a}\nabla u_h\right)_{\Omega_h} \\ &- \left\langle v_h \boldsymbol{n}, \{\{\boldsymbol{a} u_h\}\} + \frac{\alpha}{2}|\boldsymbol{n}\cdot\boldsymbol{a}|\,[\![u_h]\!] - \frac{1}{2}\boldsymbol{a} u_h\right\rangle_{\partial\Omega_h} + (v_h, b)_{\Omega_h},\end{aligned} \tag{3}$$

and the stationary Poisson problem discretized with the symmetric interior penalty discontinuous Galerkin method,

$$\begin{aligned}(\nabla v_h, \nabla u_h)_{\Omega_h} - \langle v_h \boldsymbol{n}, \{\{\nabla u_h\}\}\rangle_{\partial\Omega_h} - \left\langle \nabla v_h, \frac{1}{2}\,[\![u_h]\!]\right\rangle_{\partial\Omega_h} \\ + \left\langle v_h \boldsymbol{n}, \tau\,[\![u_h]\!]\right\rangle_{\partial\Omega_h} = (v_h, b)_{\Omega_h}.\end{aligned} \tag{4}$$

In these equations, the solution u_h is found as the function in the space

$$\mathcal{U}_h = \left\{u_h \in L_2(\Omega_h) \,:\, u_h|_{\Omega_e} \in \mathcal{P}_k \circ \boldsymbol{\Phi}_e^{-1}\right\} \tag{5}$$

that satisfies either (3) or (4) for all test functions $v_h \in \mathcal{U}_h$. In both equations, the right-hand side b is a source term. The bilinear form $(\cdot,\cdot)_{\Omega_h}$ denotes the integral over the product of the two arguments on the computational domain Ω_h, realized by the sum of n element integrals computed in reference coordinates for the reference element $\hat{\Omega} = [-1, 1]^d$. The bilinear form $\langle\cdot,\cdot\rangle_{\partial\Omega_h}$ represents the sum of the integrals along all the interior and boundary surfaces of the computational mesh, called *faces*. Along the interior faces, the two solution values from the two sides are denoted by u_h^- and u_h^+, with $\{\{v_h\}\} = \frac{v_h^- + v_h^+}{2}$ denoting the average of the given argument and $[\![v_h]\!] = \boldsymbol{n}^-(v_h^- - v_h^+)$ the jump along the direction $\boldsymbol{n}^-$ from v_h^- to v_h^+. In the interior penalty discretization, the factor $\tau = \frac{(k+1)^2}{h}$ is a penalty factor to make the discrete approximation coercive. Along the boundary, exterior values u_h^+ and ∇u_h^+ are defined in terms of the boundary conditions, e.g., via the mirror principle such as $u_h^+ = 2g_\mathrm{D} - u_h^-$ for a given Dirichlet boundary data g_D. For the derivation and background of the equations and the ingredients, we refer to the standard DG literature or the first chapter of this book.

For a computer implementation of the discontinuous Galerkin method, the integrals in the weak forms (3) or (4) are computed by a summation, the numerical quadrature. To give an example, consider the term $\left(\nabla\varphi_i^{(e)}, \boldsymbol{a}u_h^{(e)}\right)_{\Omega_e}$ in (3) with given coefficients $u_j^{(e)}$, tested by the function $\varphi_i^{(e)}$ with $i = 1, \ldots, n_\mathrm{p}$, at a fixed time t (neglected in the arguments),

$$\int_{\Omega_e} \nabla\varphi_i^{(e)}(\boldsymbol{x}) \cdot \boldsymbol{a}(\boldsymbol{x}) u_h^{(e)}(\boldsymbol{x})\,\mathrm{d}\boldsymbol{x} \approx$$

$$\sum_{q=1}^{n_\mathrm{c}} \left(\boldsymbol{J}_e^{-T} \hat{\nabla}\varphi_i(\hat{\boldsymbol{x}}_q)\right) \cdot \boldsymbol{a}\left(\Phi_e(\hat{\boldsymbol{x}}_q)\right) \underbrace{\left(\sum_{j=1}^{n_\mathrm{p}} u_j^{(e)} \varphi_j(\hat{\boldsymbol{x}}_q)\right)}_{u_h^{(e)}(\hat{\boldsymbol{x}}_q)} \det\left(\boldsymbol{J}_e(\hat{\boldsymbol{x}}_q)\right) w_q, \tag{6}$$

where $\boldsymbol{J}_e^{-T}$ denotes the inverse and transpose of the Jacobian of the transformation from reference to real coordinates, $\boldsymbol{J}_e = \hat{\nabla}\boldsymbol{\Phi}_e$, or some other suitable representation of the contravariant metric term (Kopriva 2006), $\hat{\boldsymbol{x}}_q$ the position of the quadrature points in reference space $[-1, 1]^d$, and w_q the associated quadrature weight. The most common quadrature formula is the Gauss–Legendre formula, but the exact definition is immaterial to this text as long as the integrals are computed with enough accuracy. Formula (6) is used as an abstraction in this chapter—whenever a bilinear form $(\cdot, \cdot)_{\Omega_e}$ or an integral over Ω_e is stated, it is implied to be evaluated numerically in reference coordinates by a summation.

Figure 2.1 gives an illustration of the position of the nodes as well as element and face quadrature points for the case $k = 4$ with $n_{\mathrm{c,1D}} = 6$ points per direction, giving $n_\mathrm{c} = 36$ points per element in total.

Application Efficiency

Higher order discontinuous Galerkin methods with polynomial degree $k \geq 2$ are attractive in a variety of applications because they possess a good *sequential* efficiency, enabling more accurate simulations in terms of dispersion and dissipation as

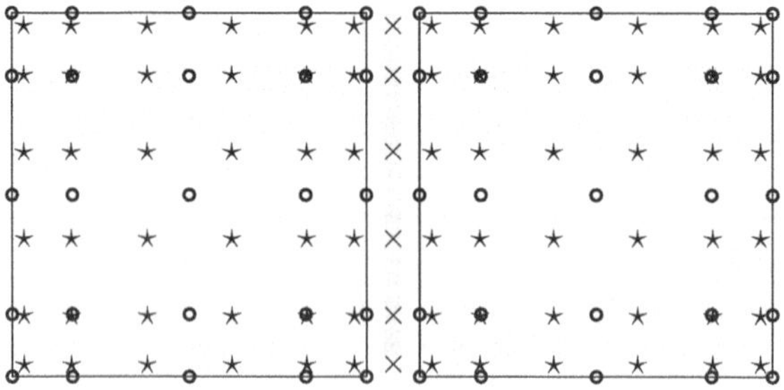

Fig. 2.1 Illustration of node values (circles), element quadrature points (stars) and face quadrature points (crosses) of a DG scheme of polynomial degree $k = 4$ with 6^2 Gaussian quadrature points on two quadrilateral elements

compared to low-order counterparts for the same number of unknowns. A review on the use of high-order methods in the fluid dynamics context can be found in Wang et al. (2013), see also Keyes et al. (2013) for the broader class of multiphysics problems. However, accuracy alone does not fully describe the capabilities of a numerical scheme, because the final application metric is accuracy delivered for a specific computational expense. Following Fehn et al. (2018b), we split the efficiency of an algorithm into two parts,

$$\text{efficiency} = \frac{\text{accuracy}}{\text{cost}} = \underbrace{\frac{\text{accuracy}}{\text{DoFs} \cdot \text{time steps}}}_{\text{discretization}} \cdot \underbrace{\frac{\text{DoFs} \cdot \text{time steps}}{\text{cost}}}_{\text{implementation}} . \quad (7)$$

The discretization efficiency refers to the quality of the spatial scheme and employed time stepping. An algorithm is considered efficient if it can deliver a particular accuracy with a minimum number of spatial degrees of freedom (DoFs) and a minimum number of time steps. Here, a time step refers to the effective work and needs to be understood as the number of Runge–Kutta stages in an explicit scheme. In case of implicit time stepping, the cost of the linear/nonlinear solvers also plays a crucial role, suggesting that the metric of time steps also needs to reflect the actual solver cost, see Fehn et al. (2018b) and Arndt et al. (2020c) for details.

The implementation efficiency measures the *throughput* of the algorithmic realization in terms of how quickly a given number of unknowns and time steps can be processed. On the one hand, it depends on the algorithm choice as well as parameters of the discretization, such as the polynomial degree, the type of elements (e.g., hexahedral versus tetrahedrals) and their deformation (affine versus curvilinear mappings $\mathbf{\Phi}_e$). On the other hand, the efficiency strongly depends on the chosen computer hardware, which is subject to continuous evolution. The exact meaning of the computational cost depends on the context—it can refer to the number of compute hours accounted for by a supercomputing grant, but it can also be the absolute wall time to run a code (time-to-solution) if resources are no limit. The hardware and implementation aspects of DG methods are the subject of high-performance computing (HPC) and covered in this book chapter.

Formula (7) suggests that a balance must be found between possibly conflicting goals. Sophisticated discretizations that decrease the number of unknowns might still be less efficient in reaching a desired accuracy if their implementation is much slower than a simpler algorithm with a higher throughput. The contribution by Chang et al. (2018) provides theoretical background for these metrics, albeit with relatively simple implementations only.

ALGORITHM 1: Evaluation of advection right hand side (1.32)

Input: Solution vector $\boldsymbol{U}_h$ defining solution u_h via (1.10)
Output: Vector $\boldsymbol{Y} = \mathcal{L}_h(\boldsymbol{U}_h, t)$ holding the value of all integrals on the right hand side of (1.32), tested by all test functions $\varphi_i^{(e)}$

If run in parallel, import section of data vector from neighboring processor to handle face integrals below
For each element $e = 1, \ldots, n$ **do**

(i) Read the solution coefficients $u_j^{(e)}$ from $\boldsymbol{U}_h$ corresponding to element Ω_e for $j = 1, \ldots, n_{\mathrm{p}}$
(ii) Compute the element integrals $Y_i^{(e)} = \frac{1}{2}\left(\nabla\varphi_i^{(e)}, \boldsymbol{a}u_h^{(e)}\right)_{\Omega_e} - \frac{1}{2}\left(\varphi_i^{(e)}, \boldsymbol{a}\cdot\nabla u_h^{(e)}\right)_{\Omega_e} + \left(\varphi_i^{(e)}, b\right)_{\Omega_e}$ with quadrature like (1.26) for $i = 1, \ldots, n_{\mathrm{p}}$
(iii) **For each** face $f = 1, \ldots, 2d$ of Ω_e **do**

(a) Interpolate the interior solution approximation $u_h^{(e)}$ to quadrature points at the respective face, defining u_h^- for quadrature point indices q
(b) **If** interior face, read the solution coefficients $u_j^{(e^+)}$ from $\boldsymbol{U}_h$ corresponding to neighboring element Ω_{e+} along the face index f, and interpolate solution to quadrature points as u_h^+
(c) **Else** at the boundary, construct the value u_h^+ from boundary condition at position $\boldsymbol{x}_q = \boldsymbol{\Phi}_e(\hat{\boldsymbol{x}}_q)$ and/or interior value u_h^-
(d) At each quadrature point, compute $\boldsymbol{n}\cdot\left(-\{\{\boldsymbol{a}u_h\}\} - \frac{\alpha}{2}|\boldsymbol{n}\cdot\boldsymbol{a}|\,[\![u_h]\!] + \frac{1}{2}\boldsymbol{a}u_h\right)$, multiply by test function $\varphi_i^{(e)}$, sum over quadrature points, and add the result to $Y_i^{(e)}$ for $i = 1, \ldots, n_{\mathrm{p}}$

(iv) Write result $Y_i^{(e)}$ for $i = 1, \ldots, n_{\mathrm{p}}$ into result vector $\boldsymbol{Y}$

Computational Characterization of DG Schemes

With explicit time integration, discontinuous Galerkin methods require either the evaluation of integrals like (6) on the cells and the faces, or some quadrature-free variant with precomputed matrices that embeds the action of the operation, see, e.g., Hesthaven and Warburton (2008). In this text, the focus is on the case of quadrature according to the algorithm structure laid out in Algorithm 1, denoting the operator by $\mathcal{L}_h(\boldsymbol{U}_h, t)$. For different variants of DG algorithms, see also Kronbichler and Kormann (2019) and references therein.

In the integration-based DG implementation of Algorithm 1 with data layout according to Fig. 2.1, the element integrals only access the unknowns of the element under consideration. The face integrals involve both the data on the element and that of immediate neighbors sharing a face for evaluating the numerical flux, such as $(\boldsymbol{a}u)^* = \{\{\boldsymbol{u}_h\}\} + \frac{|\boldsymbol{n}\cdot\boldsymbol{a}|}{2}\,[\![u_h]\!]$ for the advection Eq. (3). This represents a *compact* point-to-point communication between nearest neighbors in the computational mesh.

Figure 2.2 shows the data access pattern for computing the face integrals of the advection Eq. (3). In order to compute u_h^+, all the unknowns on a neighboring element

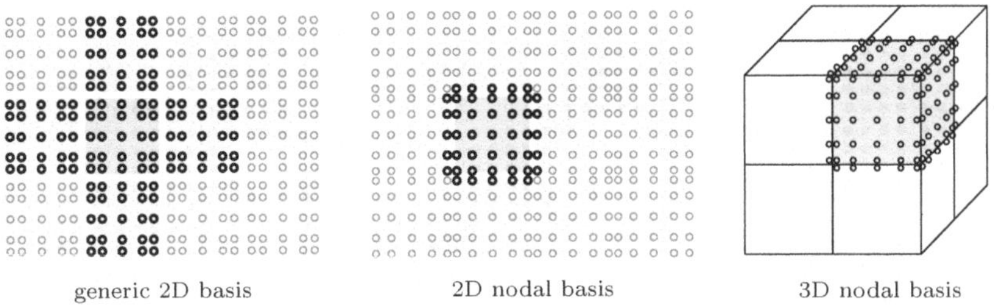

Fig. 2.2 Illustration of unknowns of a discontinuous Galerkin discretization of degree $k = 4$ for advection (3) with $k = 4$. Each circle represents a node point with an associated DG solution coefficient. The circles printed in black represent the data accessed for computing all cell and face integrals pertaining to the element shaded in red

are needed in case the polynomial basis does not have a particular structure as shown in the left panel of the figure. However, if the basis is *nodal* with nodes placed on the element boundary, like with Lagrange polynomials in the Gauss–Lobatto points, only $(k+1)^{d-1}$ out of the $(k+1)^d$ polynomials evaluate to non-zero at a face of the element. As a consequence, the data access becomes a single layer of unknowns around the element, as illustrated by the middle and right panel of Fig. 2.2. In case of a pure upwind flux with $\alpha = 1$, the exterior value on element faces in downstream direction drops out in Eq. (3), so the access of exterior information u_h could be skipped altogether. However, this optimization is rarely included in codes because it makes specific assumptions on the flow direction.

It is instructive to compare the data access of a DG method with a finite difference stencil. For the latter, unknowns with offset ± 1 in each direction around each point are accessed, e.g., the left, right, lower, and upper neighbor in a 2D grid. Figure 2.2 suggests a similar pattern in DG, if the left, right, lower, and upper neighbors are interpreted as the collective of all the unknowns on an element. Within the elements, DG schemes have a dense coupling between the unknowns, resolved by additional local computations. As a consequence, a DG method can *never* achieve a higher throughput, measured as the number of unknowns processed in a given time unit, than a well-implemented second-order finite difference method accessing a single layer of points, see, e.g., Hager and Wellein (2011) for stencil implementations.

For implicit time integration or stationary systems, a linear system associated with weak forms containing u_h in (3) or (4) needs to be solved. In that case, the data access defines a pattern of non-zero entries in a matrix. Figure 2.3 presents this pattern for the advection and Poisson problems in 1D for generic $\alpha \neq 1$. The data access pattern from Fig. 2.2 translates to both non-zero rows and columns in

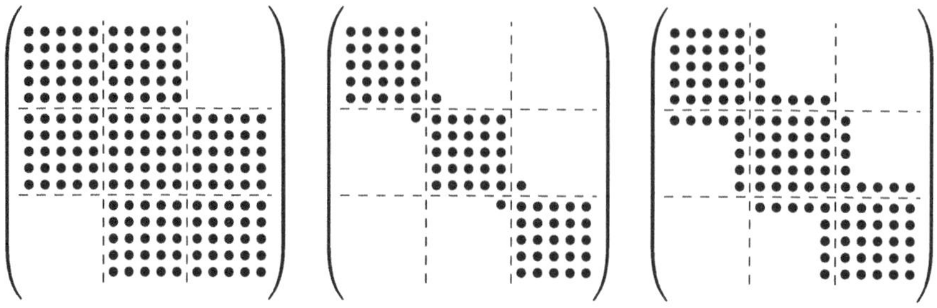

Fig. 2.3 Non-zero entries in the sparse matrix representing the 1D advection operation (3) or Poisson operation (4) with degree $k = 4$ on $n = 3$ elements. Dashed lines indicate the matrix rows and columns belonging to each element

the matrix. The matrix structure with whole blocks of non-zero entries can have favorable influence on the performance of some direct solvers like Pardiso (Schenk and Gärtner 2004) or UMFPACK (Davis 2004), or for the storage format of sparse matrices. Section "Solving Linear Systems" below provides a different matrix-free view on solvers for the linear systems associated with high-order DG discretizations.

Background on Computer Architecture

Up to around 2005, the combination of Dennard scaling (smaller transistors have a similar power density as bigger transistors; smaller transistors in turn allow to increase the frequency of integrated circuits) with Moore's law (the number of transistors in integrated circuits doubles approximately every two years) gave rise to drastic performance increases of essentially all computer programs. As Dennard scaling started fading, hardware architects continued to improve processors by adding more cores to chips in the form of multi-core processors. This increased capability, however, only translates to higher performance for codes where the predominant part can run in parallel, a consequence of Amdahl's law (Amdahl 1967). In the numerical simulation community, faster computers have been predominantly used to approach larger problems, thereby increasing the parallel portion of a program and circumventing Amdahl's law (Gustafson 1988). As of 2020, structure sizes in integrated circuits have reached structure sizes below 10 nm (a few dozens of silicon atoms). It is therefore generally expected that the classical miniaturization according to Moore's law is going to halt within the next decade for the current silicon technology. The modern hardware design tries to balance between power consumption, the amount of parallelism, and the degree of specialization, driven by market demands.

Modern supercomputers expose two main levels of parallelism, up to a few tens of thousands of *compute nodes* connected with a high-performance network fabric such as Infiniband, with each node employing parallel hardware, such as multi-core or many-core processors. Typical networks have a point-to-point latency in the range of 0.3–2 μs with a throughput between two links of 5–20 GB/s. This number can be compared to the properties within a node, which as of 2020 offer a memory bandwidth between 250 GB/s (dual-socket processor with 2×6 DDR4 memory channels) and 1.5 TB/s (Nvidia Ampere graphics processor) as well as memory latencies in the range of 50–150 ns. The current efforts of the community target exascale hardware, i.e., systems performing 10^{18} floating point operations per second (1 ExaFlop/s) aggregated over a single machine.

The classical design of multi-purpose processors (central processing units, CPUs) aims for a fast execution of latency-sensitive tasks by sophisticated techniques such as out-of-order execution, branch prediction, and speculative execution as well as deep memory hierarchies with data and instruction prefetching. In the last decades, hardware has gained a considerable amount of parallelism also within a single core, such as pipelining and instruction-level parallelism to execute several machine instructions with independent data in a single clock cycle, or by using short-length vectorization through the single-instruction/multiple data (SIMD) paradigm (Patterson and Hennessy 2013; Hager and Wellein 2011). Vectorization is the most visible feature at the programming or compilation level, aiming to execute operations such as floating point multiplications on an array of values by a single instruction. Contemporary CPUs such as the Intel Skylake or Fujitsu A64FX processors[1] contain two SIMD units that each can issue instructions for eight double-precision variables. Combined with the latency of floating point operations of 4–6 clock cycles, this means that algorithms need to supply almost 100 independent floating point operations at a time to fully utilize a modern compute core. Here, independent means that the input arguments to, e.g., a floating point multiplication must not depend on the result of any other ongoing operation. Modern out-of-order CPU cores can hold up to around 200 instructions in flight at a specific time, necessitating that those independent operations come from instructions very near in the program execution. All of that must happens while the core also loads arguments from memory, guesses on the outcome of branches that depend on some results, and stores results back to memory.

During the last decade, these classical CPUs have been complemented by hardware more specialized at throughput tasks. The most widespread accelerator architecture are graphics processing units (GPUs), which reduce the fraction of transistors spent on control logic in favor of more compute units that are fed by an even higher degree of parallelism. Combined with a separate high-bandwidth memory of size up to 32 GB (about a tenth of the memory in a node), GPUs can provide a 2–4× higher throughput per watt for workloads with enough parallelism, regular data access and no challenging caching needs, which is a good fraction of HPC loads.

[1]As of June 2020, the fastest machine listed on the Top-500 list, https://top500.org, consists of 152,064 nodes with 48-core A64FX CPUs.

From Application Code to Machine Instructions

Processors execute programs in the form of a stream of machine instructions. Machine instructions cover a few hundreds of basic operations, like loading or storing data from/to memory, arithmetic operations on integer and floating point numbers, as well as branch operations that jump to certain positions in the instruction stream in case certain conditions are met. Machine code or its human readable version, assembly code, is too low-level to allow for productive programming. Higher level languages, such as the C++ programming language used for the experiments in this book chapter, are translated to machine code by a compiler.

Compilers execute a large number of optimization steps that try to identify the most beneficial machine code representation. For example, expressions that are constant within a loop can be computed before the loop, or more expensive operations like multiplication by the integer 8 to translate an index offset of a double array into the address given in bytes are replaced by a left-shift of 3, because $8 = 2^3$ and a left-shift of an integer represented in binary format corresponds to multiplication by 2. Similarly, the compiler can try to replace scalar operations by vectorized operations using SIMD instructions. For floating point numbers, certain optimizations are only possible when breaking associativity rules, given that in floating point arithmetic it generally holds that $(a + b) + c \neq a + (b + c)$.

Given the vast amount of available code optimizations on the one hand, and restrictions of a programming language like C++ on the other hand, the finite element community has also started to develop abstractions on the level of PDEs. The Unified Form Language (UFL) (Alnæs et al. 2014) is one successful project, developed in the context of the FEniCS (Alnæs et al. 2015) and Firedrake (Rathgeber et al. 2017) projects, for example.

The Memory Wall

A general trend in computer architecture, affecting both CPUs and GPUs, has been the increasing gap between the performance of arithmetic operations and the access of data stored in memory, the so-called memory wall. For example, a contemporary Intel Xeon Platinum 8280 (released in 2019) with 28 cores running at a clock frequency of 2.7 GHz can perform up to 2.42 double-precision TFlop/s (trillion floating point operations per second) when using fused multiply-add (FMA) operations, i.e., operations of the form $a \leftarrow a + b * c$ or some permutations of the arguments. This number is computed as

$$\begin{aligned} &28 \text{ [cores]} \times 2.7 \cdot 10^9 \text{ [clock cycles/s]} \times 2 \text{ [execution units]} \\ &\times 8 \text{ [SIMD vector lanes]} \times 2 \text{ [operations / FMA]}. \end{aligned}$$

At the same time, the interface from random access memory (RAM), also called main memory, allows for up to 141 GB/s of transfer, computed by 6 memory channels with a transfer width of 8 bytes and a frequency of 2.93 GHz. As a consequence, for each byte loaded from main memory, 17.2 double-precision floating point operations can be done, a number called the machine balance. Given that a double number occupies 8 bytes of memory, 137 floating point operations can be performed for each double loaded from memory. The ratio is not substantially smaller for the NVIDIA A100 GPU released in 2020 with the latest high-bandwidth memory interface, which supports 9.7 TFlop/s with 1.56 TB/s of transfer, giving a machine balance of 6.2 Flop/byte. By comparison, the Intel Xeon W5590 from 2009 with 4 cores running at 3.33 GHz had a machine balance of 1.7-Flop/byte.

The growing gap has been partly mitigated by a deep memory hierarchy, i.e., a combination of caches of different sizes and speeds. Caches are used to keep a copy of the data from memory or temporary results closer to the processing units to allow for a faster access for repeatedly used data. Typical level-1 caches have a size between 32 and 128 kB with a high throughput of reading up to 128 bytes per clock cycle, giving a balance of 0.25 Flop/byte. Level-2 and possible level-3 caches have sizes up to a few megabytes, but they are slower in terms of throughput and latency. Caches are usually organized by the hardware with sophisticated heuristics. Data that got recently accessed (temporal locality) or data accessed nearby (spatial locality) is held in caches, whereas new data needs to be fetched from main memory. On many GPUs, also user managed scratch pad memory with similar performance as caches is available.

Dense linear algebra operations such as matrix–matrix multiplications or dense matrix factorizations can be cached well and reach a high percentage of the floating point peak performance. However, linear algebra operations relevant for PDEs almost exclusively belong to the BLAS-1 (vector-vector) or BLAS-2 (matrix-vector) category, which come with a very low data re-use. Take for example the inner product between two vectors of a few million entries. They are too big to fit into caches, so both vectors must come predominantly from main memory. In an inner product, the two floating point values or 16 byte of loaded data are used for a single multiplication and addition, which gives an *arithmetic intensity* of only 1/8 Flop/byte, much lower than typical machine balances. Similarly, the multiplication of a sparse matrix with a vector has an arithmetic intensity between 0.15 and 0.25 (Patterson and Hennessy 2013). Given the gap between hardware capabilities and algorithm characteristics, one of the main challenges in high-performance computing for PDEs is to find algorithms that do additional operations on data once loaded into the processor, i.e., to increase cache locality. In terms of the inner product, the goal would be to perform additional operations on the vector entries while the data in caches is still hot, i.e., before the inner product has been completed.

Due to the limited data re-use of many classical numerical algorithms, the community has started to consider algorithms that trade memory transfer for additional computations. This paradigm, loosely coined "flops are for free" (see, e.g., Keyes et al. 2013 and references therein), is the central pillar making discontinuous Galerkin algorithms particularly attractive on modern hardware, given the algorithm demands identified in section "Computational Characterization of DG Schemes" and evaluated extensively in the sections below.

Whereas the HPC community has traditionally focused mostly on double-precision numbers for accuracy reasons, even higher performance can be obtained by using lower precision numbers, such as single-precision, half-precision, or even 8-bit integer numbers. Lower precision numbers have the advantage of not only reducing the number of transistors needed to perform operations, but in particular to also reduce the memory transfer. In order to not affect the result quality, lower precision numbers need to be combined with higher precision corrections in the context of PDEs, see, e.g., the work by Göddeke et al. (2007).

Parallelization of DG Algorithms

Like many mesh-based discretization schemes for PDEs, discontinuous Galerkin methods can be parallelized by splitting the computational domain between the participating processors, where the term *processor* refers to a suitable compute unit to the algorithm at hand. This task is called domain partitioning and is handled either through graph partitioning algorithms such as METIS (Karypis and Kumar 1998) or by space filling curves (Burstedde et al. 2011; Bangerth et al. 2011; Weinzierl 2019). If the DG method is run in parallel on the processor-local partition, some data exchange is necessary as indicated by Algorithm 1. If the 4×3 mesh in the left two panels of Fig. 2.2 is split into two partitions, assigning the left 2×3 elements to the first partition and the right 2×3 elements to the second partition, the three faces at the interface need access to data of a foreign processor, called *ghost* or *halo data*. The data access of a DG scheme to neighbors identified in Fig. 2.2 defines the *data communication* for a parallel execution of the DG scheme. For polynomial bases without nodes at the element boundary, or with non-nodal bases, all data of the respective elements needs to be imported, whereas a single layer of points suffices for nodal polynomials. Implementations sometimes include an additional interpolation step of unknowns to the surface, in order to reduce the data exchange to the data minimally needed, see, e.g., Hindenlang et al. (2012) or Kronbichler et al. (2017). Parallelization of DG algorithms is also discussed in Kronbichler and Kormann (2019) and Kronbichler and Allalen (2018).

The DG algorithm communicates between nearest neighbors in the computational grid only, which fits well with the performance characteristics of modern Infiniband networks in terms of their bandwidth. As seen above, the network bandwidth between two nodes is between $\frac{1}{10}$ and $\frac{1}{50}$ of the memory bandwidth. At the same time, for a discretization of 100^3 unknowns per node, the exchange of the surface amounts to $6 \cdot 100^2$ data items or 6% of the traffic to access a single vector. Given that at least two local vector accesses are necessary, often even significantly more, the network access is not the dominant cost. If the number of unknowns per node increases, a more beneficial volume-to-surface ratio reduces the transfer even further in proportion. The ghost data exchange is only needed for the computation of numerical fluxes on faces located at the interface between processors, giving ample possibilities to overlap the latency in the communication phase with computations on processor-local data (Brightwell et al. 2005).

In software implementations, the message passing interface (MPI) is the dominant paradigm for data exchange between the different nodes of a parallel computer. MPI runs separate programs on each processor and data is exchanged by explicit send and receive operations. Often, MPI is also used for addressing multiple cores within a compute node. As an alternative to the classical MPI-only paradigm, MPI+X approaches make explicit use of the shared memory within a node and avoid sending messages in favor of direct access into the unknowns of a neighboring subdomain. Common paradigms are OpenMP, MPI-3 shared memory, or accelerator-specific interfaces such as CUDA on Nvidia GPUs.

Linear solvers discussed in section "Solving Linear Systems" below might add also other communication patterns. Given the data exchange property of DG and the fact that the network topology of HPC systems has been relatively steady during the last decade, the primary driver for performance of DG solvers is the node-level performance and memory access in particular, discussed in section "Fast Computation of Integrals with Sum Factorization" below. This is confirmed by exascale projections made in Ibeid et al. (2019).

Identifying the Performance Limit

In order to assess the performance of an algorithm, the limiting resource(s) need to be identified, which is done for one algorithmic component, called kernel, at a time. A kernel can refer to the loop over elements in Algorithm 1 when embedded into additional operations such as the vector operations in an explicit Runge–Kutta method. However, a kernel can also represent a sub-step in the algorithm when analyzing the phases of operations with different characters. Assuming that workloads can be parallelized to run on all cores and utilize vector units, the performance limits can be roughly divided into

- floating point operations in vector units or the immediate data transfer from/to the level-1 cache,
- integer operations like array/table lookup, search operations, or similar non-floating point work,
- data transfer from a higher cache level,
- data transfer from main memory, or
- data transfer over the Infiniband network.

For each category, the performance can be limited by throughput, which means that enough independent items are available and it is the sheer stream of operations that is overwhelming, or by latency. A latency limit is present when the execution of operations needs to be delayed not because the relevant resource is already exhaused, but because of the time it takes between the start and the end of the operation, like for memory lookup or when waiting for a previous floating point operation. Given that HPC workloads often process many data items in parallel with similar operations, the throughput is the more common limit, except for possibly the network data transfer where latency can play a role, and for unoptimized inner loop kernels where latency of in-core operations can be substantial. Given the importance of data transfer and especially the transfer from main memory on the one hand and floating point operations on the other hand, the *roofline performance model* (Williams et al. 2009) predicts performance as

$$\text{execution time} = \frac{n_{\text{fp}}}{\min\left(\text{bandwidth} * I_{\text{a}}, \text{arithmetic throughput}\right)}, \tag{8}$$

with n_{fp} representing the number of floating point operations in an algorithm and I_{a} the arithmetic intensity. This model gives a good first indication of the behavior of a computational kernel. The quality of a roofline prediction depends on the accuracy of the ingredients. For example, the arithmetic throughput is not only limited by the maximum peak performance of the architecture, but rather by the actual instructions in the kernel, like non-vectorized operations or non-floating point operations. Likewise, memory transfer is more accurately represented by measured values rather than peak values, and the roofline model can also take the bandwidth of caches into account depending on the data access pattern (Williams et al. 2009). Other performance models like the execution-cache-memory model (Treibig and Hager 2010; Hager et al. 2016) also represent limitations in the data transfer between different cache levels.

What to Measure

A key step in identifying a fast algorithm is to optimize for the right metric. Given the considerations about the performance limits in terms of memory bandwidth, arithmetic throughput, and network, one might be tempted to consider high GFlop/s or high GB/s as the primary target for code tuning. While these two metrics give useful insight for a fixed algorithm, they are not sufficient. One can easily see that doing superfluous arithmetic operations would likely increase the floating point performance measured in GFlop/s, without giving benefit to the application, see also Hoefler and Belli (2015) and references therein on how to measure and report results. From the application point of view, an unbiased metric is often the *throughput* as given in Eq. (7), such as the number of DoFs processed per second on a certain hardware. A more extensive discussion of this topic is found in Hager and Wellein (2011). In order to achieve that goal, it is important to both consider equivalent mathematical expressions in the algorithm that reduce the arithmetic work or memory access, and to utilize the hardware well with high GFlop/s and GB/s numbers.

To give an example of the tradeoffs in this process, we consider the kernel that computes the data to be multiplied by the test functions at the quadrature points for the face integral in the weak form of the advection Eq. (3),

$$\begin{aligned} &\boldsymbol{n}\cdot\left(\frac{\boldsymbol{a}u_h^- + \boldsymbol{a}u_h^+}{2} + \frac{\alpha}{2}|\boldsymbol{n}\cdot\boldsymbol{a}|\boldsymbol{n}(u_h^- - u_h^+) - \frac{1}{2}\boldsymbol{a}u_h^-\right)|\boldsymbol{t}_1\times\boldsymbol{t}_2|w_q \\ &= \left(\boldsymbol{n}\cdot\boldsymbol{a}\frac{u_h^+}{2} + \frac{\alpha}{2}|\boldsymbol{n}\cdot\boldsymbol{a}|(u_h^- - u_h^+)\right)|\boldsymbol{t}_1\times\boldsymbol{t}_2|w_q, \end{aligned} \tag{9}$$

where $|\boldsymbol{t}_1 \times \boldsymbol{t}_2|$ denotes L_2 norm of the cross product of the two 3D tangential vectors to produce the surface area contribution of the integral transformation, and w_q is the quadrature weight.

In case the kernel is called many times for the same $\boldsymbol{n}$ and $\boldsymbol{a}$ with different u_h, the number of operations is minimized by pre-computing the scalar factors $\frac{1}{2}\boldsymbol{n}\cdot\boldsymbol{a}\,|\boldsymbol{t}_1\times\boldsymbol{t}_2|w_q$ and $\frac{\alpha}{2}|\boldsymbol{n}\cdot\boldsymbol{a}|\,|\boldsymbol{t}_1\times\boldsymbol{t}_2|w_q$. Then, at each quadrature one addition for $u_h^- - u_h^+$, one multiplication with $\frac{\alpha}{2}|\boldsymbol{n}\cdot\boldsymbol{a}|\,|\boldsymbol{t}_1\times\boldsymbol{t}_2|w_q$, and one fused multiply-add operation to multiply u_h^+ by the second factor $\frac{1}{2}\boldsymbol{n}\cdot\boldsymbol{a}\,|\boldsymbol{t}_1\times\boldsymbol{t}_2|w_q$ and add the result to the remaining part is needed. For these four floating point operations, four variables need to be loaded from memory, u_h^-, u_h^+ as well as the two factors. With an arithmetic intensity of $\frac{1}{8}$ Flop/byte, this operation is memory limited on all architectures and for all levels of caches.

Since α is a constant with the same values at all quadrature points and on all faces, and since the modulus of a number ξ is a cheap operation, it suffices to only load the factor $\xi = \frac{1}{2}\boldsymbol{n}\cdot\boldsymbol{a}\,|\boldsymbol{t}_1\times\boldsymbol{t}_2|w_q$ from memory and compute $\alpha|\xi|$ *on the fly* from the available operands. The arithmetic balance of this variant is 5 Flop for 3 doubles loaded from memory or $\frac{5}{24}$ Flop/byte. This kernel will be $\frac{4}{3}$ times faster than the original version, despite performing additional floating point operations. The advantage grows to a factor of two in case the solution coefficients u_h^- and u_h^+ already sit in fast cache memory when the operation at quadrature points is invoked, because now only one double variable for ξ needs to be loaded per quadrature point, compared to two in the original setup.

The situation becomes even more interesting if the speed $\boldsymbol{a}$ and the geometry data $\boldsymbol{n}$ and $|\boldsymbol{t}_1 \times \boldsymbol{t}_2|$ are constant throughout all quadrature points of a face, or can be computed without much additional memory access. Then, only the quadrature weight w_q differs from one quadrature point to the next. Assuming in turn that w_q can be served from cache because the same weights occur on every face in the computational mesh, separating w_q from the stored factors allows to increase the arithmetic intensity even more and gain additional performance.

This simple example shows that investing some additional floating point operations can increase performance. The paradigm of trading memory transfer for additional operations to re-compute necessary input arguments over and over again has become a corner stone in high-performance computing. Note that the number of arithmetic operations and the memory transfer have changed during the algorithmic modifications, showing the importance of an absolute metric that reflects how many integrals have been computed.

However, computing more information over and over again with the goal to transfer less is only useful as long as the memory transfer is the bottleneck. If, conversely, the speed $\boldsymbol{a}$ in (9) depends on the spatial coordinates $\boldsymbol{x}$, e.g., as $\sin(\pi x_1)\sin(\pi x_2)\sin(\pi x_3)$, then computing $\boldsymbol{a}$ on the fly is typically more expensive than merely loading $\boldsymbol{a}\cdot\boldsymbol{n}$ from memory. This is because each trigonometric function evaluation involves many arithmetic operations and also involves overhead due to latencies. If we assume that 50 arithmetic operations are needed for each trigonometric function that perfectly fill the arithmetic units (which would not be the case in reality), computing the term (9) takes around 160 arithmetic operations. If we assume that one double is loaded from memory in that kernel to get u_h^- (cached from the other side u_h^+), and one double is written to memory for the result, the arithmetic intensity of 10 Flop/byte suggests that memory transfer is advantageous over computing those expensive terms on the fly, as long as the machine balance is below 10 Flop/byte. If additional arithmetic operations are done on cached data, the threshold goes to even higher Flop/byte ratios.

Fast Computation of Integrals with Sum Factorization

In this section, the implementation of a DG algorithm will be matched with the capabilities of hardware. If we compute the integrals in Eq. (3) with quadrature as specified in Eq. (6), we observe that the multiplication of $u_h^{(e)}$ with the transport speed $\boldsymbol{a}$ and the metric terms $\boldsymbol{J}_e^{-1}$ as well as $\det(\boldsymbol{J}_e(\hat{\boldsymbol{x}}_q))w_q$ only depend on the quadrature point index but not on the index i of the test functions. Thus, they can be computed once per point before the final multiplication by $\hat{\nabla}\varphi_i$ for $i = 1, \ldots, n_{\mathrm{p}}$. Likewise, the summation defining $u_h^{(e)}$ and $\hat{\nabla}u_h^{(e)}$ according to Eq. (1) for a specific quadrature point index does not depend on the metric term.

The algorithm pattern for computing this particular integral is generic to a large class of weak forms, such as the other integrals in advection (3) or for the Poisson operator (4), and can be described by the following steps:

(i) interpolation from the input vector coefficients $u_j^{(e)}$ to the value $u_h^{(e)}(\hat{\boldsymbol{x}}_q)$ or gradient $\hat{\nabla}u_h^{(e)}(\hat{\boldsymbol{x}}_q)$ (in reference coordinates) at all quadrature points $\hat{\boldsymbol{x}}_q$,
(ii) at each quadrature point, transform the gradients from reference to real space by using $\boldsymbol{J}_e^{-T}$, evaluate the flux term such as $\boldsymbol{f}(u_h)$ based on $u_h^{(e)}(\hat{\boldsymbol{x}}_q)$, multiply with integration factor $\det(\boldsymbol{J}_e(\hat{\boldsymbol{x}}_q))w_q$ and possibly multiply with the metric term $\boldsymbol{J}_e^{-1}$ factored out from the test functions,
(iii) and multiplication by test function φ_i or gradient $\hat{\nabla}\varphi_i$ in reference coordinates and summation over all quadrature points.

While this concept appears in almost any finite element computation, the algorithmic split into these three components, done for one or a few elements at a time to achieve good performance on modern computer hardware, appears to be rather recent and evolved in Brown (2010), Kronbichler and Kormann (2012, 2019) and Knepley et al. (2013).

The first step in this subdivision applies an interpolation operation on a part of the solution vector $\boldsymbol{U}_h$ and generates temporary results at quadrature points. In the second step, these temporary results are combined with additional quantities, the metric terms and variable coefficients, to again create $n_c \times d$ temporary results. In the third step, these results are further processed and written to a part of the result vector $\boldsymbol{Y}_h$.

Naive Interpolation and Differentiation

The interpolation from $(k+1)^d$ solution values to n_c quadrature points as illustrated in Fig. 2.1 in step (i) of the above procedure is represented by a small matrix-vector multiplication with a dense matrix $\mathcal{S} \in \mathbb{R}^{n_c \times (k+1)^d}$. Gradients $\hat{\nabla} u_h^{(e)}$ at all quadrature points are the result of a matrix-vector multiplication with a differentiation matrix $\mathcal{D} \in \mathbb{R}^{dn_c \times (k+1)^d}$. Similarly, the summation over quadrature points for all test function gradients in Step (iii) for the integral (6) corresponds to a multiplication of the $d\, n_c$ quantities at quadrature points by the dense matrix $\mathcal{D}^T$. In a naive realization of this algorithm, both Steps (i) and (iii) are of quadratic complexity in the number of unknowns for $n_c \sim (k+1)^d$, whereas the operation at quadrature points and the data access are linear in the number of unknowns. For higher polynomial degrees k and especially in $d = 3$ dimensions, the quadratic complexity parts would soon become the bottleneck.

For the example of $k = 5$ in 3D, the number of floating point operations to compute the integral (6) on $n_c = (k+1)^3$ quadrature points is $2 \cdot 6^6 = 93{,}300$ for the interpolation of u_h, around 4,100 for the work at quadrature points assuming a precomputed $\boldsymbol{J}_e^{-1}$, and 280,000 for the multiplication by the gradient of all test functions and summation over quadrature points. Assuming that only $u_j^{(e)}$ needs to be read from memory and $Y_j^{(e)}$ is written to memory, this gives an arithmetic intensiy of around 109 Flop/byte. If also $\boldsymbol{J}_e^{-1}$ and $\det(\boldsymbol{J}_e(\hat{\boldsymbol{x}}_q))w_q$ are read from memory, the arithmetic intensity would be 18.2 Flop/byte. Doing the second element integral of Eq. (3) and the face integrals on the same data would further increase the arithmetic intensity. Hence, the arithmetic throughput would be the limiting resource on all contemporary hardware. Given enough effort on implementation, such a variant has the potential to reach a high fraction of the floating point performance of a modern processor, but the code is not necessarily fast. While arithmetic operations might be cheap, they are definitely not for free if they are the factor limiting performance.

Utilizing the Tensor Product

For tensor product shape functions (2) that are integrated on a tensor product quadrature formula, the interpolation matrix $\mathcal{S}$ and the derivative matrix $\mathcal{D}$ have additional structure. If we denote the matrix $\mathcal{S}$ to hold the evaluation of the jth basis function at the qth quadrature point in the qj entry, $\mathcal{S}_{qj} = \varphi_j(\hat{\boldsymbol{x}}_q)$, it is given by the Kronecker product of d one-dimensional matrices,

$$\mathcal{S} = \mathcal{S}_d^{1\mathrm{D}} \otimes \cdots \otimes \mathcal{S}_2^{1\mathrm{D}} \otimes \mathcal{S}_1^{1\mathrm{D}}.$$

Here, $\mathcal{S}_1^{1\mathrm{D}}$ denotes the one-dimensional interpolation matrix obtained by evaluating $\varphi_{i_1}^{1\mathrm{D}}(\hat{x}_{q,1})$ at all one-dimensional quadrature points. Similarly, the matrix $\mathcal{S}_2^{1\mathrm{D}}$ denotes the respective matrix in the second direction. In this formula, the number of quadrature points and polynomials needs not be the same, and different numbers of points in different directions are allowed. The only restriction is the tensor product form (2) of the shape functions and a tensor product quadrature formula. Similarly, the derivative matrix $\mathcal{D}$ is given by

$$\mathcal{D} = \begin{bmatrix} \mathcal{S}_d^{1\mathrm{D}} \otimes \cdots \otimes \mathcal{S}_2^{1\mathrm{D}} \otimes \mathcal{D}_1^{1\mathrm{D}} \\ \mathcal{S}_d^{1\mathrm{D}} \otimes \cdots \otimes \mathcal{D}_2^{1\mathrm{D}} \otimes \mathcal{S}_1^{1\mathrm{D}} \\ \vdots \\ \mathcal{D}_d^{1\mathrm{D}} \otimes \cdots \otimes \mathcal{S}_2^{1\mathrm{D}} \otimes \mathcal{S}_1^{1\mathrm{D}} \end{bmatrix},$$

with $\mathcal{D}_1^{1\mathrm{D}} = \frac{\mathrm{d}}{\mathrm{d}x_1}\varphi_{i_1}^{1\mathrm{D}}(\hat{x}_{q,1})$ denoting the derivative of all 1D shape functions evaluated at all points.

The multiplication of matrices forming a Kronecker product by a vector can be implemented by sequentially going through the one-dimensional interpolations, see, e.g., Buis and Dyksen (1996). In two dimensions, the operation can be visualized by a matrix–matrix product of small matrices,

$$\left(\mathcal{S}_2^{1\mathrm{D}} \otimes \mathcal{S}_1^{1\mathrm{D}}\right) \boldsymbol{U}^{(e)} = \mathrm{vec}\left(\mathcal{S}_1^{1\mathrm{D}} \mathrm{mat}\left(\boldsymbol{U}^{(e)}\right) \left(\mathcal{S}_2^{1\mathrm{D}}\right)^T\right),$$

where $\mathrm{mat}(\boldsymbol{U}^{(e)})$ denotes the interpretation of the $(k+1)^2$ vector $\boldsymbol{U}^{(e)}$ as a $(k+1) \times (k+1)$ matrix with the indices in x_1 direction running in columns and indices in x_2 running in rows, and $\mathrm{vec}(\cdot)$ the interpretation of the resulting matrix as a vector.

The mechanisms of this tensorial evaluation can also be seen by looking at the interpolation for all points $q = (q_1, q_2)$,

$$u_h(\hat{x}_{q_1,1}, \hat{x}_{q_2,2}) = \sum_{j_2=0}^{k} \sum_{j_1=0}^{k} \varphi_{j_1}(\hat{x}_{q_1,1}) \varphi_{j_2}(\hat{x}_{q_2,2}) u_{j_1,j_2}^{(e)}.$$

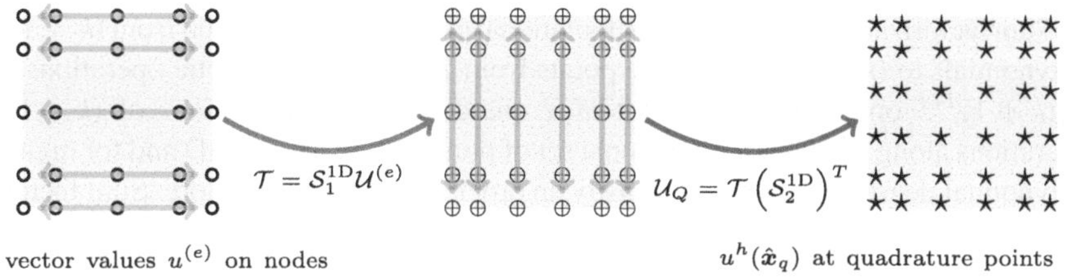

Fig. 2.4 Illustration of sum factorization for interpolation from node values on the left to the values at quadrature points (right) for $k = 4$ and $n_c = 6^2$

The multiplication by $\varphi_{j_2}(\hat{x}_{q_2,2})$ is independent of the summation over the first index j_1, so all sums $\sum_{j_1=0}^{k} \varphi_{j_1}(\hat{x}_{q_1,1})u_{j_1,j_2}^{(e)}$ for the point indices q_1 and q_2 can be computed and stored in a temporary array of $(k+1) \times n_{c,1D}$ unknowns, before performing a second pass through the data to perform the interpolation in the second direction. Factoring out the interpolation from other directions while performing the summation in one direction motivates the name for this algorithm, *sum factorization*. Figure 2.4 gives a graphical interpretation of the two summation steps in sum factorization for the interpolation operation and a corresponding matrix–matrix multiplication with some temporary matrix $\mathcal{T}$.

Sum factorization has its origin in the spectral element community (Orszag 1980; Patera 1984; Fischer and Patera 1991) and is described in the text books on high-order methods by Deville et al. (2002), Karniadakis and Sherwin (2005), and Kopriva (2009). Sum factorization is nowadays widely adopted in finite element software packages such as deal.II (Arndt et al. 2020b), DUNE (Bastian et al. 2020), Firedrake (Rathgeber et al. 2017; Sun et al. 2020), mfem (Anderson et al. 2020), Nek5000 (Fischer et al. 2020), Nektar++ (Cantwell et al. 2015; Moxey et al. 2020b), or NGSolve (Schöberl 2014), via optimized kernels like libCEED (Barra et al. 2020) or OCCA (Remacle et al. 2016), with support from loop optimizers such as Loopy (Klöckner 2014), as well as in application codes such as the compressible flow solver framework Flexi (Krais et al. 2020), pTatin3D (May et al. 2014), SPECFEM3D (Komatitsch et al. 2015) or the work by Huismann et al. (2020).

Sum factorization similar to the above can be applied for the summation over quadrature points with the matrix $\mathcal{D}^T$, for the cell integral $\frac{1}{2}\left(\varphi_i^{(e)}, \boldsymbol{a} \cdot \nabla u_h^{(e)}\right)_{\Omega_e}$ as well as the interpolation of finite element basis functions on the faces of an element for face integrals, see Kronbichler and Kormann (2019) for details on the algorithms involved in the DG context. In the spectral element and spectral DG context where node points coincide with quadrature points, the interpolation matrix $\mathcal{S}^{1D}$ becomes a unit matrix, leaving only the operations of the derivative matrix. As described in Kronbichler and Kormann (2019) and Fischer et al. (2020), a change of basis into the Lagrange basis in the quadrature points is the fastest evaluation option for non-trivial $\mathcal{S}$, as long as the number of quadrature points is not much higher than the number of polynomials.

Sum factorization reduces the computational effort for interpolation from $(k+1)^d$ polynomials to $(k+1)^d$ quadrature points from $2(k+1)^{2d}$ arithmetic operations to $2d(k+1)^{d+1}$ operations, the result of d one-dimensional operations of $(k+1)^2$ operations along each one-dimensional set of points. Especially in 3D and for higher polynomial degrees, the saving is easily an order of magnitude or more. Sum factorization represents matrix-matrix multiplication of small matrices with most dimensions equal $k+1$. Intermediate results are often small enough to fit into caches of CPUs (Kronbichler and Kormann 2019) or registers of GPUs (Świrydowicz et al. 2019; Kronbichler and Ljungkvist 2019). Thus, a high utilization of modern hardware can be obtained as long as the data for $u_h^{(e)}$ can be loaded quickly enough.

Given the challenge in applying vectorization within the different stages of sum factorization due to the low dimensions of the involved matrices for typical $2 \leq k \leq 8$, the most straight-forward way to identify SIMD-friendly loops is to do the same operations of several elements at once (Kronbichler and Kormann 2012, 2019; Anderson et al. 2020; Moxey et al. 2020a; Sun et al. 2020), i.e., vectorization across elements. In that case, all the operations within the integration loop can be done in parallel, except for the access to the portions of the solution in the result vector for face integrals. Modern C++ implementation typically rely on code around intrinsics to ensure vectorized SIMD instructions, which can be made user-friendly by operator overloading (Arndt et al. 2020b). Auto-vectorization by the compiler is generally not effective because vectorization goes over an outer loop with indirect data access and code indirections in between, see also the analysis in Sun et al. (2020). Algorithms for vectorization within elements are described in Müthing et al. (2017) and Kempf et al. (2018).

Operation Counts and Measured Throughput with Sum Factorization

We analyze the cost per degree of freedom for the evaluation of the advection Eq. (3) in an explicit time integration setting for the 2D case (quadrilateral elements) in the left panel and for the 3D case (hexahedral elements) in the right panel of Fig. 2.5. In order to make the work of high degrees more easily comparable to $k=1$, the figure presents the work done per unknown, i.e., the cost per element divided by $(k+1)^d$. For the cell term, a total of $4d$ sum factorization sweeps, i.e., operations applying one 1D interpolation on all unknowns of a cell, are needed, with d sweeps for the interpolation $u_h^{(e)}(\hat{\boldsymbol{x}}_q)$, d for the gradient of $u_h^{(e)}$, d for the multiplication by the gradient of the test function and d for the value of the test function. The setup assumes precomputed coefficients and metric terms and periodic boundary conditions to ignore the evaluation of boundary data. For sum factorization, the symmetry of the basis functions with respect to the domain center is used to cut the arithmetic operations down by almost a factor of two using the so-called even-odd decomposition as proposed by Solomonoff (1992) and evaluated in Kronbichler and

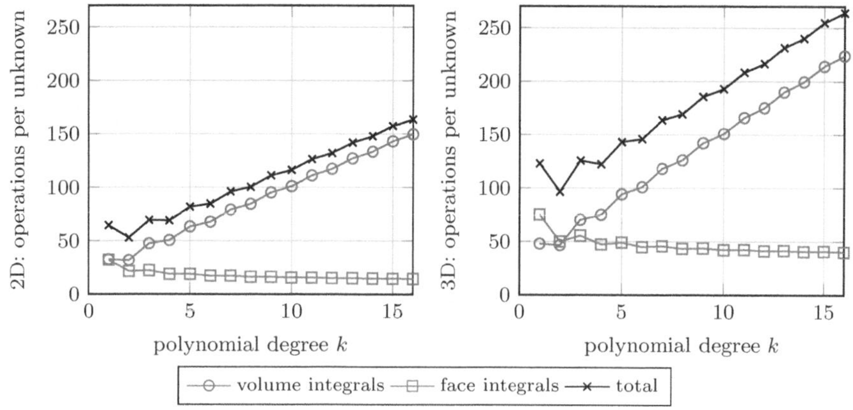

Fig. 2.5 Cost per element for evaluation of advection term with sum factorization, normalized by the number of unknowns

Kormann (2019) and Fischer et al. (2020). This represents the cheapest currently known evaluation strategy for low to moderate polynomial degrees with possibly variant coefficients.

In 3D, the face integrals involve more operations than the volume integrals for polynomial degree $k = 1$. The work for face integrals decreases as the polynomial degree increases because of a lower surface-to-volume ratio for higher order elements. Thus, the operations to be performed for cubic basis functions appear similar as for linear ones, computed relative to the number of degrees of freedom. Due to the even-odd decomposition, there is a stair-case effect in the operation counts with some zeros for even polynomial degrees or odd numbers of unknowns per direction due to the co-location of the quadrature point and node point in the center of the cells. For very large polynomial degrees, the asymptotic behavior with work proportional to the polynomial degree becomes apparent.

In terms of the arithmetic intensity discussed in section "Naive Interpolation and Differentiation", the implementation with sum factorization in 3D when only vectors are accessed from memory comes with an arithmetic intensity of around 7 for $k = 2$ (97 arithmetic operations per unknown for 2×8 bytes of transfer in an idealized situation), reaching the value of 17 Flop/byte for $k = 15$. As explained in section "Roofline performance evaluation" below, the actual memory transfer is higher due to non-perfect caching, leading to somewhat lower arithmetic intensities.

Finally, we look at a high-performance implementation of the sum factorization techniques, implemented in the deal.II finite element library (Arndt et al. 2020b; Alzetta et al. 2018) with so-called face-centric loops to minimize operation counts (Kronbichler and Kormann 2019). The source code is available online at https://github.com/kronbichler/advection_miniapp.[2] It is derived from measuring the computational time on a workstation with an Intel Xeon E5-2687W v4 proces-

[2]Commit 385c588, retrieved on August 21, 2020.

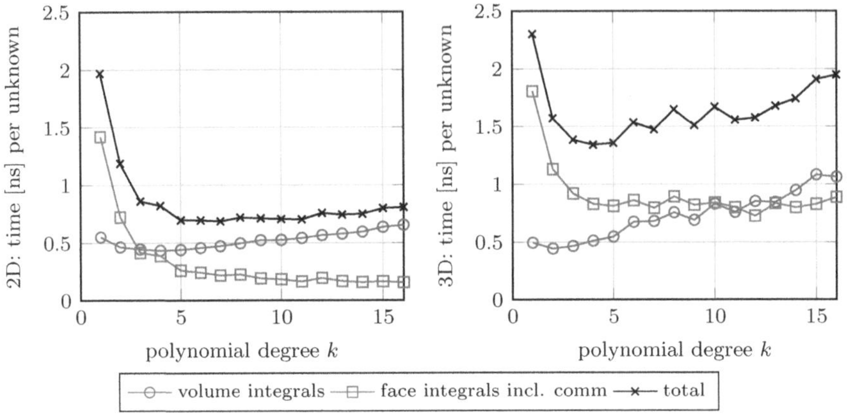

Fig. 2.6 Measured run time of evaluation of advection on a 12-core processor with optimized sum factorization routines from the deal.II library

sor with 4-wide vectorization (AVX2) when using all 12 cores for an experiment with approximately 15 million degrees of freedom. The code is parallelized with MPI. Figure 2.6 reports the computational time spent per unknown, expressed in nanoseconds. A value of 1.0 nanosecond in the figure means that it has taken 0.01 s to evaluate the right-hand side for a problem with 10 million unknowns.

Since the computation of cell and face integrals is overlapped to utilize data that is already in caches, the times reported for the face integrals in Fig. 2.6 are computed indirectly by the difference between a run with only the cell integrals and another one with both cell and face integrals. As a result, the cost of face integrals would be higher than reported in the figure if they were run alone. Nonetheless, the face integrals appear more expensive in the implementation than what the operation counts suggest. This is because the data access of face integrals into the neighbors is less regular than for the volume integrals, slowing down execution on the CPU. In addition, the parallel execution needs to exchange data between the different cores of the CPU via MPI messages, which is added to the cost of face integrals. Finally, for lower degrees the arithmetic work done on each face is relatively low, especially in 2D. In this case, the loop overhead of the C++ implementation and some metadata of the implementation also plays a role. On the other hand, the expected linear increase in run time with the degree is less pronounced than what would be expected from the operation counts alone. This is because higher degrees can utilize the computer hardware better with a higher arithmetic intensity, and because the even-odd decomposition (Kronbichler and Kormann 2019) becomes more heavy on fused multiply-add operations for higher degrees, speeding up execution. Only for very high degrees $k > 10$, the element integrals start to become dominant. On systems of equations or with nonlinear operators that are using over-integration with $n_c > (k+1)^d$, like for the compressible Navier–Stokes equations, also the aspect of fitting all temporary data of sum factorization in cache with vectorization across cells can set a limit, leading to a decay in performance for large $k > 8$ (Fehn et al. 2019b).

Scaling Within a Compute Node

The considerations in the previous subsection reveal that state-of-the-art implementations with sum factorization have arithmetic intensities of around 10 Flop/byte when only considering the vector access. This means that code optimizations need to both address the arithmetic performance and the memory transfer to achieve optimal performance. To better understand the behavior of Algorithm 1 in terms of data access and the attractive properties of DG, we compare the DG algorithm to an optimized finite difference implementation of the transport term with the stencil $\frac{1}{2h}(a_1(u_{i+1,j,k} - u_{i-1,j,k}) + a_2(u_{i,j+1,k} - u_{i,j-1,k}) + a_3(u_{i,j,k+1} - u_{i,j,k-1}))$ for constant factors a_1, a_2, a_3. Ideally, this code needs to load the input vector $\boldsymbol{u}$ once and write the result to a second vector. This is achieved with spatial blocking (loop tiling) to keep all data along layers $u_{i,j\pm1,k\pm1}$ in the cache according to Hager and Wellein (2011) as the grid is traversed. When measuring the actual memory transfer of an optimized implementation with the LIKWID tool (Treibig et al. 2010), a read transfer of 8.7 byte per unknown and writes of 8.0 byte per unknown are recorded when using special streaming (non-temporal) store instructions to avoid the read-for-ownership data transfer (Hager and Wellein 2011).

Here and in the following, experiments are done on a dual-socket Intel Xeon Platinum 8174 compute node.[3] Each core runs at a fixed clock frequency of 2.3 GHz, which gives an arithmetic peak performance of 3.53 TFlop/s. The memory bandwidth measured with the STREAM triad benchmark is 205 GB/s. In this subsection, all algorithms are parallelized in shared memory with OpenMP using the "spread" affinity policy to fill cores on different CPU sockets alternately, whereas later subsections also use MPI. The experiments are performed for problem sizes of a few tens to a few hundreds of millions of unknowns per node, running the experiment in a loop with 100 repetitions, and compute the throughput as

$$\text{throughput} = \frac{\text{problem size} \times \text{repetitions}}{\text{measured time}}.$$

The experiments run for at least a few seconds, which is long enough to report steady-state performance on the chosen hardware and its cooling system. In order to reduce the influence of noise, the experiment is run five times and the minimum out of these experiments is reported. Standard deviations are not more than a few percent, which is why they are not separately reported in this text.

[3] SuperMUC-NG supercomputer, https://doku.lrz.de/display/PUBLIC/SuperMUC-NG, retrieved on July 27, 2020.

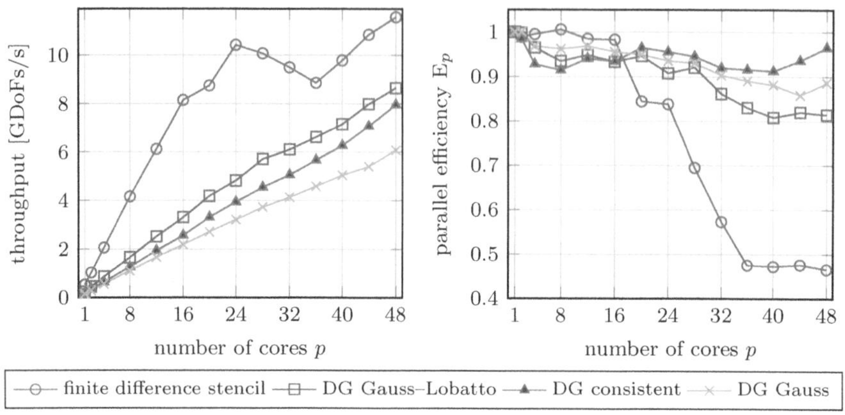

Fig. 2.7 Comparison of throughput of DG method run in 3D with optimized sum factorization implement for an affine cell geometry and constant transport speed $\boldsymbol{a}$ against an optimized finite difference stencil. The only access to memory is from reading the input vector $\boldsymbol{U}_h$ and writing the result $\boldsymbol{Y}$

Figure 2.7 presents the throughput of the difference stencil on a problem with 216 million unknowns of a 600^3 grid when executed on 1, 2, 4, 8, 12, 16, ..., 48 cores of the Intel Xeon Platinum system. The DG code is executed with 32.8 million unknowns on a 64^3 mesh with polynomial degree $k = 4$. In both cases, the size of two vectors in memory is much larger than available caches (combined L2 and L3 cache size is around 114 MB), so the predominant part of the vector data needs to be served from the main memory interface. Three variants of polynomial bases and quadrature points are considered for the DG case. The standard case, labeled "DG consistent", uses Lagrange polynomials integrated with the consistent 5^3 point Gauss–Legendre quadrature. The case "DG Gauss–Lobatto" uses the same polynomials but the spectral Gauss–Lobatto quadrature, resulting in an identity operation for the interpolation $\mathcal{S}$. Finally, the "DG Gauss spectral" cases use the consistent Gaussian integration together with polynomials in the same points. The latter two cases involve only around 0.58× the operation count of the "DG consistent" case due to the spectral (collocation) property.

The operation counts of the three DG variants do not accurately present the achieved throughput, though. On 48 cores, the "DG consistent" implementation runs with 8.0 GDoFs/s (billion unknowns per second), the variant with Gauss–Lobatto quadrature with 8.6 GDoFs/s, and the spectral Gauss quadrature with 6.1 GDoFs/s. These numbers are the result of an arithmetic performance of 972 GFlop/s, 647 GFlop/s and 468 GFlop/s, respectively. The measured memory transfer with the LIKWID tool (Treibig et al. 2010) is relatively similar between the three schemes with 176, 189, and 183 GB/s, respectively. For the spectral "DG Gauss" approach, the access to all unknowns of the neighboring elements according to the left panel of Fig. 2.2 leads to significantly more main memory access because not all data can be kept in caches.

The comparison to the finite difference scheme reveals how the heavy arithmetic work in the DG scheme benefits from more cores: While the finite difference stencil runs with 2.6× the throughput of DG when only 24 cores are used (10.3 GDoFs/s versus 4.0 GDoFs/s), the advantage on 48 cores is 1.4× against the consistent Gauss integration (11.6 GDoFs/s versus 8.0 GDoFs). The reason is that the finite difference code is completely limited by the memory bandwidth, with 193 GB/s measured transfer. In the discontinuous Galerkin scheme, additional arithmetic operations are performed while the data is transferred from and to memory. The higher arithmetic intensity is one of the pillars that make high-order DG attractive for current and future hardware. The fifth order DG scheme produces a much higher accuracy than the second-order finite difference method for the same number of unknowns, or can run with much fewer unknowns (and time steps) than the difference stencil. Given that the gap in throughput is not as pronounced as the operation counts would suggest, the application efficiency model (7) suggests a lower cost. Note that similar considerations would also apply to higher order difference stencils, which, however, are not as easily applied to complex geometries.

Figure 2.7 also reports the parallel efficiency with p cores computed as

$$\mathrm{E}_p = \frac{\text{time (1 core)}}{\text{time } (p \text{ cores})} \frac{1}{p},$$

with an ideal parallel efficiency of $\mathrm{E}_p = 1$. For the finite difference stencil, the efficiency drops significantly beyond 16 cores. This is the point when the memory interface shared between the cores is saturated. Conversely, the DG schemes are primarily limited by the in-core execution and show a much higher parallel efficiency. The spectral variant with the Gauss–Lobatto integration also drops in efficiency for more than 30 cores as the kernels get closer to saturating the memory bandwidth.

The parallel efficiency plot of Fig. 2.7 highlights the importance of absolute run times in high performance computing: When only looking at the parallel efficiency, the "DG consistent" algorithms look like the best algorithm. However, the graph is misleading because the consistent DG scheme actually achieves a lower throughput than the finite difference stencil and the "DG Gauss–Lobatto" algorithm. Incidentally, the difference stencil displays the worst parallel efficiency despite delivering the highest throughput. Hence, comparing relative quantities between different algorithms is only useful when also taking absolute run times into account. The picture gets even more multi-faceted when additionally considering accuracy, in which case the "DG consistent" approach is most efficient again.

The throughput of the high-order DG schemes in Fig. 2.7 with a constant velocity and constant metric terms is, despite being slightly slower than the difference stencil, nonetheless very high when considering the generality of the method. However, as soon as the velocity depends on the spatial variables or the elements are deformed, more data needs to be loaded for the evaluation of the integrals. Figure 2.8 considers the case where a variable transformed velocity $\boldsymbol{J}_e^{-1}\boldsymbol{a}$ (three doubles) needs to be loaded for each quadrature point of the element integrals, and the normal part of the velocity $\boldsymbol{n} \cdot \boldsymbol{a}$ at all the quadrature points of the face integrals (one double). Here, the

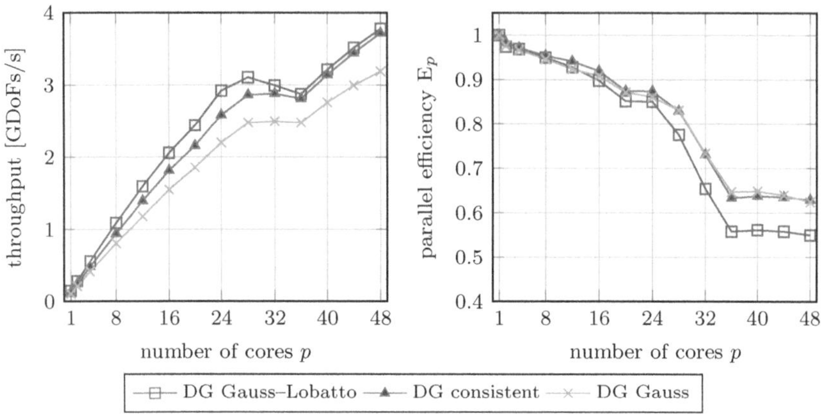

Fig. 2.8 Throughput of DG method with variable speed, reading tabulated values for $\boldsymbol{J}_e^{-1}\boldsymbol{a}$ at cell quadrature points and $\boldsymbol{n} \cdot \boldsymbol{a}$ at face quadrature points as a function of the number of cores

memory transfer increases by about 4 doubles per unknown in 3D, and the code is now operated in the completely memory bandwidth limited regime. As a result, the parallel efficiency figures look similar as for the finite difference stencil. Figure 2.8 reports essentially the same throughput for the "DG consistent" and "DG Gauss–Lobatto" schemes for high core counts because they have the same memory access patterns, and all additional computations of the consistent integration can be hidden behind the memory transfer. Note that the drop in performance between 24 and 36 cores observed both for the difference stencil in Fig. 2.7 and the DG scheme with variable speed in Fig. 2.8 is due to the hardware properties of the Intel Xeon Platinum and the particular way OpenMP schedules threads to the CPU cores.

In case the speed $\boldsymbol{a}$ is not the same in multiple evaluations of the DG right-hand side, a precomputed factor $\boldsymbol{J}_e^{-1}\boldsymbol{a}$ does not make sense, and instead both $\boldsymbol{J}_e^{-1}$ and $\boldsymbol{a}$ need to accessed from memory (or $\boldsymbol{a}$ needs to be computed from the spatial coordinates $\boldsymbol{x}(\hat{\boldsymbol{x}}_q)$). This leads to even lower throughput due to the increased memory transfer.

Roofline performance evaluation

The roofline performance model (Williams et al. 2009) can be represented visually in terms of the achievable GFlop/s rate. An upper horizontal line displays the limit of the arithmetic units, whereas the memory bandwidth limit in Eq. (8) appears as a diagonal line with slope according to the product of the arithmetic intensity and the available memory bandwidth. A given kernel cannot exceed the performance *roof* described by these two limits. The publication by Williams et al. (2009) shows that

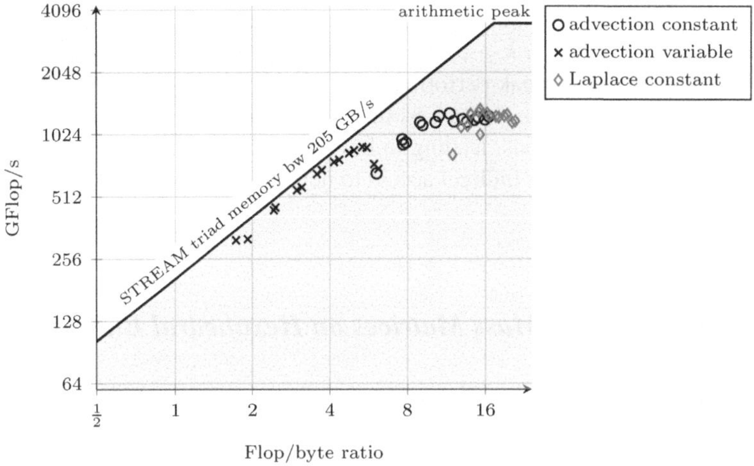

Fig. 2.9 Roofline model for the evaluation of the 3D advection operation (3) and the 3D Laplace operation (4) with constant coefficients and affine geometry as well as advection with variable coefficient on 2×24 Intel Xeon Platinum 8174 cores. Each experiments includes 16 data points, corresponding to polynomial degrees between $k = 1$ and $k = 16$. The memory transfer is based on the theoretical estimates of 16 bytes per unknwon (8 read, 8 write) for the affine geometry and $40 + 8 \cdot 6/(k + 1)$ bytes per unknown for the variable speed case

the model is flexible and additional limits can be inserted into the graph, such as the bandwidth from caches.

Figure 2.9 presents a roofline analysis of the 3D advection operation with constant and variable coefficients as well as the 3D Laplace operator discretized with the symmetric interior penalty method, i.e., the left-hand side of Eq. (4), using the techniques described in Kronbichler et al. (2019). The analysis covers all polynomial degrees from 1 to 16 for large problem sizes that access the global vectors from main (RAM) memory. All experiments are run with 2-way hyperthreading, i.e., 96 OpenMP threads. The roofline data is based on the best-case memory transfer and theoretical operation counts based on the measured DoF/s throughput. The assumption is that only the vectors and the variable velocity need to be loaded from memory, whereas all other data sits in caches, including the exterior solution values u_h^+ that are needed for the numerical flux. In reality, the transfer is higher. For example, for the constant-coefficient advection case, the actual transfer is between 21 and 25 bytes per unknown, rather than the theoretical transfer of 16 bytes. When computing the arithmetic intensity from the measured memory transfer, constant-coefficient advection is within 15% of the memory bandwidth limit up to degree $k \leq 6$. The roofline performance evaluation shows that the memory access is the main performance limit of the presented DG algorithms, especially on deformed meshes and on affine meshes for $k \leq 5$. To further increase performance, this analysis suggests that hardware with higher memory bandwidth or algorithmic variants to lower the memory access would be most profitable.

Figure 2.9 suggests that both the constant-coefficient advection case for $k \geq 7$ and the Laplace operator for $k \geq 5$ reach a limit of around 1.4 TFlop/s that is much lower than the arithmetic peak performance. The analysis in Kronbichler and Kormann (2019) lists some causes, such as the instruction mix (not all operations are fused multiply-add operations, leading to lower occupancy), latency in some local computations, the strided or indirect access to neighbor data for u_h^+, and the level-2 cache bandwidth.

Fast Inversion of DG Mass Matrices on Hexahedral Elements

Explicit time integration like the advection scheme (3) also needs the action of the inverse of the mass matrix $\mathcal{M}$, computed as

$$\mathcal{M}_{ij}^{(e)} = \left(\varphi_i^{(e)}, \varphi_j^{(e)}\right)_{\Omega_e}, \quad e = 1, \ldots, n,$$

on the integrals of the right-hand side. Due to separate basis functions for each element, $\mathcal{M}$ is easily inverted by inverting the matrices on each element. However, the application of a block-diagonal matrix (e.g., via an array of LU factors), stored with separate matrices for each element, would lead to a memory transfer of $(k+1)^d \times (k+1)^d$ values for $(k+1)^d$ unknowns. Since the evaluation of integrals with sum factorization is close in cost to the access of the vectors only, the inverse mass matrix would be several times more expensive than the integrals.

A large part of the DG community is therefore either considering the case where the same mass matrix (scaled by a single number) can be used on all elements like for affine element shapes of triangles and tetrahedra (Hesthaven and Warburton 2008), or bases where the mass matrix is indeed diagonal. Examples for the latter are the L_2-orthogonal Legendre basis using hierarchical polynomials or Lagrange polynomials on the points of the quadrature, such as Gauss points (consistent integration on affine elements) or Gauss–Lobatto points (inexact integration). While the diagonal property breaks down for deformed elements, the error made by assuming a diagonal mass matrix for Gauss quadrature does not alter discretization accuracy, as seen, e.g., by the experiments in the first chapter of this book.

The property of a diagonal mass matrix in L_2-orthogonal bases can also be used for other bases. The idea, proposed with the goal of avoiding the performance bottleneck of non-diagonal bases in Kronbichler et al. (2016), is to perform a change of the basis. The mass matrix on Ω_e is computed by the formula $\mathcal{M}^{(e)} = \mathcal{S}^T \mathcal{W}^{(e)} \mathcal{S}$, with $\mathcal{S}$ the interpolation matrix and $\mathcal{W}^{(e)}$ the diagonal matrix with the determinant of the Jacobian times the quadrature weight, $\det(\boldsymbol{J}_e(\hat{\boldsymbol{x}}_q))w_q$, as entries. If the number of polynomials equals the number of quadrature points, all matrices in $\mathcal{S}^T \mathcal{W}^{(e)} \mathcal{S}$ are square, and also the ingredients to $\mathcal{S}$ in the Kronecker product are square. Thus, one can invert each matrix to form the overall inverse, e.g., in two dimensions as

$$\left(\mathcal{M}^{(e)}\right)^{-1} = \left(\mathcal{S}^{\mathrm{1D}}\right)^{-1} \otimes \left(\mathcal{S}^{\mathrm{1D}}\right)^{-1} \left(\mathcal{W}^{(e)}\right)^{-1} \left(\mathcal{S}^{\mathrm{1D}}\right)^{-T} \otimes \left(\mathcal{S}^{\mathrm{1D}}\right)^{-T}.$$

This formula has exactly the same structure as the evaluation of integrals, with a different matrix, the projection matrix from the chosen basis into the Lagrange basis at the (Gaussian) quadrature points. Thus, the application of the inverse mass matrix on a vector can be computed with sum factorization techniques as described above.

With this procedure, the cost of applying non-diagonal mass matrices is essentially the same as for diagonal ones, given that the $2d$ sum factorization sweeps are cheaper than the transfer of vectors from memory, see the experiments in Schoeder et al. (2018a) and Kronbichler et al. (2019). This means that one is free to choose the basis that minimizes costs in other parts of the code, for example, by choosing Lagrange polynomials with nodes at the boundary of the reference domain as shown in the previous subsection.

Massively Parallel Computations

The DG method with explicit time integration runs well on large-scale parallel computers due to the nearest-neighbor communication. Figure 2.10 presents the results of three strong scaling experiments with the advection problem (3) on up to 256 nodes of SuperMUC-NG (12,288 cores), with 6.7 million, 53 million, and 426 million spatial unknowns, respectively. An MPI-only implementation is used. The time integration is based on an explicit Runge–Kutta scheme of order four with five stages, using a low-storage implementation according to Kennedy et al. (2000). In this case, a velocity that varies in space and time is tested, such that both $\boldsymbol{a}$ and $\boldsymbol{J}_e^{-1}$ need to be loaded at each quadrature point. The polynomial degree is $k = 4$.

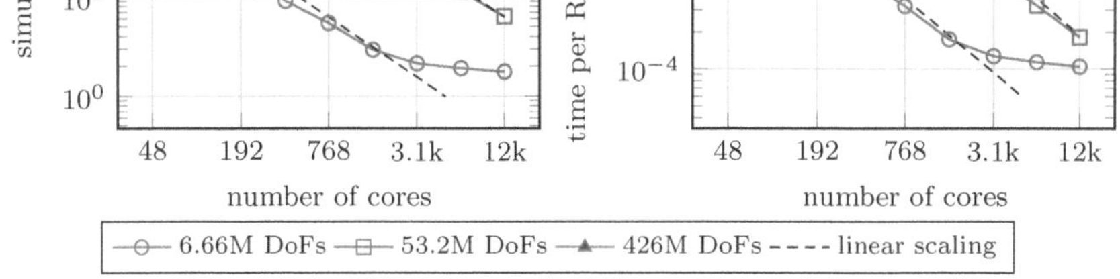

Fig. 2.10 Strong scaling of explicit time stepping of the advection operation on up 256 nodes of SuperMUC-NG

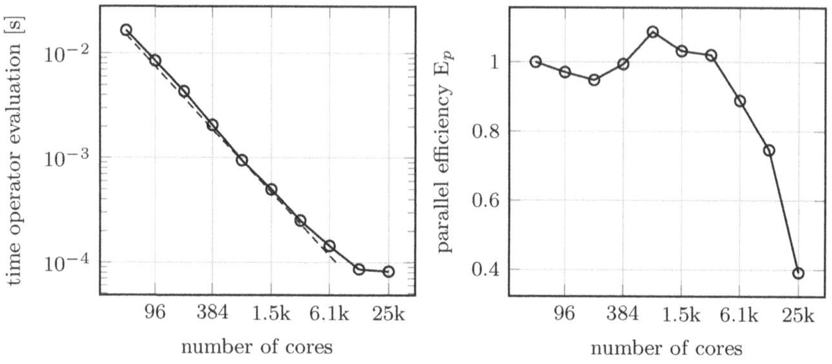

Fig. 2.11 Strong scaling of evaluation of the Laplace operator of Eq. (4) on an affine mesh with 524k elements and degree $k = 4$, giving 65.5 million unknowns

In the strong scaling experiment, the simulation run time goes down as the number of cores increases. For the smallest problem size, the scaling saturates above 3,072 cores. The reason for the saturation can be better identified in the right panel of Fig. 2.10, with a scaling limit of 10^{-4} s for one Runge–Kutta stage. One Runge–Kutta stage comprises the evaluation of the advection right-hand side, subsequent evaluation of the inverse mass matrix, and the Runge–Kutta vector updates. This limit can be related to the time it takes for a single point-to-point communication with MPI (a few microseconds), aggregated over 6–8 messages that are needed on the chosen mesh, see also Raffenetti et al. (2017) and experiments in Kronbichler and Wall (2018), Arndt et al. (2020c), and Fischer et al. (2020). Note that the lack of further scaling is not missing parallelism, as with 3,072 cores each core is responsible for approximately 17 cells and 2200 unknowns.

For run times per Runge–Kutta stage of around 10^{-3} s, Fig. 2.10 reveals a super-linear speedup, which can be seen by a deflection of the measured time below the dashed curve representing ideal scaling. This is due to a cache effect: As the local problem size gets smaller, the active data set (vectors, metric terms, velocities) eventually fits into the level-2 and level-3 cache of the processors, relaxing the memory bandwidth limit and making the code run faster, see also the graphs displaying the throughput over the computational time in Arndt et al. (2020c, Fig. 6).

Figure 2.11 presents a similar strong scaling experiment for the constant-coefficient Laplace operator discretized with the symmetric interior penalty method with a polynomial degree $k = 4$. A 3D mesh with 2^{19} elements of affine shape (parallelepipeds) is chosen, where the Jacobians $\boldsymbol{J}_e$ are constant over the mesh but have non-zero entries in all components. The strong scaling is very good with parallel efficiency above 95% all the way to $3 \cdot 10^{-4}$ s. The efficiency starts to decline once the absolute times reach 10^{-4} s, as in the advection experiment. The parallel efficiency goes above one when the data fits into cache, albeit in a less pronounced way than for the advection. Additional large-scale experiments with the infrastructure in the deal.II library up to the full size of the SuperMUC-NG machine are found in Arndt et al. (2020c), Fischer et al. (2020), and Arndt et al. (2020b).

Trends and Perspectives

The experiments in this section summarize the state of the art in the field which has found that high-performance implementations of sum factorization make the evaluation of DG (and high-order finite element) operators come close in throughput to simply copying the input vector to the output vector. This has implications on how to write code around these operators:

- All the integrals on cells and faces should be computed within the same loop in order to use data that sits in caches for architectures that have relatively slow memory access and good caches, like CPUs. Hence, large auxiliary data structures projecting solution data to, e.g., all faces in the mesh, like used in traditional DG implementations as in Kronbichler et al. (2017), need to be avoided. Likewise, operators composed of multiple integrals are more efficiently evaluated by a single loop combined at the quadrature points.
- The data access patterns have a strong influence on performance. Polynomial basis functions with a compact data access, like the nodal basis on the nodes of the Gauss–Lobatto quadrature formula proposed in Fig. 2.2, are beneficial.
- The performance documented for the case of a constant coefficient and affine geometry in Fig. 2.7 is not achievable for deformed geometries or variable speeds. It is an ongoing challenge to find algorithms to cheaply store some information and compute other on the fly to bring performance closer to the affine case.
- The evolution of hardware with computations getting cheaper allows, e.g., to change the basis in the inverse matrix evaluation without affecting throughput. Hence, collocation of quadrature and node positions of polynomials is not a must when targeting high throughput, as observed in Schoeder et al. (2018a).
- Given that the evaluation of a DG operator has similar costs as simple BLAS-1 operations, a holistic algorithm design is necessary, e.g., fusing the loops for the operator evaluation with vector updates in the Runge–Kutta scheme or the vector operations in an iterative solver like used in Arndt et al. (2020c).
- In order to best utilize the capabilities of SIMD units in CPUs or to achieve good performance on GPUs, hardware-specific implementations are required. The community is working on separating the formulation of the PDE or other equation-specific contributions from the hardware, e.g., by domain-specific languages, C++ template classes for different hardware targets, or libraries such as libCEED (Barra et al. 2020).
- The relevant application metric to assess performance is the throughput measured as DoFs/s, including the MPI communication. The community has started to develop well-defined benchmarks, such as the CEED bakeoff problems for high-order continuous finite element and spectral element implementations

(Fischer et al. 2020), following the principle "what gets measured gets improved". Nonetheless, cross-project performance comparisons are still rare and many scientific contributions fall short on comparing algorithmic variants against the *best-performing* baseline, a prerequisite for scientific progress (Hoefler and Belli 2015). Especially evolving architectures such as GPUs are sometimes compared against poor baseline implementations, leading to unrealistically high claims of performance, rather than the more realistic factor 2–4 observed, e.g., in Kronbichler and Ljungkvist (2019).

The Euler Equations

The high-performance DG ingredients presented in the previous section straightforwardly apply to explicit time integration of the multi-component systems in the fluid dynamics. In this section, the compressible Euler equations are considered as a model of fluids in the inviscid limit. Here, the velocity of the fluids very large, making viscous forces negligible compared to the inertial forces. Flow situations where this assumption is justified is for example air flow around aircrafts in the supersonic case (Ma $>$ 1), or for acoustic waves driven by density changes.

In conservative form, the compressible Euler equations are given by

$$\begin{aligned} \frac{\partial \rho}{\partial t} + \nabla \cdot (\rho \boldsymbol{u}) &= 0, \\ \frac{\partial \rho \boldsymbol{u}}{\partial t} + \nabla \cdot (\rho \boldsymbol{u} \otimes \boldsymbol{u} + p \boldsymbol{I}) &= \rho \boldsymbol{g}, \\ \frac{\partial E}{\partial t} + \nabla \cdot (\boldsymbol{u}(E + p)) &= \rho \boldsymbol{u} \cdot \boldsymbol{g}, \end{aligned} \tag{10}$$

where ρ denotes the fluid density, $\boldsymbol{u} = (u_1, u_2, \ldots, u_d)$ denotes the fluid velocity, and E the internal energy. The fluid pressure p is related to the other quantities by an equation of state, e.g., for an ideal gas,

$$p = (\gamma - 1)\left(E - \frac{1}{2}\rho \boldsymbol{u} \cdot \boldsymbol{u}\right),$$

with gas constant $\gamma = 1.4$. The rank-2 tensor $\rho \boldsymbol{u} \otimes \boldsymbol{u}$ represents the outer product between two vectors and $\boldsymbol{I}$ the $d \times d$ identity tensor. The vector of gravity is denoted by $\boldsymbol{g}$.

To ease notation, the Euler equations are written in concise vector form

$$\frac{\partial \boldsymbol{w}}{\partial t} + \nabla \cdot \boldsymbol{F}(\boldsymbol{w}) = \boldsymbol{G},$$

using the quantities

$$\boldsymbol{w} = [\rho, \rho\boldsymbol{u}, E]^T, \quad \boldsymbol{F}(\boldsymbol{w}) = \begin{bmatrix} \rho\boldsymbol{u} \\ \rho\boldsymbol{u} \otimes \boldsymbol{u} + p\boldsymbol{I} \\ \boldsymbol{u}(E+p) \end{bmatrix}, \quad \boldsymbol{G}(\boldsymbol{w}) = [0, \rho\boldsymbol{g}, \rho\boldsymbol{u}\cdot\boldsymbol{g}]^T.$$

For the DG discretization, polynomial approximations of the form (1) are applied for each component of the equation. Note that the polynomial approximation is set up for each component of the momentum $\rho\boldsymbol{u}$, rather than the velocity $\boldsymbol{u}$. The equations are multiplied by test functions $\boldsymbol{v}_h \in \mathcal{U}_h^{d+2}$, integrated over each element, and integrated by parts. Summing up over the elements gives the final weak form of the Euler equations,

$$\left(\boldsymbol{v}_h, \frac{\partial \boldsymbol{w}_h}{\partial t}\right)_{\Omega_h} - (\nabla \boldsymbol{v}_h, \boldsymbol{F}(\boldsymbol{w}_h))_{\Omega_h} + \left\langle \boldsymbol{v}_h, \boldsymbol{n}\cdot\boldsymbol{F}^*(\boldsymbol{w}_h)\right\rangle_{\partial\Omega_h} = (\boldsymbol{v}_h, \boldsymbol{G}(\boldsymbol{w}_h))_{\Omega_h}. \quad (11)$$

There exist many approximate Riemann solvers to define the numerical flux $\boldsymbol{F}^*$ for the Euler equations (LeVeque 2002; Hesthaven and Warburton 2008), with the most common variants being the basic local Lax–Friedrichs flux, the Roe flux, or the Harten–Lax–van Leer (contact) flux.

Equation (11) can be straight-forwardly propagated in time by an explicit time integrator, using the techniques from section "Fast Computation of Integrals with Sum Factorization" to compute the integrals and apply the inverse mass matrix. However, a basic DG scheme lacks robustness for higher Mach numbers. This is because the Euler equations develop shocks in the solution. The high-order nature of the polynomials leads to oscillations near shocks and as a result, the computation of derived quantities, such as the velocity $\boldsymbol{u}$ from the momentum $\rho\boldsymbol{u}$, with a division by the density ρ, will break down once ρ becomes zero or negative due to oscillations. Therefore, the need to find appropriate limiters is an active topic of research in the DG community, especially in the context of fast sum factorization implementations. Furthermore, the form (11) is not energy-conserving because integrals contain rational expressions and cannot be computed exactly. Nonetheless, over-integration with more quadrature points can sometimes help to relax the problems (Winters et al. 2018). We refer to the chapter by Winters, Kopriva, Gassner, and Hindenlang later in this book for split-form variants.

A Modern C++ Implementation—The Deal.II Step-67 Tutorial Program

For certain regimes of parameters, the Euler equations describe wave-like phenomena that are transported by a background fluid flow. For this scenario, the simple form (11) is sufficient. The `step-67` tutorial program part of the open-source finite element library deal.II (Arndt et al. 2020b, a) presents a modern implementation, written in the C++ programming language. The tutorial program can be studied online at https://

www.dealii.org/9.2.0/doxygen/deal.II/step_67.html

The code is based on a separation of the sum factorization algorithm (vectorized and highly optimized, inherited from deal.II) from the application-specific code. Algorithm 2 shows the implementation of the operation at quadrature points. All physics are queried through the `euler_flux` function that takes the state at the quadrature point with index `q`. The latter are defined in a compact way by free functions for the derived quantities such as $\boldsymbol{u}$, p, and speed of sound $c = \sqrt{\frac{\gamma p}{\rho}}$ or for the physical flux $\boldsymbol{F}$ (see Algorithm 3) and the numerical flux $\boldsymbol{F}^*$.

ALGORITHM 2: C++ implementation of the operation at quadrature points of cells for the Euler equations

```
const auto w_q = evaluator.get_value(q);
evaluator.submit_gradient(euler_flux<dim>(w_q), q);
if (body_force.get() != nullptr)
  evaluator.submit_value(compute_body_force(*body_force,
                             evaluator.quadrature_point(q), w_q), q);
```

ALGORITHM 3: C++ implementation computing the flux $\boldsymbol{F}(\boldsymbol{w}_q)$ from a vector of conserved variables for the Euler equations

```
template <int dim, typename Number>
inline Tensor<1, dim+2, Tensor<1, dim, Number>>
euler_flux(const Tensor<1, dim+2, Number> &conserved_variables)
{
  const auto velocity   = euler_velocity<dim>(conserved_variables);
  const Number pressure = euler_pressure<dim>(conserved_variables);
  Tensor<1, dim+2, Tensor<1, dim, Number>> flux;
  for (unsigned int d = 0; d < dim; ++d) {
    flux[0][d] = conserved_variables[1 + d];
    for (unsigned int e = 0; e < dim; ++e)
      flux[e + 1][d] = conserved_variables[e + 1] * velocity[d];
    flux[d + 1][d] += pressure;
    flux[dim + 1][d] =
      velocity[d] * (conserved_variables[dim + 1] + pressure);
  }
  return flux;
}
```

In Algorithm 3, quantities are defined in terms of the abstract (templated) number type `Number`. It can refer to a simple `double` variable, but in the code it is also used as a placeholder for an array in terms of the SIMD execution, called `VectorizedArray` in deal.II. By an expressive interface in terms of C++ operator overloading, the array can be used similarly to classical numbers, yet be directly translated to SIMD-vectorized instructions by the compiler. Furthermore, the space dimension is specified by a template argument `dim`, which allows to use the same code in 1D, 2D, and 3D without a performance penalty: Optimizing compilers unroll the loop over the derivative components. Also, `Tensor` is a fixed-size array on the stack without memory allocation. Note that the velocity and pressure are computed by the separate functions `euler_velocity` and `euler_pressure`. In order to avoid repeated computation of derived quantities, such as $\boldsymbol{u}$ needed both in $\boldsymbol{F}$ and for p, the code proposes to inline the functions such that code optimization passes of the compiler remove the redundant computations. This combines readable code and high-performance execution.

Besides sum factorization, the code makes use of a large range of other capabilities from the deal.II library, such as handling of parallel meshes, definition of curved cells, the definition of polynomials, or MPI communication. The program can also easily be extended to adaptive mesh refinement. Altogether, the user-visible code is around 2360 lines of C++ code, a good fraction of which in-code comments and postprocessor facilities.

Acoustic Wave Equation

An important specialization of the Euler equations is the acoustic wave equation, which describes the propagation of sound waves. It is the result of linearization of the Euler equations about a background density ρ_0 and speed of sound c in a resting fluid. The resulting equation is

$$\begin{aligned} \rho_0 \frac{\partial \boldsymbol{v}}{\partial t} + \nabla p = 0, \\ \frac{1}{c^2}\frac{\partial p}{\partial t} + \rho_0 \nabla \cdot \boldsymbol{v} = 0. \end{aligned} \tag{12}$$

Here, $\boldsymbol{v}$ denotes the particle velocity and p the acoustic pressure. This first-order system can be discretized with DG similar to the Euler equations, as detailed in Schoeder et al. (2018a, b), and available as an open-source project called ExWave at https://github.com/kronbichler/exwave[4] (Schoeder et al. 2019). When using a numerical flux with upwinding contributions, the discrete scheme inherits the dissipative properties of high-order DG schemes for barely resolved waves, preventing spurious contributions from growing. An alternative is to treat the wave equation in the

[4]Retrieved on July 27, 2020.

second-order form and apply DG on that system with techniques like the symmetric interior penalty method (4). This gives a smaller system, at the price of somewhat worse dissipative and dispersive properties.

Assuming constant c and ρ_0, the DG discretization of the first-order system (12) is derived by multiplication by test functions $(\boldsymbol{w}_h, q_h)$, integration over the elements, and integration by parts,

$$\begin{aligned} \left(\boldsymbol{w}_h, \frac{\partial \boldsymbol{v}_h}{\partial t}\right)_{\Omega_h} - \left(\nabla\cdot\boldsymbol{w}_h, \frac{1}{\rho_0} p_h\right)_{\Omega_h} + \left\langle \boldsymbol{w}_h, \frac{1}{\rho_0}\boldsymbol{n} p^*\right\rangle_{\partial\Omega_h} &= 0, \\ \left(q_h, \frac{\partial p}{\partial t}\right)_{\Omega_h} - \left(\nabla q_h, \rho_0 c^2 \boldsymbol{v}_h\right)_{\Omega_e} + \left\langle q_h \boldsymbol{n}, \rho_0 c^2 \boldsymbol{v}^*\right\rangle_{\partial\Omega_h} &= 0. \end{aligned} \tag{13}$$

An accurate numerical flux is the scheme originally proposed for the hybridizable discontinuous Galerkin flux by Nguyen et al. (2011) and used for explicit time stepping in Stanglmeier et al. (2016) and Kronbichler et al. (2016),

$$\begin{aligned} \frac{1}{\rho_0} p^* &= \frac{1}{2\tau}\left(\boldsymbol{v}_h^- - \boldsymbol{v}_h^+\right)\cdot\boldsymbol{n} + \frac{1}{2\rho_0}(p_h^- + p_h^+), \\ \rho_0 c^2 \boldsymbol{v}^* &= \rho_0 c^2 \frac{\boldsymbol{v}_h^- + \boldsymbol{v}_h^+}{2} + \frac{c^2\tau}{2}\boldsymbol{n}(p_h^- - p_h^+), \end{aligned}$$

where τ is a stabilization parameter typically chosen as $\frac{1}{c}$ (Nguyen et al. 2011). This flux can provide more accurate results than the local Lax–Friedrichs flux due to super-convergent postprocessing (Stanglmeier et al. 2016; Schoeder et al. 2018b).

Following the considerations of the previous sections, an efficient realization can be achieved by representing each component by a Lagrange polynomial basis on the Gauss–Lobatto points with a nodal property at the boundary, integrated with the consistent Gauss quadrature on $k+1$ points. In terms of arithmetic work, the spatial integral in the first equation of the weak form (13) involves the divergence of a vector-valued quantity, which requires $2d^2$ sum factorization sweeps, $2d$ for each of the d components (Kronbichler and Kormann 2019) and d sweeps for the solution interpolation, giving $2d^2 + d$ sweeps in total. Conversely, the spatial integral on the second line involves d^2 sum factorization sweeps for evaluating $\boldsymbol{v}_h$ at quadrature points and $2d$ sweeps for the gradient of the (scalar) test function, i.e., $d^2 + 2d$ sweeps. Thus, the spatial integral in the first equation is integrated by parts, resulting in the equation

$$\begin{aligned} \left(\boldsymbol{w}_h, \frac{\partial \boldsymbol{v}_h}{\partial t}\right)_{\Omega_h} + \left(\boldsymbol{w}_h, \frac{1}{\rho_0}\nabla p_h\right)_{\Omega_h} + \left\langle \boldsymbol{w}_h, \frac{1}{\rho_0}\boldsymbol{n}(p^* - p_h)\right\rangle_{\partial\Omega_h} &= 0, \\ \left(q_h, \frac{\partial p}{\partial t}\right)_{\Omega_h} - \left(\nabla q_h, \rho_0 c^2 \boldsymbol{v}_h\right)_{\Omega_e} + \left\langle q_h \boldsymbol{n}, \rho_0 c^2 \boldsymbol{v}^*\right\rangle_{\partial\Omega_h} &= 0. \end{aligned} \tag{14}$$

This form has the additional advantage of being skew-symmetric, which gives energy stability also with inexact integration.

Computational Challenges for Wave Propagation

In the context of wave propagation, like the Euler equations away from shocks, the linearized Euler equations, or the acoustic wave equation, the challenge is often to transport moderate to high-frequency content over long distances. When operated near the resolution limit, high-order methods are significantly more accurate in terms of the dissipation and dispersion behavior than low-order methods. Furthermore, the reflection of waves at material interfaces or exterior boundaries such as walls necessitates a high-order curved representation of interfaces, which the DG method can handle well by using high-order mappings.

Solving Linear Systems

The techniques presented in the previous sections allow for a highly efficient implementation of explicit time integration. Additional ingredients are needed when solving linear systems of stationary problems or with implicit time integration. For direct solvers, fast evaluation methods are not directly applicable, because a matrix needs to be explicitly set up before it can be factored with some sparse solver (Schenk and Gärtner 2004; Davis 2004). The hybridizable discontinuous Galerkin method, presented in the chapter by Fernández-Méndez of this book, offers a way to reduce the size of the associated system. For larger three-dimensional problems, iterative solvers, such as the conjugate gradient (CG) or generalized minimum residual (GMRES) method (Saad 2003), need to be employed. Iterative methods only need the *action* of the matrix on a vector. Then, sum factorization algorithms can be used to implement a *matrix-free evaluation* of the underlying discrete operator, using the spectral element (Orszag 1980; Patera 1984; Fischer and Patera 1991; Deville et al. 2002; Karniadakis and Sherwin 2005; Kopriva 2009) and high-order finite element background (Brown 2010; Cantwell et al. 2011; Kronbichler and Kormann 2012; Kronbichler and Wall 2018). In the context of nonlinear problems, this leads to the concept of Jacobian-free Newton–Krylov methods (Knoll and Keyes 2004).

Matrix-free operator evaluation inherits the algorithmic advantages presented in section "Fast Computation of Integrals with Sum Factorization" in terms of a high throughput, and is beneficial for iterative solvers which are dominated by the matrix-vector product. Compared to conventional sparse matrix-vector products, matrix-free implementations can give speedups of an order of magnitude or more for polynomial degrees between $k = 2$ and $k = 5$. This speedup has its reasons both in the higher arithmetic intensity and in the complexity of operator evaluation as $\mathcal{O}(k)$ per unknown, as compared to $\mathcal{O}(k^d)$ for traditional sparse matrices or $\mathcal{O}(k^{d-2})$ for methods relying on static condensation or hybridized discontinuous Galerkin approaches, see Kronbichler and Wall (2018) for a recent performance evaluation. Other research has approximated high-order schemes with sparser stencils, such as the work by Persson (2013). Matrix-free evaluation with sum factorization on hexahedra is also several times faster per unknown than a sparse matrix-vector product resulting from linear finite elements (Arndt et al. 2020c), and close in throughput to stencil representations of low-order finite elements according to Gmeiner et al. (2015) or difference stencils. Matrix-free methods allow to avoid the high memory consumption of the densely coupled matrices of high-order DG schemes, see, e.g., the contributions by Bassi et al. (2020) or Yan et al. (2020) using memory consumption as the primary driver rather than performance.

Preconditioners

The cost of iterative solvers is proportional to the number of iterations, which in turn depends on the condition number of the underlying system matrix. Approximate inverses of the operator that are cheap to compute are then used to accelerate the convergence, called preconditioners, as elaborated in the chapter by Persson later in this book. For linear systems of low to moderate condition numbers that are efficiently solved with either no preconditioner or the diagonal of the matrix (Jacobi preconditioning), matrix-free methods provide an immediate benefit because all operations apart from the matrix-vector product only involve BLAS-1 type vector operations. This is the case for discretizations of time-dependent problems with implicit time stepping and small enough time steps, such as the advection system (3) with $\Delta t \lesssim \frac{h}{|a|}$, or discretizations of time-dependent heat conduction of the form $\frac{\partial u}{\partial t} - \epsilon \nabla^2 u$ with time step size $\Delta t \lesssim \frac{h^2}{\epsilon}$.

However, many of the successful preconditioners established by the linear algebra community, such as incomplete factorizations (Saad 2003), sparse approximate inverses (Grote and Huckle 1997), or algebraic multigrid (Ruge and Stüben 1987; Vaněk et al. 1996), explicitly rely on matrix entries. For relatively low polynomial degrees $k = 2, 3$, matrices with some coupling dropped have been used to define incomplete factorizations or Gauss–Seidel type operations (Kronbichler et al. 2018; Yan et al. 2020). Also, the spectral equivalence of a high-order method to a discretization with linear shape functions on a refined mesh with the same number of unknowns is frequently used (Orszag 1980; Olson 2007; Pazner 2019; Franco et al. 2020).

Alternative preconditioners are based on the fast diagonalization method originally proposed by Lynch et al. (1964) and used for the approximate inversion of an overlapping Schwarz setup for spectral element methods by Lottes and Fischer (2005) and Huismann et al. (2017). Block-Jacobi schemes on the unknowns of an element using approximate tensor product inverses have been pursued in Pazner and Persson (2018), Diosady and Murman (2019), and Witte et al. (2019).

Multigrid Solvers and Preconditioners

For operators with a dominating elliptic contribution, the condition number scales as $\mathcal{O}(h^{-2})$. Basic preconditioners act only locally and can therefore not provide robust iteration counts for large-scale elliptic operators, meaning that the number of iteration grows as the problem size increases. One of the most powerful solution methods with optimal complexity $\mathcal{O}(N)$ for N unknowns are multigrid methods, see also Gholami et al. (2016) and Ibeid et al. (2019) for an evaluation of fast Poisson solvers. Multigrid methods combine simple iterative schemes, like the Jacobi or Gauss–Seidel iteration or incomplete factorizations, on a hierarchy of coarser problem representations. The iterative schemes need to be effective in reducing the high-frequency content of the error, which explains their name *smoothers*. The low-frequency content turns into higher frequencies on coarser grids, which means that the recursive application of simple schemes can effectively reduce all error components with optimal methods. The process of traversing through the grid levels with a direct solver on the coarsest level is called a V-cycle for the classical approach of visiting each level once; there are also variants like the W-cycle or F-cycle with additional operations on coarser levels (Trottenberg et al. 2001).

In the classical geometric multigrid method, the coarsening hierarchy is generated by discretizing the problem also on coarser meshes. Transfer operators between the mesh levels, called restriction (fine to coarse) and prolongation (coarse to fine), can be created by the natural embedding operators. Algebraic multigrid methods generate the coarsening solely based on the connectivity in the matrix. For high-order discontinuous Galerkin methods which have many unknowns within the elements, additional coarsening options are available, such as p-multigrid or the transfer to continuous finite element methods (auxiliary space methods), see the chapter by Colombo, Ghidoni, Noventa, and Rebay later in this book as well as Fehn et al. (2020b) for an experimental evaluation in the HPC context.

With respect to the smoother selection, similar considerations as for preconditioners apply when aiming for fast matrix-free implementations. Often, there is a conflict between the cost of a preconditioner application and the effectiveness of the smoothers. A popular choice is the Chebyshev iteration around the Jacobi method (Adams et al. 2003), which combines the inverse of the matrix diagonal or any other cheap preconditioner with matrix-vector products with parameters that damp high frequencies effectively. It was used in a matrix-free context, e.g., in May et al. (2014, 2015), Kronbichler and Wall (2018), Kronbichler and Ljungkvist (2019), or Fehn et al. (2020b). Block-Jacobi smoothers with local iterative solvers (Bastian et al. 2019) or overlapping Schwarz methods relying on (approximate) tensor product inverses (Lottes and Fischer 2005; Huismann et al. 2019; Witte et al. 2019) are also widespread.

Multigrid methods can be used as standalone solvers or as preconditioners in standard iterative schemes like CG or GMRES. The use as preconditioner provides additional robustness especially for variable coefficients and non-uniform meshes. On the other hand, additional efficiency can be gained with multigrid solvers in case of high-quality smoothers and level transfer operations. The most successful variant is the full multigrid method or its nonlinear variant, the full approximation scheme (Trottenberg et al. 2001). Here, the solution is recursively approximated starting from the coarser levels, such that only a single V-cycle is needed on the finest level to reach errors on the order of the discretization accuracy.

Research Trends

The main trend in the context of linear solvers for high-order DG schemes is the development of cheap yet effective preconditioners and smoothers that offer similar performance per application as the matrix-free operator evaluation presented in section "Fast Computation of Integrals with Sum Factorization". The main ingredient is approximate inversions based on the fast diagonalization method or local iterative solvers. In this context, the topic of textbook multigrid efficiency is pursued, which means that solving the linear system should cost not more than ten matrix-vector products (Thomas et al. 2003; Gmeiner et al. 2015). As indicated by the efficiency model (7), a more expensive iterative solver needs to allow for significantly larger time steps to improve the time-to-solution.

One of the major challenges of massively parallel application of iterative solvers is the strong scaling limit. Multigrid V-cycles, with parallelization by domain partitioning on the levels and a direct solver on the coarsest level, imply a specific form of global communication. It is related to a tree-based implementation of `MPI_Allreduce`, but using the communication hierarchy implied by the grid coarsening and level transfer operations. Furthermore, the communication tree is interleaved with computations and nearest-neighbor communication during the matrix-vector products and smoother operations. The strong scaling limit of a multigrid V-cycle is on the order of 3–20 ms (Arndt et al. 2020b, c) on modern large-scale computers with state-of-the-art implementations, a factor of around 50 higher than the scaling limit of a single matrix-vector product. There are efforts on non-nested cycles, e.g., based on the BPX scheme originally proposed by Bramble et al. (1991).

Another challenge arising with matrix-free evaluation of very high throughput is the fact that the matrix-vector product may no longer be the dominant cost, with vector updates of smoothers or transfer taking a significant share of the time. This was observed on CPUs in Kronbichler and Ljungkvist (2019) and is expected to become more pressing on future architectures (Ibeid et al. 2019). Hence, different operations within the algorithm need to be fused into a single loop, which necessitates modifications to software interfaces. Finally, the development of mixed-precision setups, such as smoothing in single-precision or even half-precision, combined with double-precision corrections, has the potential to increase throughput by less memory transfer and higher hardware capabilities, see the results by Kronbichler and Ljungkvist (2019), Fehn et al. (2020b), and Oo and Vogel (2020).

The Incompressible Navier–Stokes Equations

The compressible Navier–Stokes equations extend the Euler equations by including viscous effects. With explicit time integration, the ingredients presented for the Euler equations on the one hand and for second-derivative terms on the other hand can be applied straight forwardly, see Fehn et al. (2019b) for an example. The implementation of implicit methods for the compressible Navier–Stokes equations is an area of active research, see, e.g., Yan et al. (2020) and references therein, as the cost of solving a nonlinear system of equations needs to be weighed against the gain in time step size. Besides the ratio between inertial and viscous effects, the choice also depends on the type of mesh elements—problems involving uniformly sized elements at small viscosities are difficult to implement more efficiently with implicit time stepping than with explicit methods with the current state of the art. The picture can change for meshes with a few strongly deformed elements, a topic discussed in the chapter by Persson later in this book.

In this section, an incompressible fluid with constant fluid density ρ is allowing to skip the energy equation in favor of the fluid pressure p as a primal solution variable. The assumption of incompressibility is accurate for air with Mach numbers up to around $\mathrm{Ma} \approx 0.3$ or for most liquid flows without additional physical effects.

The incompressible Navier–Stokes equations seek to find the fluid velocity $\boldsymbol{u}$ and the (kinematic) fluid pressure p that fulfill the system of partial differential equations

$$\begin{aligned} \frac{\partial \boldsymbol{u}}{\partial t} + \nabla \cdot (\boldsymbol{u} \otimes \boldsymbol{u} + p\boldsymbol{I} - \nu \nabla \boldsymbol{u}) &= \boldsymbol{f}, \\ \nabla \cdot \boldsymbol{u} &= 0, \end{aligned} \tag{15}$$

where ν denotes the (kinematic) fluid viscosity. The equations are completed by the initial condition $\boldsymbol{u}(\cdot, t = 0) = \boldsymbol{u}_0$ and boundary conditions of Dirichlet type, $\boldsymbol{u} = \boldsymbol{g}_u$ on $\partial\Omega_\mathrm{D}$ (inflow or walls with $\boldsymbol{g}_u = \boldsymbol{0}$, and Neumann type, $(-p\boldsymbol{I} + \nu\nabla\boldsymbol{u}) \cdot \boldsymbol{n} = \boldsymbol{h}$ on $\partial\Omega_\mathrm{N}$ (outflow). The Dirichlet and Neumann portions form a partition of the boundary.

Compared to the compressible Navier–Stokes equations, the incompressible Navier–Stokes equations eliminate the time scale of acoustic wave propagation and have therefore lower resolution requirements. In addition, the spatial terms are at most quadratically nonlinear via the convective term $\nabla \cdot (\boldsymbol{u} \otimes \boldsymbol{u})$, compared to rational expressions in the compressible formulation. Finally, the incompressible Navier–Stokes equations do not develop shocks, avoiding the topic of limiting.

Similarities of flow with different viscosities ν as well as length and velocity scales are revealed by non-dimensionalizing the equations and introducing the Reynolds number as the ratio between interial and viscous effects as

$$\mathrm{Re} = \frac{|\boldsymbol{u}_\mathrm{c}| L_\mathrm{c}}{\nu}$$

with some characteristic magnitude of velocity $|\boldsymbol{u}_\mathrm{c}|$ and some characteristic length scale L_c.

Many practically relevant flows are turbulent with Reynolds numbers Re in the range $10^5 \dots 10^7$. The amount of scales contributing to the flow physics is proportional to $\mathrm{Re}^{3/4}$ per dimension, i.e., the computational work scales as Re^3 in three space dimensions plus time, which is at the edge or often beyond the capabilities of even the largest supercomputers. Here, the range of scales is defined as the ratio from the global geometrical features to be represented to the size of the smallest vortex-like structures developing in the flow. There exist techniques to simulate with lower resolution than the one required for a complete resolution of all relevant processes in a direct numerical simulation (DNS), like the large eddy simulation (LES) or the Reynolds-averaged Navier–Stokes equations (RANS).

Time Integration

For a spatial discretization with discrete solution vectors $\boldsymbol{U}_h$ and $\boldsymbol{P}_h$, to be derived in section "Discretization in Space" below, the resulting system is

$$\begin{bmatrix} \mathcal{M} & 0 \\ 0 & 0 \end{bmatrix} \begin{bmatrix} \frac{\mathrm{d}\boldsymbol{U}_h}{\mathrm{d}t} \\ \boldsymbol{0} \end{bmatrix} + \begin{bmatrix} \mathcal{C}(\boldsymbol{U}_h) + \mathcal{A} & \mathcal{G} \\ \mathcal{D} & 0 \end{bmatrix} \begin{bmatrix} \boldsymbol{U}_h \\ \boldsymbol{P}_h \end{bmatrix} = \begin{bmatrix} \boldsymbol{F}_h \\ \boldsymbol{0} \end{bmatrix}. \tag{16}$$

Here, $\mathcal{M}$ is the velocity mass matrix, $\mathcal{C}(\boldsymbol{U}_h)\boldsymbol{U}_h$ represents the nonlinear convective term, the viscous matrix is denoted by $\mathcal{A}$, the pressure gradient operator by $\mathcal{G}$, and the divergence-free condition by $\mathcal{D}$. The matrix representation reveals a saddle point structure with a zero matrix on the diagonal $(2, 2)$ block. It is a consequence of the fact that only the fluid velocity $\boldsymbol{u}$ is subject to a time derivative, whereas the pressure p is an instantaneous variable that is defined from the constraint to make the velocity divergence-free, $\nabla \cdot \boldsymbol{u} = 0$. As a result, fully explicit time stepping of the incompressible Navier–Stokes equations is not immediately possible, and at least the pressure needs to be determined implicitly. In terms of linear algebra, the choice of appropriate preconditioners for system (16) has remained an active research topic (Elman et al. 2005).

There are two alternatives which avoid the nonlinear saddle point system. The first option, which is not discussed here, is to introduce an artificial compressibility as a numerical parameter, see, e.g., Noventa et al. (2016), Loppi et al. (2018), Guermond and Minev (2019), or Manzanero et al. (2020). The second option are projection methods, where each time step is divided into several sub-steps with a simpler structure. A general overview of projection methods is given by Guermond et al. (2006). The present text focuses on the *dual-splitting method* introduced by Karniadakis et al. (1991) using the backward differentiation formula (BDF) family of time integrators. For the order $J = 1, 2, 3$, a time step is split into the four phases as follows:

- Explicit convective step,

$$\gamma_0 \hat{\boldsymbol{u}} = \sum_{i=0}^{J-1} \left(\alpha_i \boldsymbol{u}^{n-i} - \Delta t \beta_i \nabla \cdot (\boldsymbol{u}^{n-i} \otimes \boldsymbol{u}^{n-i}) \right) + \Delta t \boldsymbol{f}^{n+1},$$

where γ_0 and α_i are coefficients of the BDF-J time integrator and β_i the respective coefficients for Jth order extrapolation to the new time level t^{n+1}. For $J = 2$ and constant time steps, the coefficients are $\gamma_0 = \frac{3}{2}$, $\alpha_0 = 2$, $\alpha_1 = -\frac{1}{2}$ and $\beta_0 = 2$, $\beta_1 = -1$.

- Implicit pressure step involving the Poisson equation

$$-\nabla^2 p^{n+1} = -\frac{\gamma_0}{\Delta t} \nabla \cdot \hat{\boldsymbol{u}}$$

with a Neumann boundary condition on Dirichlet portion $\partial\Omega_{\mathrm{D}}$ of the boundary,

$$\nabla p^{n+1} \cdot \boldsymbol{n} = -\Bigg[\frac{\partial \boldsymbol{g}_u(t_{n+1})}{\partial t} + \sum_{i=0}^{J_p-1} \beta_i \left(\nabla \cdot (\boldsymbol{u} \otimes \boldsymbol{u})^{n-i} + \nu \nabla \times (\nabla \times \boldsymbol{u}^{n-i}) \right) - \boldsymbol{f}(t_{n+1}) \Bigg] \cdot \boldsymbol{n}.$$

This condition was proposed in Karniadakis et al. (1991) as the remainder term in the momentum equation with the rotational form of the viscous term in order to reduce the divergence errors near the boundary. In the definition of the boundary extrapolation, $J_p \leq J$ steps are used. The splitting is unconditionally stable only for $J_p \leq 2$ with time error $\mathcal{O}(\Delta t^3)$ in the velocity and $\mathcal{O}(\Delta t^{5/2})$ for the pressure. On the Neumann boundary $\partial\Omega_{\mathrm{N}}$ of the fluid, a Dirichlet value $p^{n+1} = g_p(t_{n+1})$ is imposed based on a decomposition of the fluid Neumann condition $\boldsymbol{h} = \boldsymbol{h}_u - g_p \boldsymbol{n}$ into a viscous contribution $\boldsymbol{h}_u$ for defining $(\nu \nabla \boldsymbol{u}) \cdot \boldsymbol{n} = \boldsymbol{h}_u$ and a pressure contribution g_p.

- Projection step,

$$\hat{\hat{\boldsymbol{u}}} = \hat{\boldsymbol{u}} - \frac{\Delta t}{\gamma_0} \nabla p^{n+1}.$$

- Implicit viscous step,

$$\frac{\gamma_0}{\Delta t} \left(\boldsymbol{u}^{n+1} - \hat{\hat{\boldsymbol{u}}} \right) = \nu \nabla \cdot \nabla \boldsymbol{u}^{n+1},$$

using the boundary data $\boldsymbol{g}_u$ on Dirichlet boundaries and $\boldsymbol{h}_u$ on Neumann boundaries.

Note that the explicit treatment of the convective term imposes a time step limit similar to explicit time integration for linear transport. The settings $J = 2$ and $J = 3$ are most common, combining stability of the underlying BDF integrator and extrapolation with the stability of the pressure Neumann condition. The time step limit of the convective term usually renders the time step size small enough to make the time error negligible compared to the errors of the spatial discretizations.

Discretization in Space

The DG approximate solutions $\boldsymbol{u}_h$ and p_h are defined in terms of polynomial expansions according to (1). The weak form is derived by multiplying the momentum equation by vector-valued test functions $\boldsymbol{v}_h$ and the continuity equation by scalar test functions q_h and subsequent integration. The discrete representation is done term-by-term as follows, see Fehn et al. (2017) for details,

- Mass operator:
$$M_h^{(e)}(\boldsymbol{v}_h, \boldsymbol{u}_h) = (\boldsymbol{v}_h, \boldsymbol{u}_h)_{\Omega_e}\,.$$
- For the convective operator, we integrate by parts
$$C_h^{(e)}(\boldsymbol{v}_h, \boldsymbol{u}_h; \boldsymbol{g}_{\boldsymbol{u}}) = -\,(\nabla \boldsymbol{v}_h, \boldsymbol{u}_h \otimes \boldsymbol{u}_h)_{\Omega_e} + \left\langle \boldsymbol{v}_h \otimes \boldsymbol{n}, (\boldsymbol{u} \otimes \boldsymbol{u})^* \right\rangle_{\partial\Omega_e},$$
with a common numerical flux given by the local Lax–Friedrichs flux,
$$(\boldsymbol{u} \otimes \boldsymbol{u})^* = \{\{\boldsymbol{u} \otimes \boldsymbol{u}\}\} + \frac{\max(|\boldsymbol{u}_h^-|, |\boldsymbol{u}_h^+|)}{2} [\![\boldsymbol{u}]\!]\,.$$
Along the boundary, the outer values $\boldsymbol{u}_h^+ = -\boldsymbol{u}_h^- + 2\boldsymbol{g}_{\boldsymbol{u}}$ on the Dirichlet portion $\partial\Omega_{\mathrm{D}}$ and $\boldsymbol{u}_h^+ = \boldsymbol{u}_h^-$ on the Neumann portion $\partial\Omega_{\mathrm{N}}$ of the boundary according to the mirror principle are used.
- The velocity divergence on the right-hand side of the pressure Poisson equation is represented by integration by parts as
$$D_h^{(e)}(q_h, \boldsymbol{u}_h; \boldsymbol{g}_{\hat{\boldsymbol{u}}}) = -\,(\nabla q_h, \boldsymbol{u}_h)_{\Omega_e} + \langle q_h \boldsymbol{n}, \{\{\boldsymbol{u}_h\}\}\rangle_{\partial\Omega_e},$$
using a central flux $\boldsymbol{u}^* = \{\{\boldsymbol{u}_h\}\}$. As the divergence operator is applied on the intermediate-step velocity $\hat{\boldsymbol{u}}$, care is needed for the definition of the boundary condition. The papers by Fehn et al. (2017) and Fehn et al. (2020a) propose to use a value extrapolated from the momentum equation at old times steps in order to obtain optimal convergence rates. Note that the numerical flux $\boldsymbol{u}^*$ is essential to avoid instabilities for small time steps (Fehn et al. 2017).
- The pressure Poisson operator is discretized using the symmetric interior penalty method according to Eq. (4), with outside values p_h^+ and ∇p_h^+ according to the dual-splitting boundary condition defined in section "Time Integration".
- The pressure gradient term in the projection step is defined as
$$G_h^{(e)}(\boldsymbol{v}_h, p_h; g_p) = -\,(\nabla \cdot \boldsymbol{v}_h, p_h)_{\Omega_e} + \langle \boldsymbol{v}_h \cdot \boldsymbol{n}, \{\{p_h\}\}\rangle_{\partial\Omega_e},$$

using $p_h^+ = p_h^-$ on $\partial\Omega_{\mathrm{D}}$ and $p_h^+ = -p_h^- + 2g_p$ on $\partial\Omega_{\mathrm{N}}$.

- The viscous operator is discretized with the interior penalty method applied to each velocity component,

$$V_h^{(e)}(\boldsymbol{v}_h, \boldsymbol{u}_h; \boldsymbol{g}_u, \boldsymbol{h}_u) = (\nabla \boldsymbol{v}_h, \nu \nabla \boldsymbol{u}_h)_{\Omega_e} - \left\langle \nabla \boldsymbol{v}_h, \nu \frac{1}{2} [\![\boldsymbol{u}_h]\!] \right\rangle_{\partial\Omega_e}$$
$$- \langle \boldsymbol{v}_h \boldsymbol{n}, \nu \{\{\nabla \boldsymbol{u}_h\}\} \rangle_{\partial\Omega_e} + \langle \boldsymbol{v}_h \boldsymbol{n}, \tau\nu [\![\boldsymbol{u}_h]\!] \rangle_{\partial\Omega_e},$$

again using the mirror principle and boundary conditions for the definition of the outside values along the boundary.

The convective step and the projection step only involve a mass matrix on the left-hand side, which is cheap to apply, as are the right-hand sides. On the other hand, the pressure step and the viscous step involve the solution of coupled linear systems. For the pressure Poisson equation, the most common solver type are multigrid algorithms outlined in section "Multigrid Solvers and Preconditioners". For the viscous operator, simpler solvers like the conjugate gradient method preconditioned by the inverse mass matrix are the most efficient choice as long as $\nu \Delta t \lesssim h^2$, which is often the case for moderate to high-Reynolds numbers due to the CFL limit (Fehn et al. 2018b).

Stability

The first ingredient to stability is the compatibility of the velocity and pressure ansatz spaces related to the inf–sup condition. In analogy to the inf–sup stable Taylor–Hood element pair, Fehn et al. (2017) proposed to use mixed orders with degree $k+1$ for the velocity and degree k for the pressure, as equal-order pairs with the above fluxes violate the inf–sup condition for certain examples.

The second ingredient is an energy estimate of the numerical solution. We consider the spatial discretization of the original saddle point system (15), insert the spatial operators defined above, and keep the time continuous. Other time stepping schemes are discussed in Fehn et al. (2020a). Furthermore, assume a domain with periodic boundaries and no forcing, $\boldsymbol{f} = \boldsymbol{0}$. The resulting weak form is

$$M_h\left(\boldsymbol{v}_h, \frac{\partial \boldsymbol{u}_h}{\partial t}\right) + C_h(\boldsymbol{v}_h, \boldsymbol{u}_h) + V_h(\boldsymbol{v}_h, \boldsymbol{u}_h) - G_h(\boldsymbol{v}_h, p_h) = 0, \tag{17}$$

$$D_h(\boldsymbol{q}_h, \boldsymbol{u}_h) = 0. \tag{18}$$

With the energy method, we monitor the time derivative

$$\frac{\mathrm{d}}{\mathrm{d}t} E_{u_h}(t) = \frac{1}{2}\frac{\mathrm{d}}{\mathrm{d}t} M_h(\boldsymbol{u}_h(\cdot,t), \boldsymbol{u}_h(\cdot,t)) = M_h\left(\boldsymbol{u}_h(\cdot,t), \frac{\partial \boldsymbol{u}_h(\cdot,t)}{\partial t}\right)$$
$$= -C_h(\boldsymbol{u}_h, \boldsymbol{u}_h) - V_h(\boldsymbol{u}_h, \boldsymbol{u}_h) + G_h(\boldsymbol{u}_h, p_h),$$

where the second line has been obtained by inserting Eq. (17). By integration by parts, one can show that $G_h(\boldsymbol{v}_h, q_h) = -D_h(q_h, \boldsymbol{v}_h)$ for all $\boldsymbol{v}_h$ and q_h. Thus, the last term on the right-hand side can be replaced by $-D_h(p_h, \boldsymbol{u}_h)$, which is zero by (18). Furthermore, the discretization of the viscous operator is coercive, $V_h(\boldsymbol{u}_h, \boldsymbol{u}_h) \geq c\|\boldsymbol{u}_h\|^2$ for some constant c that does not depend on the solution $\boldsymbol{u}_h$ but only the parameters of the discretization. As a consequence,

$$\frac{\mathrm{d}}{\mathrm{d}t} E_{\boldsymbol{u}_h} \leq -C_h(\boldsymbol{u}_h, \boldsymbol{u}_h). \tag{19}$$

In the analytical setting, one can show that the convective term is neutral in terms of energy. A stable numerical scheme replicates the exact energy behavior, i.e., the numerical energy must be non-increasing. For the element integral in C_h on element Ω_h, it holds that

$$(\nabla \boldsymbol{u}_h, \boldsymbol{u}_h \otimes \boldsymbol{u}_h)_{\Omega_e} = -\frac{1}{2} (\boldsymbol{u}_h, (\nabla \cdot \boldsymbol{u}_h)\boldsymbol{u}_h)_{\Omega_e} + \frac{1}{2} \langle \boldsymbol{u}_h, (\boldsymbol{u}_h \otimes \boldsymbol{u}_h) \cdot \boldsymbol{n} \rangle_{\partial\Omega_e}, \tag{20}$$

which is the result of integration by parts as follows:

$$-(\nabla \boldsymbol{u}_h, \boldsymbol{u}_h \otimes \boldsymbol{u}_h)_{\Omega_e} = (\boldsymbol{u}_h, \nabla \cdot (\boldsymbol{u}_h \otimes \boldsymbol{u}_h))_{\Omega_e} - \langle \boldsymbol{u}_h \otimes \boldsymbol{n}, \boldsymbol{u}_h \otimes \boldsymbol{u}_h \rangle_{\partial\Omega_e},$$

applying the product rule $\nabla \cdot (\boldsymbol{u}_h \otimes \boldsymbol{u}_h) = (\nabla \cdot \boldsymbol{u}_h)\boldsymbol{u}_h + \boldsymbol{u}_h \cdot (\nabla \boldsymbol{u}_h)$ combined with the equivalences $(\boldsymbol{u}_h, \boldsymbol{u}_h \cdot (\nabla \boldsymbol{u}_h))_{\Omega_e} = (\boldsymbol{u}_h \otimes \boldsymbol{u}_h, \nabla \boldsymbol{u}_h)_{\Omega_e}$ and $\langle \boldsymbol{u}_h \otimes \boldsymbol{n}, \boldsymbol{u}_h \otimes \boldsymbol{u}_h \rangle_{\partial\Omega_e} = \langle \boldsymbol{u}_h, (\boldsymbol{u}_h \otimes \boldsymbol{u}_h) \cdot \boldsymbol{n} \rangle_{\partial\Omega_e}$. Inserting relation (20) into the equation of $C_h^{(e)}(\boldsymbol{u}_h, \boldsymbol{u}_h)$, summing up over all elements, and combing terms along the interior faces Γ_{i} yields

$$\begin{aligned} C_h(\boldsymbol{u}_h, \boldsymbol{u}_h) = \frac{1}{2} (\boldsymbol{u}_h \cdot \boldsymbol{u}_h, \nabla \cdot \boldsymbol{u}_h)_{\Omega_h} &- \frac{1}{2} \left\langle \boldsymbol{u}_h^-, \left(\boldsymbol{u}_h^+ \otimes \boldsymbol{u}_h^+\right) \cdot \boldsymbol{n}^- \right\rangle_{\Gamma_{\mathrm{i}}} \\ &- \frac{1}{2} \left\langle \boldsymbol{u}_h^+, \left(\boldsymbol{u}_h^- \otimes \boldsymbol{u}_h^-\right) \cdot \boldsymbol{n}^+ \right\rangle_{\Gamma_{\mathrm{i}}} + \left\langle [\![\boldsymbol{u}_h]\!], \frac{\Lambda}{2} [\![\boldsymbol{u}_h]\!] \right\rangle_{\Gamma_{\mathrm{i}}}, \end{aligned}$$

with the constant $\Lambda = \max(|\boldsymbol{u}_h^-|, |\boldsymbol{u}_h^+|)$. Some algebraic manipulations are applied to the second and third term on the right-hand side. Then, the expression is inserted into the energy relation (19),

$$\begin{aligned} \frac{\mathrm{d}}{\mathrm{d}t} E_{\boldsymbol{u}_h} \leq &- \frac{1}{2} (\boldsymbol{u}_h \cdot \boldsymbol{u}_h, \nabla \cdot \boldsymbol{u}_h)_{\Omega_h} + \frac{1}{2} \left\langle \left(\boldsymbol{u}_h^- - \boldsymbol{u}_h^+\right) \cdot \boldsymbol{n}^-, \boldsymbol{u}_h^- \cdot \boldsymbol{u}_h^+ \right\rangle_{\Gamma_{\mathrm{i}}} \\ &- \left\langle [\![\boldsymbol{u}_h]\!], \frac{\Lambda}{2} [\![\boldsymbol{u}_h]\!] \right\rangle_{\Gamma_{\mathrm{i}}}. \end{aligned} \tag{21}$$

The last term on the right-hand side involves the square of $[\![\boldsymbol{u}_h]\!]$ and is thus non-increasing. However, the first two terms in the energy estimate cannot be bounded a priori. While the divergence $\nabla \cdot \boldsymbol{u}_h$ should be zero due to the continuity equation, the equation is only satisfied in a weak sense and subject to a discretization error. Likewise, the directed jump in the solution $\left(\boldsymbol{u}_h^- - \boldsymbol{u}_h^+\right) \cdot \boldsymbol{n}^-$, written $[\boldsymbol{u}_h] \cdot \boldsymbol{n}$ in the following, becomes small as the numerical solution approaches the analytical solution.

As elaborated in the recent publications by Krank et al. (2017), Fehn et al. (2018a, 2019a), these two terms indeed render the numerical scheme unstable for *under-resolved* flows in the context of implicit large eddy simulation, despite optimal convergence rates for sufficiently fine meshes and high polynomial degrees. Here, under-resolved means that the numerical approximation is too coarse to resolve all vortex structures in a (turbulent) flow of a certain Reynolds number. While the jump term originating from the Lax–Friedrichs flux could in principle absorb parts of the jump in the normal direction, it becomes inefficient to choose Λ very large due to its influence on the maximal stable time step size in an explicit treatment of convection.

Pressure Robustness and H(div) Conforming Schemes

As a more robust solution, the community has developed pressure-robust discretizations in the last decade, see Fehn et al. (2019a) and references therein. From an energy point of view, the natural approximation space for the velocity would be $H(\text{div})$, a mix of fully continuous and fully discontinuous function spaces such that the velocity $\boldsymbol{u}_h$ is continuous over element faces in the normal direction but discontinuous in the other directions. This ensures that $[\boldsymbol{u}_h] \cdot \boldsymbol{n} = 0$ by construction of the function space. The normal continuity can be generalized to deformed elements by using a so-called contravariant Piola transform.

For a normal-continuous velocity space, the divergence of the velocity $\nabla \cdot \boldsymbol{u}_h$ is a conforming operation on Ω_h without the need for face terms. By choosing the local polynomial expansion of the velocity to degree $k+1$ in the continuous direction and degree k in the other directions (the Raviart–Thomas finite element space), the divergence of the velocity (seen as a global function) *exactly* maps to the discontinuous L_2-conforming space of the pressure of degree k. As a result, the weak continuity equation, $D_h(q_h, \boldsymbol{u}_h) = 0$ for all q_h induces a velocity field that is exactly divergence-free, i.e., divergence-free in all points of the computational domain. This is because $\nabla \cdot \boldsymbol{u}_h$ can be represented by a weighted sum of all test functions q_h. However, the equation is zero for all test functions by the best-approximation property in a Galerkin weak form. The $H(\text{div})$ property and the judicious choice of polynomial degrees makes both problematic terms in the energy estimate (21), the term involving $\nabla \cdot \boldsymbol{u}_h$ and the term involving the jump of the solution in normal direction, $[\boldsymbol{u}_h] \cdot \boldsymbol{n}$, disappear, giving a stable spatial discretization.

An alternative scheme, presented in Fehn et al. (2018a), is to weakly enforce the continuity of the numerical solution and the divergence-free condition by a penalization approach. This is implemented by treating the velocity $\hat{\hat{\boldsymbol{u}}}_h$ after the viscous step of the dual-splitting algorithm as intermediate (because it still violates $\left[\hat{\hat{\boldsymbol{u}}}_h\right] \cdot \boldsymbol{n} = 0$ and $\nabla \cdot \hat{\hat{\boldsymbol{u}}}_h = 0$) and perform a *stabilization* step that finds the end-of-step velocity

$\boldsymbol{u}_h^{n+1}$ as

$$\left(\boldsymbol{v}_h, \boldsymbol{u}_h^{n+1}\right)_{\Omega_h} + \left(\nabla \cdot \boldsymbol{v}_h, \tau_{\mathrm{D}} \nabla \cdot \boldsymbol{u}_h^{n+1}\right)_{\Omega_h} + \left\langle [\boldsymbol{v}_h] \cdot \boldsymbol{n}, \tau_{\mathrm{C}} \left[\boldsymbol{u}_h^{n+1}\right] \cdot \boldsymbol{n}\right\rangle_{\Gamma_{\mathrm{i}}}$$
$$= \left(\boldsymbol{v}_h, \hat{\hat{\boldsymbol{u}}}_h\right)_{\Omega_h} .$$

Here, the parameters $\tau_{\mathrm{D}} = \zeta_{\mathrm{D}} \overline{\|\boldsymbol{u}_h^n\|} h_e/(k_u + 1)$ and $\tau_{\mathrm{C}} = \zeta_{\mathrm{C}} \overline{\|\boldsymbol{u}_h^n\|}$ involve dimensionless factors ζ_{D} and ζ_{C} and are scaled by the velocities and mesh size h_e. When choosing $\boldsymbol{v}_h = \boldsymbol{u}_h$ in the energy estimate (21), they represent $|\nabla \cdot \boldsymbol{u}_h|^2$ and $|[\boldsymbol{u}_h] \cdot \boldsymbol{n}|^2$ with a negative sign when moved to the right-hand side of the energy estimate. Thus, the energy is decreased by a factor corresponding to the discretization error. By choosing ζ_{D} and ζ_{C} large enough, these terms can dominate over the problematic terms of the energy estimate (21). In terms of function spaces, we observe that the stabilization terms represent a penalty to the divergence and jump in normal velocity. Thus, the properties of an $H(\mathrm{div})$-conforming solution space are mimicked weakly. The results in Fehn et al. (2019a) show that this weak enforcement gives similar solution quality as a Raviart–Thomas velocity space with a similar number of unknowns.

Finally, we note that the energy estimate (21) has been derived under the assumption of exact integration. For curved elements with inexact integration, additional measures are necessary to make the scheme provably energy-stable, such as skew-symmetric weak forms in the spirit of split-form DG methods (Kopriva and Gassner 2016). While these schemes could avoid the need for $H(\mathrm{div})$ conformity from an energy point of view, exactly divergence-free discrete velocities $\boldsymbol{u}_h$ ensure that additional conservation laws of the Navier–Stokes equations, besides energy or some entropy underlying the split-form, can be accurately represented.

Computational Examples

The accuracy of the presented incompressible Navier–Stokes solver is first assessed on a simple benchmark setup of a two-dimensional vortex flow with the analytical velocity

$$u_1(x, y, t) = -\sin(2\pi y) \mathrm{e}^{-4\pi^2 \nu t}, \qquad u_2(x, y, t) = \sin(2\pi x) \mathrm{e}^{-4\pi^2 \nu t},$$

and the pressure

$$p(x, y, t) = -\cos(2\pi x) \cos(2\pi y) \mathrm{e}^{-8\pi^2 \nu t}.$$

The computational domain is $\Omega = [-0.5, 0.5]^2$. In this problem, inflow is along the vertical lines $[-0.5] \times (-0.5, 0)$ and $[0.5] \times (0, 0.5)$ as well as the horizontal lines $(0, 0.5) \times [-0.5]$ and $(-0.5, 0) \times [0.5]$, with Dirichlet data $\boldsymbol{g}_u$ set according to the analytical solution. On the other half of the boundary, the flow leaves the domain and

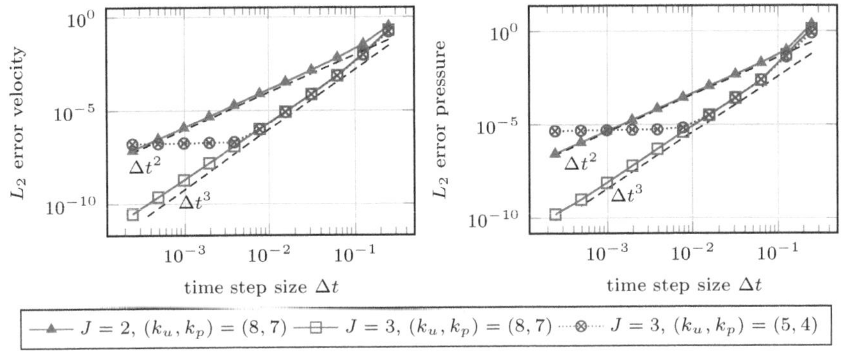

Fig. 2.12 Convergence of dual-splitting method with 8^2 elements of velocity and pressure degrees (8, 7) and (5, 4) with orders $J = 2, 3$ in time

Neumann conditions using a split into g_p and $\boldsymbol{h}_u$ according to the analytical solution are used. We consider the case with viscosity $\nu = 0.025$ and the final time $T_f = 1$. Continuity and divergence penalty stabilization with $\zeta_C = 1$ and $\zeta_D = 1$ is used.

Figure 2.12 lists the relative error recorded when choosing a very accurate spatial discretization with 8^2 elements of polynomial degree $k_u = 8$ for the velocity and $k_p = 7$ for the pressure. Optimal convergence in time according to the orders $J = 2, 3$ of the applied time stepping scheme is observed. Note that this test stresses in particular the boundary conditions and the various components in the dual-splitting scheme, whereas the spatial resolution requirements are modest. The plots in Fig. 2.12 also include a coarser spatial discretization of degree $(k_u, k_p) = (5, 4)$ that initially follows the error of the higher resolved case. However, as the time stepping becomes more accurate, the error levels off due to the spatial error.

Figure 2.13 presents a spatial convergence study. Here, a very small time step size $\Delta t = 2.44 \cdot 10^{-4}$ is chosen for all mesh sizes and polynomial degrees. Optimal rates of convergence $\mathcal{O}(h^{k_u+1})$ for the fluid velocity and $\mathcal{O}(h^{k_p+1})$ for the fluid pressure compared to the analytical solution. Note that the pressure converges with one order less than the velocity, in line with the lower degree of the pressure shape functions. There is a slight deviation of the convergence for the highest resolutions with both $k_u = 4, 5$, which is due to the influence of the error made by the time stepping.

In Fig. 2.13, higher orders deliver much more accurate results than lower orders also when using the metric of the number of spatial unknowns. With an efficient implementation according to section "Fast Computation of Integrals with Sum Factorization", it is possible to demonstrate better time to solution both for smooth solutions and for 3D turbulent flows in the pre-asymptotic regime, see Fehn et al. (2018b).

As a more challenging test, the so-called Orr–Sommerfeld stability problem is considered, see, e.g., Fischer (1997). It is a slight perturbuation of a Poiseuille flow in a channel of extent $[0, 2\pi] \times [-1, 1]$ with periodic boundary conditions in x direction and homogeneous Dirichlet boundary conditions $\boldsymbol{u} = \boldsymbol{0}$ on the walls. The

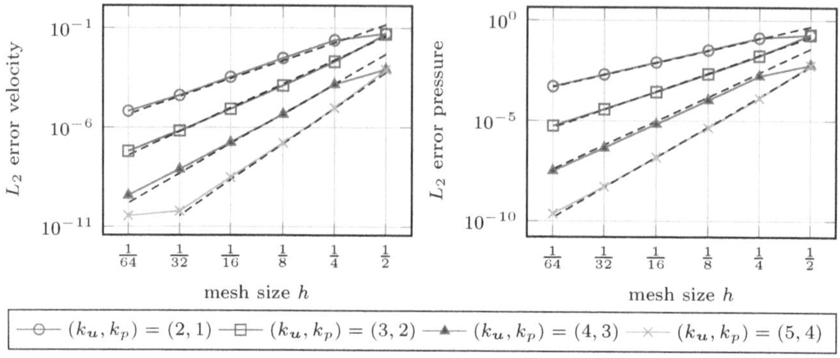

Fig. 2.13 Convergence of dual-splitting method with time step size $\Delta t = 2.44 \cdot 10^{-4}$ for various mesh sizes and polynomial degrees. Optimal rates of convergence of $\mathcal{O}(h^{k_u+1})$ for velocity and $\mathcal{O}(h^{k_p+1})$ are indicated by dashed lines

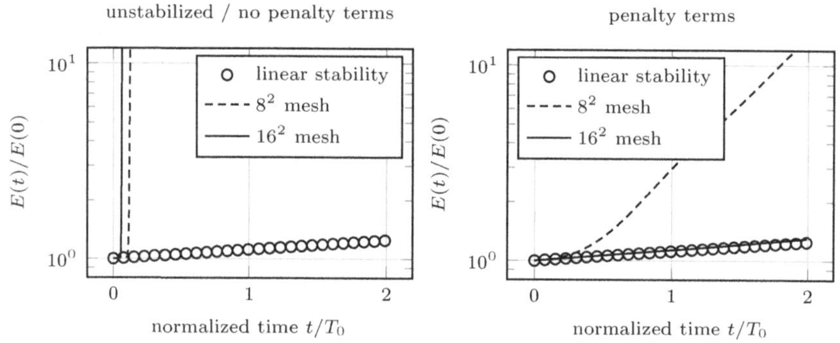

Fig. 2.14 Development of perturbation energy of Orr–Sommerfeld stability problem with plain DG solver (left) and the stabilized approach based on divergence and continuity penalty terms (right) compared to the result of linear stability theory

viscosity is set to $\nu = 1/7500$ with a unit-size velocity, which results in a Reynolds number related to half of the channel height of Re $= 7500$. As an initial condition, a perturbation velocity derived from a stream function $\varepsilon\psi(x_2)\exp(i(x_1 - \omega t))$ is added to the horizontal flow field. Here, ψ and ω are found as the unstable eigenfunction and eigenvalue to the Orr–Sommerfeld differential equation, a fourth order differential equation in the space variable y, with ω taking complex values. This results in the initial velocity field

$$
\begin{aligned}
u_1(x_1, x_2, t = 0) &= (1 - x_2^2) + \varepsilon \mathrm{Re}\left\{\frac{\mathrm{d}\psi(x_2)}{\mathrm{d}x_2} \mathrm{e}^{i(x_1 - \omega t)}\right\}, \\
u_2(x_1, x_2, t = 0) &= -\varepsilon \mathrm{Re}\left\{i\psi(x_2)\mathrm{e}^{i(x_1 - \omega t)}\right\}.
\end{aligned}
$$

The size of the perturbation is chosen as $\varepsilon = 10^{-5}$. Figure 2.14 presents the development of the perturbation energy,

$$E(t) = \int_{\Omega} \left(\boldsymbol{u}_h(\boldsymbol{x}, t) - \left((1 - x_2)^2, 0 \right)^T \right)^2 \, \mathrm{d}\boldsymbol{x},$$

as compared to the initial perturbation energy. The polynomial degree is chosen to $k_u = 3, k_p = 2$ and the time is normalized by the travel time of the perturbation waves through the domain. Clearly, the plain DG discretization without stabilization terms, which violates the energy conservation according to Eq. (21), quickly becomes unstable. On the other hand, the stabilized approach is stable for all mesh sizes and converges towards the reference result as the mesh is refined.

Perspectives

The stabilized DG discretization of the incompressible Navier–Stokes equations is compatible with fast matrix-free ingredients from sections "Fast Computation of Integrals with Sum Factorization" and "Solving Linear Systems" by design. However, similarly efficient implementations for $H(\mathrm{div})$-conforming elements with Piola transforms have not been developed yet, so a performance comparison to inform whether the stabilized approach or $H(\mathrm{div})$-conformity perform better is still an open topic. Given that sum factorization is most efficient on hexahedral meshes for which, however, mesh generation is non-trivial, practical applications need support for hex-dominant meshes, unfitted (cut), or overlapping mesh techniques, combining body-fitted boundary layer meshes with a more structured background mesh.

The HPC challenges of incompressible flow solvers are naturally the ones identified in sections "Fast Computation of Integrals with Sum Factorization" and "Solving Linear Systems", such as the strong scaling limit of pressure multigrid solvers with global communication in the V-cycle. The topic is pressing because of time step limitations due to the CFL condition from treating the convective term explicitly, leading to millions of time steps for many industrially relevant high-Reynolds number flows. As a result, sub-stepping of the convective term has been proposed by Maday et al. (1990) and applied for incompressible flow in many works, e.g., recently in Karakus et al. (2019), but a fair evaluation of performance versus accuracy has not yet been given to the best of the author's knowledge. On the other end, the community has been approaching increasingly larger problems as computer systems are becoming more powerful. This needs continuous improvements to node-level throughput where memory bandwidth is the limit. Likewise, the development of better iterative solvers to enable fully implicit schemes with better performance than the dual-splitting scheme is subject to intensive research. Finally, new trends to overcome the strong scaling limit, such as parallel-in-time methods, are slowly gaining momentum for simpler equations such as parabolic problems or wave propagation, and it needs to be determined whether extensions to the nonlinear Navier–Stokes equations are viable.

References

Adams, M., Brezina, M., Hu, J., & Tuminaro, R. (2003). Parallel multigrid smoothing: Polynomial versus Gauss-Seidel. *Journal of Computational Physics*, *188*, 593–610. https://doi.org/10.1016/S0021-9991(03)00194-3.

Alnæs, M. S., Logg, A., Ølgaard, K. B., Rognes, M. E., & Wells, G. N. (2014). Unified form language. *ACM Transactions on Mathematical Software*, *40*(2), 1–37. https://doi.org/10.1145/2566630.

Alnæs, M. S., Blechta, J., Hake, J., Johansson, A., Kehlet, B., Logg, A., et al. (2015). The FEniCS project version 1.5. *Archive of Numerical Software*, *3*(100). https://doi.org/10.11588/ans.2015.100.20553.

Alzetta, G., Arndt, D., Bangerth, W., Boddu, V., Brands, B., Davydov, D., et al. (2018). The `deal.II` library, version 9.0. *Journal of Numerical Mathematics*, *26*(4), 173–184. https://doi.org/10.1515/jnma-2018-0054.

Amdahl, G. M. (1967). Validity of the single processor approach to achieving large scale computing capabilities. In *AFIPS Conference Proceedings* (Vol. 30, pp. 483–485). https://doi.org/10.1145/1465482.1465560.

Anderson, R., Andrej, J., Barker, A., Bramwell, J., Camier, J.-S., Cerveny, J., et al. (2020). MFEM: A modular finite element methods library. *Computers and Mathematics with Applications*, in press. https://doi.org/10.1016/j.camwa.2020.06.009.

Arndt, D., Bangerth, W., Blais, B., Clevenger, T. C., Fehling, M., Grayver, A. V., et al. (2020a). The deal.II library, version 9.2. *Journal of Numerical Mathematics*, in press. https://doi.org/10.1515/jnma-2020-0043.

Arndt, D., Bangerth, W., Davydov, D., Heister, T., Heltai, L., Kronbichler, M., et al. (2020b). The deal.II finite element library: Design, features, and insights. *Computers and Mathematics with Applications*, in press. https://doi.org/10.1016/j.camwa.2020.02.022.

Arndt, D., Fehn, N., Kanschat, G., Kormann, K., Kronbichler, M., Munch, P., et al. (2020c). ExaDG – high-order discontinuous Galerkin for the exa-scale. In H.-J. Bungartz, S. Reiz, B. Uekermann, P. Neumann, & W. E. Nagel (Eds.), *Software for exascale computing – SPPEXA 2016–2019*. Lecture notes in computational science and engineering (Vol. 136, pp. 189–224). Cham: Springer International Publishing. https://doi.org/10.1007/978-3-030-47956-5_8.

Bangerth, W., Burstedde, C., Heister, T., & Kronbichler, M. (2011). Algorithms and data structures for massively parallel generic finite element codes. *ACM Transactions on Mathematical Software*, *38*(2), 14:1–14:28. https://doi.org/10.1145/2049673.2049678.

Barra, V., Beams, N., Brown, J., Camier, J.-S., Dobrev, V., Dudouit, Y., et al. (2020). libCEED development site. https://github.com/ceed/libceed.

Bassi, F., Botti, L., Colombo, A., Crivellini, A., Franciolini, M., Ghidoni, A., et al. (2020). A p-adaptive matrix-free discontinuous Galerkin method for the implicit LES of incompressible transitional flows. *Flow, Turbulence and Combustion*, *105*(2), 437–470. https://doi.org/10.1007/s10494-020-00178-2.

Bastian, P., Müller, E. H., Müthing, S., & Piatkowski, M. (2019). Matrix-free multigrid block-preconditioners for higher order discontinuous Galerkin discretisations. *Journal of Computational Physics*, *394*, 417–439. https://doi.org/10.1016/j.jcp.2019.06.001.

Bastian, P., Blatt, M., Dedner, A., Dreier, N.-A., Engwer, C., Fritze, R., et al. (2020). The DUNE framework: Basic concepts and recent developments. *Computers and Mathematics with Applications*, in press. https://doi.org/10.1016/j.camwa.2020.06.007.

Bramble, J. H., Pasciak, J. E., & Xu, J. (1991). The analysis of multigrid algorithms with nonnested spaces or noninherited quadratic forms. *Mathematics of Computation, 56*(193), 1–1. https://doi.org/10.1090/s0025-5718-1991-1052086-4.

Brightwell, R., Riesen, R., & Underwood, K. D. (2005). Analyzing the impact of overlap, offload, and independent progress for message passing interface applications. *International Journal of High Performance Computing Applications, 19*(2), 103–117. https://doi.org/10.1177/1094342005054257.

Brown, J. (2010). Efficient nonlinear solvers for nodal high-order finite elements in 3D. *Journal of Scientific Computing, 45*(1–3), 48–63.

Buis, P. E., & Dyksen, W. R. (1996). Efficient vector and parallel manipulation of tensor products. *ACM Transactions on Mathematical Software, 22*(1), 18–23.

Burstedde, C., Wilcox, L. C., & Ghattas, O. (2011). p4est: Scalable algorithms for parallel adaptive mesh refinement on forests of octrees. *SIAM Journal on Scientific Computing, 33*(3), 1103–1133. https://doi.org/10.1137/10079163, http://p4est.org.

Cantwell, C. D., Sherwin, S. J., Kirby, R. M., & Kelly, P. H. J. (2011). Form h to p efficiently: Selecting the optimal spectral/*hp* discretisation in three dimensions. *Mathematical Modelling of Natural Phenomena, 6*.

Cantwell, C. D., Moxey, D., Comerford, A., Bolis, A., Rocco, G., Mengaldo, G., et al. (2015). Nektar++: An open-source spectral/hp element framework. *Computer Physics Communications, 192*, 205–219. https://doi.org/10.1016/j.cpc.2015.02.008.

Chang, J., Fabien, M. S., Knepley, M. G., & Mills, R. T. (2018). Comparative study of finite element methods using the time-accuracy-size (TAS) spectrum analysis. *SIAM Journal on Scientific Computing, 40*(6), C779–C802. https://doi.org/10.1137/18m1172260.

Davis, T. A. (2004). Algorithm 832: UMFPACK V4.3—an unsymmetric-pattern multifrontal method. *ACM Transactions on Mathematical Software, 30*, 196–199. https://doi.org/10.1145/992200.992206.

Deville, M. O., Fischer, P. F., & Mund, E. H. (2002). *High-order methods for incompressible fluid flow* (Vol. 9). Cambridge: Cambridge University Press.

Diosady, L. T., & Murman, S. M. (2019). Scalable tensor-product preconditioners for high-order finite-element methods: Scalar equations. *Journal of Computational Physics, 394*, 759–776. https://doi.org/10.1016/j.jcp.2019.04.047.

Elman, H., Silvester, D., & Wathen, A. (2005). *Finite elements and fast iterative solvers with applications in incompressible fluid dynamics*. Oxford: Oxford Science Publications.

Fehn, N., Wall, W. A., & Kronbichler, M. (2017). On the stability of projection methods for the incompressible Navier–Stokes equations based on high-order discontinuous Galerkin discretizations. *Journal of Computational Physics, 351*, 392–421. https://doi.org/10.1016/j.jcp.2017.09.031.

Fehn, N., Wall, W. A., & Kronbichler, M. (2018a). Robust and efficient discontinuous Galerkin methods for under-resolved turbulent incompressible flows. *Journal of Computational Physics, 372*, 667–693. https://doi.org/10.1016/j.jcp.2018.06.037.

Fehn, N., Wall, W. A., & Kronbichler, M. (2018b). Efficiency of high-performance discontinuous Galerkin spectral element methods for under-resolved turbulent incompressible flows. *International Journal for Numerical Methods in Fluids, 88*(1), 32–54. https://doi.org/10.1002/fld.4511.

Fehn, N., Kronbichler, M., Lehrenfeld, C., Lube, G., & Schroeder, P. W. (2019a). High-order DG solvers for under-resolved turbulent incompressible flows: A comparison of L^2 and $H(\mathrm{div})$ methods. *International Journal for Numerical Methods in Fluids, 91*(11), 533–556. https://doi.org/10.1002/fld.4763.

Fehn, N., Wall, W. A., & Kronbichler, M. (2019b). A matrix-free high-order discontinuous Galerkin compressible Navier–Stokes solver: A performance comparison of compressible and incompressible formulations for turbulent incompressible flows. *International Journal for Numerical Methods in Fluids, 89*(3), 71–102. https://doi.org/10.1002/fld.4683.

Fehn, N., Heinz, J., Wall, W. A., & Kronbichler, M. (2020a). High-order arbitrary Lagrangian-Eulerian discontinuous Galerkin methods for the incompressible Navier–Stokes equations. Technical report. arXiv:2003.07166.
Fehn, N., Munch, P., Wall, W. A., & Kronbichler, M. (2020b). Hybrid multigrid methods for high-order discontinuous Galerkin discretizations. *Journal of Computational Physics*, *415*, 109538. https://doi.org/10.1016/j.jcp.2020.109538.
Fischer, P., Min, M., Rathnayake, T., Dutta, S., Kolev, T., Dobrev, V., et al. (2020). Scalability of high-performance PDE solvers. *International Journal of High Performance Computing Applications*, *34*(5), 562–586. https://doi.org/10.1177/1094342020915762.
Fischer, P. F. (1997). An overlapping Schwarz method for spectral element solution of the incompressible Navier–Stokes equations. *Journal of Computational Physics*, *133*(1), 84–101. https://doi.org/10.1006/jcph.1997.5651.
Fischer, P. F., & Patera, A. T. (1991). Parallel spectral element solution of the Stokes problem. *Journal of Computational Physics*, *92*(2), 380–421. https://doi.org/10.1016/0021-9991(91)90216-8.
Fischer, P. F., Kerkemeier, S., et al. (2020). Nek5000 Web page. https://nek5000.mcs.anl.gov.
Franco, M., Camier, J.-S., Andrej, J., & Pazner, W. (2020). High-order matrix-free incompressible flow solvers with GPU acceleration and low-order refined preconditioners. *Computers and Fluids*, *203*, 104541. https://doi.org/10.1016/j.compfluid.2020.104541.
Gholami, A., Malhotra, D., Sundar, H., & Biros, G. (2016). FFT, FMM, or multigrid? A comparative study of state-of-the-art Poisson solvers for uniform and nonuniform grids in the unit cube. *SIAM Journal on Scientific Computing*, *38*(3), C280–C306. https://doi.org/10.1137/15M1010798.
Gmeiner, B., Rüde, U., Stengel, H., Waluga, C., & Wohlmuth, B. (2015). Towards textbook efficiency for parallel multigrid. *Numerical Mathematics-Theory, Methods and Applications*, *8*(1), 22–46.
Göddeke, D., Strzodka, R., & Turek, S. (2007). Performance and accuracy of hardware-oriented native-, emulated-and mixed-precision solvers in FEM simulations. *International Journal of Parallel, Emergent and Distributed Systems*, *22*(4), 221–256. https://doi.org/10.1080/17445760601122076.
Grote, M. J., & Huckle, T. (1997). Parallel preconditioning with sparse approximate inverses. *SIAM Journal on Scientific Computing*, *18*(3), 838–853. https://doi.org/10.1137/s1064827594276552.
Guermond, J.-L., & Minev, P. (2019). High-order adaptive time stepping for the incompressible Navier–Stokes equations. *SIAM Journal on Scientific Computing*, *41*(2), A770–A788. https://doi.org/10.1137/18m1209301.
Guermond, J. L., Minev, P., & Shen, J. (2006). An overview of projection methods for incompressible flows. *Computer Methods in Applied Mechanics and Engineering*, *195*(44–47), 6011–6045. https://doi.org/10.1016/j.cma.2005.10.010.
Gustafson, J. L. (1988). Reevaluating Amdahl's law. *Communications of the ACM*, *31*(5), 532–533. https://doi.org/10.1145/42411.42415.
Hager, G., & Wellein, G. (2011). *Introduction to high performance computing for scientists and engineers*. Boca Raton: CRC Press.
Hager, G., Treibig, J., Habich, J., & Wellein, G. (2016). Exploring performance and power properties of modern multi-core chips via simple machine models. *Concurrency and Computation*, *28*(2), 189–210. https://doi.org/10.1002/cpe.3180.
Hesthaven, J. S., & Warburton, T. (2008). *Nodal discontinuous Galerkin methods: Algorithms, analysis, and applications*. Berlin: Springer. https://doi.org/10.1007/978-0-387-72067-8.
Hindenlang, F., Gassner, G., Altmann, C., Beck, A., Staudenmaier, M., & Munz, C.-D. (2012). Explicit discontinuous Galerkin methods for unsteady problems. *Computers and Fluids*, *61*, 86–93.
Hoefler, T., & Belli, R. (2015). Scientific benchmarking of parallel computing systems. In *Proceedings of the International Conference for High Performance Computing, Networking, Storage and Analysis on - SC'15*. ACM Press. https://doi.org/10.1145/2807591.2807644.

Huismann, I., Stiller, J., & Fröhlich, J. (2017). Factorizing the factorization - a spectral-element solver for elliptic equations with linear operation count. *Journal of Computational Physics*, *346*, 437–448. https://doi.org/10.1016/j.jcp.2017.06.012.

Huismann, I., Stiller, J., & Fröhlich, J. (2019). Scaling to the stars – a linearly scaling elliptic solver for p-multigrid. *Journal of Computational Physics*, *398*, 108868. https://doi.org/10.1016/j.jcp.2019.108868.

Huismann, I., Stiller, J., & Fröhlich, J. (2020). Efficient high-order spectral element discretizations for building block operators of CFD. *Computers and Fluids*, *197*, 104386. https://doi.org/10.1016/j.compfluid.2019.104386.

Ibeid, H., Olson, L., & Gropp, W. (2019). FFT, FMM, and multigrid on the road to exascale: Performance challenges and opportunities. *Journal of Parallel and Distributed Computing*, *136*, 63–74. https://doi.org/10.1016/j.jpdc.2019.09.014.

Karakus, A., Chalmers, N., Świrydowicz, K., & Warburton, T. (2019). A GPU accelerated discontinuous Galerkin incompressible flow solver. *Journal of Computational Physics*, *390*, 380–404. https://doi.org/10.1016/j.jcp.2019.04.010.

Karniadakis, G., & Sherwin, S. J. (2005). *Spectral/hp element methods for computational fluid dynamics* (2nd ed.). Oxford: Oxford University Press. https://doi.org/10.1093/acprof:oso/9780198528692.001.0001.

Karniadakis, G. E., Israeli, M., & Orszag, S. A. (1991). High-order splitting methods for the incompressible Navier–Stokes equations. *Journal of Computational Physics*, *97*(2), 414–443. https://doi.org/10.1016/0021-9991(91)90007-8.

Karypis, G., & Kumar, V. (1998). A fast and high quality multilevel scheme for partitioning irregular graphs. *SIAM Journal on Scientific Computing*, *20*(1), 359–392. https://doi.org/10.1137/S1064827595287997.

Kempf, D., Hess, R., Müthing, S., & Bastian, P. (2018). Automatic code generation for high-performance discontinuous Galerkin methods on modern architectures. Technical report. arXiv:1812.08075.

Kennedy, C. A., Carpenter, M. H., & Lewis, R. M. (2000). Low-storage, explicit Runge–Kutta schemes for the compressible Navier–Stokes equations. *Applied Numerical Mathematics*, *35*(3), 177–219. https://doi.org/10.1016/s0168-9274(99)00141-5.

Keyes, D. E., McInnes, L. C., Woodward, C., Gropp, W., Myra, E., Pernice, M., et al. (2013). Multiphysics simulations: Challenges and opportunities. *International Journal of High Performance Computing Applications*, *27*(1), 4–83. https://doi.org/10.1177/1094342012468181.

Klöckner, A. (2014). Loo.py: Transformation-based code generation for GPUs and CPUs. In *Proceedings of ARRAY '14: ACM SIGPLAN Workshop on Libraries, Languages, and Compilers for Array Programming, Edinburgh, Scotland, 2014*. Association for Computing Machinery. https://doi.org/10.1145/2627373.2627387.

Knepley, M. G., Brown, J., Rupp, K., Smith, B. F. (2013). Achieving high performance with unified residual evaluation. Technical report. arXiv:1309.1204.

Knoll, D. A., & Keyes, D. E. (2004). Jacobian-free Newton–Krylov methods: A survey of approaches and applications. *Journal of Computational Physics*, *193*(2), 357–397. https://doi.org/10.1016/j.jcp.2003.08.010.

Komatitsch, D., et al. (2015). SPECFEM 3D cartesian user manual. Technical report, Computational Infrastructure for Geodynamics, Princeton University, CNRS and University of Marseille, and ETH Zürich.

Kopriva, D. A. (2006). Metric identities and the discontinuous spectral element method on curvilinear meshes. *Journal of Scientific Computing*, *26*(3), 301–327. https://doi.org/10.1007/s10915-005-9070-8.

Kopriva, D. A. (2009). *Implementing spectral methods for partial differential equations*. Berlin: Springer.

Kopriva, D. A., & Gassner, G. J. (2016). Geometry effects in nodal discontinuous Galerkin methods on curved elements that are provably stable. *Applied Mathematics and Computation*, *272*, 274–290. https://doi.org/10.1016/j.amc.2015.08.047.

Krais, N., Beck, A., Bolemann, T., Frank, H., Flad, D., Gassner, G., et al. (2020). FLEXI: A high order discontinuous Galerkin framework for hyperbolic–parabolic conservation laws. *Computers and Mathematics with Applications*. https://doi.org/10.1016/j.camwa.2020.05.004.
Krank, B., Fehn, N., Wall, W. A., & Kronbichler, M. (2017). A high-order semi-explicit discontinuous Galerkin solver for 3D incompressible flow with application to DNS and LES of turbulent channel flow. *Journal of Computational Physics*, *348*, 634–659. https://doi.org/10.1016/j.jcp.2017.07.039.
Kronbichler, M., & Allalen, M. (2018). Efficient high-order discontinuous Galerkin finite elements with matrix-free implementations. In H.-J. Bungartz, D. Kranzlmüller, V. Weinberg, J. Weismüller, & V. Wohlgemuth (Eds.), *Advances and new trends in environmental informatics* (pp. 89–110). Berlin: Springer. https://doi.org/10.1007/978-3-319-99654-7_7.
Kronbichler, M., & Kormann, K. (2012). A generic interface for parallel cell-based finite element operator application. *Computers and Fluids*, *63*, 135–147.
Kronbichler, M., & Kormann, K. (2019). Fast matrix-free evaluation of discontinuous Galerkin finite element operators. *ACM Transactions on Mathematical Software*, *45*(3), 29:1–29:40. https://doi.org/10.1145/3325864.
Kronbichler, M., & Ljungkvist, K. (2019). Multigrid for matrix-free high-order finite element computations on graphics processors. *ACM Transactions on Parallel Computing*, *6*(1), 2:1–2:32. https://doi.org/10.1145/3322813.
Kronbichler, M., & Wall, W. A. (2018). A performance comparison of continuous and discontinuous Galerkin methods with fast multigrid solvers. *SIAM Journal on Scientific Computing*, *40*(5), A3423–A3448. https://doi.org/10.1137/16M110455X.
Kronbichler, M., Schoeder, S., Müller, C., & Wall, W. A. (2016). Comparison of implicit and explicit hybridizable discontinuous Galerkin methods for the acoustic wave equation. *International Journal for Numerical Methods in Engineering*, *106*(9), 712–739. https://doi.org/10.1002/nme.5137.
Kronbichler, M., Kormann, K., Pasichnyk, I., & Allalen, M. (2017). Fast matrix-free discontinuous Galerkin kernels on modern computer architectures. In J. M. Kunkel, R. Yokota, P. Balaji, & D. E. Keyes (Eds.), *ISC high performance 2017*. LNCS (Vol. 10266, pp. 237–255). https://doi.org/10.1007/978-3-319-58667-0_13.
Kronbichler, M., Diagne, A., & Holmgren, H. (2018). A fast massively parallel two-phase flow solver for microfluidic chip simulation. *International Journal of High Performance Computing Applications*, *32*(2), 266–287.
Kronbichler, M., Kormann, K., Fehn, N., Munch, P., Witte, J. (2019). A Hermite-like basis for faster matrix-free evaluation of interior penalty discontinuous Galerkin operators. Technical report. arXiv:1907.08492.
LeVeque, R. J. (2002). *Finite volume methods for hyperbolic problems*. Cambridge texts in applied mathematics. Cambridge.
Loppi, N. A., Witherden, F. D., Jameson, A., & Vincent, P. E. (2018). A high-order cross-platform incompressible Navier–Stokes solver via artificial compressibility with application to a turbulent jet. *Computer Physics Communications*, *233*, 193–205. https://doi.org/10.1016/j.cpc.2018.06.016.
Lottes, J. W., & Fischer, P. F. (2005). Hybrid multigrid-Schwarz algorithms for the spectral element method. *Journal of Scientific Computing*, *24*, 613–646. https://doi.org/10.1007/s10915-004-4787-3.
Lynch, R. E., Rice, J. R., & Thomas, D. H. (1964). Direct solution of partial difference equations by tensor product methods. *Numerische Mathematik*, *6*, 185–199. https://doi.org/10.1007/BF01386067.
Maday, Y., Patera, A. T., & Rønquist, E. M. (1990). An operator-integration-factor splitting method for time-dependent problems: Application to incompressible fluid flow. *Journal of Scientific Computing*, *5*(4), 263–292. https://doi.org/10.1007/bf01063118.
Manzanero, J., Rubio, G., Kopriva, D. A., Ferrer, E., & Valero, E. (2020). An entropy–stable discontinuous Galerkin approximation for the incompressible Navier–Stokes equations with variable

density and artificial compressibility. *Journal of Computational Physics*, *408*, 109241. https://doi.org/10.1016/j.jcp.2020.109241.
May, D. A., Brown, J., & Le Pourhiet, L. (2014). pTatin3D: High-performance methods for long-term lithospheric dynamics. In J. M. Kunkel, T. Ludwig, & H. W. Meuer (Eds.), *Supercomputing (SC14), New Orleans* (pp. 1–11). https://doi.org/10.1109/SC.2014.28.
May, D. A., Brown, J., & Le Pourhiet, L. (2015). A scalable, matrix-free multigrid preconditioner for finite element discretizations of heterogeneous Stokes flow. *Computer Methods in Applied Mechanics and Engineering*, *290*, 496–523.
Moxey, D., Amici, R., & Kirby, M. (2020a). Efficient matrix-free high-order finite element evaluation for simplicial elements. *SIAM Journal on Scientific Computing*, *42*(3), C97–C123. https://doi.org/10.1137/19m1246523.
Moxey, D., Cantwell, C. D., Bao, Y., Cassinelli, A., Castiglioni, G., Chun, S., et al. (2020b). Nektar++: Enhancing the capability and application of high-fidelity spectral/hp element methods. *Computer Physics Communications*, *249*, 107110. https://doi.org/10.1016/j.cpc.2019.107110.
Müthing, S., Piatkowski, M., & Bastian, P. (2017). High-performance implementation of matrix-free high-order discontinuous Galerkin methods. Technical report. arXiv:1711.10885.
Nguyen, N. C., Peraire, J., & Cockburn, B. (2011). High-order implicit hybridizable discontinuous Galerkin methods for acoustics and elastodynamics. *Journal of Computational Physics*, *230*, 3695–3718. https://doi.org/10.1016/j.jcp.2011.01.035.
Noventa, G., Massa, F., Bassi, F., Colombo, A., Franchina, N., & Ghidoni, A. (2016). A high-order discontinuous Galerkin solver for unsteady incompressible turbulent flows. *Computers and Fluids*, *139*, 248–260. https://doi.org/10.1016/j.compfluid.2016.03.007.
Olson, L. (2007). Algebraic multigrid preconditioning of high-order spectral elements for elliptic problems on a simplicial mesh. *SIAM Journal on Scientific Computing*, *29*(5), 2189–2209. https://doi.org/10.1137/060663465.
Oo, K. L., & Vogel, A. (2020). Accelerating geometric multigrid preconditioning with half-precision arithmetic on GPUs. Technical report. arxiv:2007.07539.
Orszag, S. A. (1980). Spectral methods for problems in complex geometries. *Journal of Computational Physics*, *37*, 70–92.
Patera, A. T. (1984). A spectral element method for fluid dynamics: Laminar flow in a channel expansion. *Journal of Computational Physics*, *54*(3), 468–488. https://doi.org/10.1016/0021-9991(84)90128-1.
Patterson, D. A., & Hennessy, J. L. (2013). *Computer organization and design: The hardware/software interface* (5th ed.). Burlington: Morgan Kaufmann.
Pazner, W. (2019). Efficient low-order refined preconditioners for high-order matrix-free continuous and discontinuous Galerkin methods. Technical report. arXiv:1908.07071.
Pazner, W., & Persson, P.-O. (2018). Approximate tensor-product preconditioners for very high order discontinuous Galerkin methods. *Journal of Computational Physics*, *354*, 344–369. https://doi.org/10.1016/j.jcp.2017.10.030.
Persson, P. O. (2013). A sparse and high-order accurate line-based discontinuous Galerkin method for unstructured meshes. *Journal of Computational Physics*, *233*, 414–429. https://doi.org/10.1016/j.jcp.2012.09.008.
Raffenetti, K., Amer, A., Oden, L., Archer, C., Bland, W., Fujita, H., et al. (2017). Why is MPI so slow?: Analyzing the fundamental limits in implementing MPI-3.1. In *Proceedings of the International Conference for High Performance Computing, Networking, Storage and Analysis, SC'17, New York, NY, USA, 2017* (pp. 62:1–62:12). ACM. https://doi.org/10.1145/3126908.3126963. ISBN 978-1-4503-5114-0.
Rathgeber, F., Ham, D. A., Mitchell, L., Lange, M., Luporini, F., McRae, A. T. T., et al. (2017). Firedrake: Automating the finite element method by composing abstractions. *ACM Transactions on Mathematical Software*, *43*(3), 24:1–24:27. https://doi.org/10.1145/2998441.
Remacle, J.-F., Gandham, R., & Warburton, T. (2016). GPU accelerated spectral finite elements on all-hex meshes. *Journal of Computational Physics*, *324*, 246–257. https://doi.org/10.1016/j.jcp.2016.08.005.

Ruge, J. W., & Stüben, K. (1987). Algebraic multigrid (AMG). In *Multigrid methods* (pp. 73–130). Philadelphia: Society for Industrial and Applied Mathematics. https://doi.org/10.1137/1.9781611971057.ch4.

Saad, Y. (2003). *Iterative methods for sparse linear systems* (2nd ed.). Philadelphia: SIAM.

Schenk, O., & Gärtner, K. (2004). Solving unsymmetric sparse systems of linear equations with PARDISO. *Future Generation Computer Systems*, *20*(3), 475–487. https://doi.org/10.1016/j.future.2003.07.011, https://www.pardiso-project.org/.

Schöberl, J. (2014). C++11 implementation of finite elements in NGSolve. Technical report ASC Report No. 30/2014, Vienna University of Technology.

Schoeder, S., Kormann, K., Wall, W. A., & Kronbichler, M. (2018a). Efficient explicit time stepping of high order discontinuous Galerkin schemes for waves. *SIAM Journal on Scientific Computing*, *40*(6), C803–C826. https://doi.org/10.1137/18M1185399.

Schoeder, S., Kronbichler, M., & Wall, W. A. (2018b). Arbitrary high-order explicit hybridizable discontinuous Galerkin methods for the acoustic wave equation. *Journal of Scientific Computing*, *76*, 969–1006. https://doi.org/10.1007/s10915-018-0649-2.

Schoeder, S., Wall, W. A., & Kronbichler, M. (2019). ExWave: A high performance discontinuous Galerkin solver for the acoustic wave equation. *SoftwareX*, *9*, 49–54. https://doi.org/10.1016/j.softx.2019.01.001.

Solomonoff, A. (1992). A fast algorithm for spectral differentiation. *Journal of Computational Physics*, *98*(1), 174–177. https://doi.org/10.1016/0021-9991(92)90182-X.

Stanglmeier, M., Nguyen, N. C., Peraire, J., & Cockburn, B. (2016). An explicit hybridizable discontinuous Galerkin method for the acoustic wave equation. *Computer Methods in Applied Mechanics and Engineering*, *300*, 748–769. https://doi.org/10.1016/j.cma.2015.12.003.

Sun, T., Mitchell, L., Kulkarni, K., Klöckner, A., Ham, D. A., & Kelly, P. H. J. (2020). A study of vectorization for matrix-free finite element methods. *International Journal of High Performance Computing Applications*, page in press. https://doi.org/10.1177/1094342020945005.

Świrydowicz, K., Chalmers, N., Karakus, A., & Warburton, T. (2019). Acceleration of tensor-product operations for high-order finite element methods. *International Journal of High Performance Computing Applications*, *33*(4), 735–757. https://doi.org/10.1177/1094342018816368.

Thomas, J. L., Diskin, B., & Brandt, A. (2003). Textbook multigrid efficiency for fluid simulations. *Annual Reviews of Fluid Mechanics*, *35*, 317–340. https://doi.org/10.1146/annurev.fluid.35.101101.161209.

Treibig, J., & Hager, G. (2010). Introducing a performance model for bandwidth-limited loop kernels. In R. Wyrzykowski, J. Dongarra, K. Karczewski, & J. Wasniewski (Eds.), *Parallel Processing and Applied Mathematics: 8th International Conference, PPAM 2009, Wroclaw, Poland, 13–16 September 2009. Revised Selected Papers, Part I* (pp. 615–624). Berlin, Heidelberg: Springer. https://doi.org/10.1007/978-3-642-14390-8_64.

Treibig, J., Hager, G., & Wellein, G. (2020). LIKWID: A lightweight performance-oriented tool suite for x86 multicore environments. In *Proceedings of PSTI2010, the First International Workshop on Parallel Software Tools and Tool Infrastructures, San Diego, CA*. https://doi.org/10.1109/ICPPW.2010.38, https://github.com/RRZE-HPC/likwid. Retrieved 27 July 2020.

Trottenberg, U., Oosterlee, C., & Schüller, A. (2001). *Multigrid*. London: Elsevier/Academic.

Vaněk, P., Mandel, J., & Brezina, M. (1996). Algebraic multigrid by smoothed aggregation for second and fourth order elliptic problems. *Computing*, *56*(3), 179–196. https://doi.org/10.1007/bf02238511.

Wang, Z. J., Fidkowski, K., Abgrall, R., Bassi, F., Caraeni, D., Cary, A., et al. (2013). High-order CFD methods: Current status and perspective. *International Journal for Numerical Methods in Fluids*, *72*(8), 811–845. https://doi.org/10.1002/fld.3767.

Weinzierl, T. (2019). The Peano software—parallel, automaton-based, dynamically adaptive grid traversals. *ACM Transactions on Mathematical Software*, *45*(2), 14:1–14:41. https://doi.org/10.1145/3319797.

Williams, S., Waterman, A., & Patterson, D. (2009). Roofline: An insightful visual performance model for multicore architectures. *Communications of the ACM*, *52*(4), 65–76. https://doi.org/10.1145/1498765.1498785.

Winters, A. R., Moura, R. C., Mengaldo, G., Gassner, G. J., Walch, S., Peiró, J., et al. (2018). A comparative study on polynomial dealiasing and split form discontinuous Galerkin schemes for under-resolved turbulence computations. *Journal of Computational Physics*, *372*, 1–21. https://doi.org/10.1016/j.jcp.2018.06.016.

Witte, J., Arndt, D., & Kanschat, G. (2019). Fast tensor product Schwarz smoothers for high-order discontinuous Galerkin methods. Technical report. arXiv:1910.11239.

Yan, Z.-G., Pan, Y., Castiglioni, G., Hillewaert, K., Peiró, J., Moxey, D., et al. (2020). Nektar++: Design and implementation of an implicit, spectral/hp element, compressible flow solver using a Jacobian-free Newton Krylov approach. *Computers and Mathematics with Applications*, in press. https://doi.org/10.1016/j.camwa.2020.03.009.

Chapter 3
Construction of Modern Robust Nodal Discontinuous Galerkin Spectral Element Methods for the Compressible Navier–Stokes Equations

Andrew R. Winters, David A. Kopriva, Gregor J. Gassner, and Florian Hindenlang

Abstract Discontinuous Galerkin (DG) methods have a long history in computational physics and engineering to approximate solutions of partial differential equations due to their high-order accuracy and geometric flexibility. However, DG is not perfect and there remain some issues. Concerning robustness, DG has undergone an extensive transformation over the past seven years into its modern form that provides statements on solution boundedness for linear and nonlinear problems. This chapter takes a constructive approach to introduce a modern incarnation of the DG spectral element method for the compressible Navier–Stokes equations in a three-dimensional curvilinear context. The groundwork of the numerical scheme comes from classic principles of spectral methods including polynomial approximations and Gauss-type quadratures. We identify aliasing as one underlying cause of the robustness issues for classical DG spectral methods. Removing said aliasing errors requires a particular differentiation matrix and careful discretization of the advective flux terms in the governing equations.

A. R. Winters
Linköping University, Linköping, Sweden

D. A. Kopriva
The Florida State University, FL, USA

San Diego State University, San Diego, CA, USA

G. J. Gassner (✉)
University of Cologne, Cologne, Germany
e-mail: ggassner@math.uni-koeln.de

F. Hindenlang
Max Planck Institute for Plasma Physics, Garching, Germany

M. Kronbichler and P.-O. Persson (eds.), *Efficient High-Order Discretizations for Computational Fluid Dynamics*, CISM International Centre for Mechanical Sciences 602, https://doi.org/10.1007/978-3-030-60610-7_3

Prologue

The discontinuous Galerkin (DG) method dates back to the work of Nitsche (1971) for the solution of elliptic problems and to the work of Reed and Hill (1973) for the solution of linear hyperbolic advection problems. However, it was the work by Cockburn, Shu, and others starting in 1989, e.g., Cockburn et al. (1990), Cockburn and Shu (1998a), Cockburn and Shu (1991), Cockburn and Shu (1998b), that initiated a substantial interest in DG methods for the approximation of nonlinear hyperbolic and mixed hyperbolic-parabolic conservation laws. Bassi and Rebay (1997) were among the first who applied the DG method to the compressible Navier–Stokes equations.

As time went on, the DG methodology gained more and more traction in many different applications fields, such as compressible flows (Black 1999, 2000; Rasetarinera and Hussaini 2001), electromagnetics and optics (Kopriva et al. 2000, 2002; Deng 2007; Deng et al. 2004), acoustics (Chan et al. 2017; Rasetarinera et al. 2001; Stanescu et al. 2002a, b; Wilcox et al. 2010), meteorology (Giraldo et al. 2002; Giraldo and Restelli 2008; Restelli and Giraldo 2009; Bonev et al. 2018), and geophysics (Fagherazzi et al. 2004a, b). Nowadays, DG is applied in almost all sciences where high fidelity computational approximations of differential equations is necessary.

The first available book on DG methods was published in 1999 Cockburn et al. (2000). This book was a collection of proceedings articles and hence still left many practical issues related to an actual implementation unanswered. The situation changed however in 2005, when Karniadakis and Sherwin (2005) released their book, which not only includes the theory but also provided detailed explanations of the scheme and the algorithms. Focus of that early work was mostly on the modal variant of the DG scheme on hybrid mixed meshes.

Whereas the mathematical formulation of the discontinuous Galerkin scheme is agnostic to the particular choice of element types and basis function type, the actual scheme, i.e., the algorithms and the numerical properties such as efficiency and accuracy depend strongly on the choice of basis functions (and many other choices such as the type of discrete integration and type of element shapes). Later, Hesthaven and Warburton (2008) published a DG book with focus on nodal basis function on simplex shaped elements, while Kopriva (2009) published a book on nodal DG methods on quadrilateral (and hexahedral) elements. These three books cover the vast majority of commonly used DG variants and somewhat represented the state of the art, at least at their publication times.

The textbooks together with the promising theoretical properties of DG methods, such as high dispersion accuracy and low dissipation errors, e.g., Hu et al. (1999), Gassner and Kopriva (2010), high potential for parallel computing, e.g., Altmann et al. (2013), Stanescu et al. (2002a), Baggag et al. (2000), and natural stabilization for advection dominated problems via (approximate) Riemann solvers, certainly helped to attract more and more interest in the application of DG in research and industry.

However, it became clear that additional development and advancement in the state of the art was needed to make DG methods really competitive, as, for instance, documented in large European research collaborations ADIGMA[1] and IDIHOM.[2] The main issues that hold back high-order DG are (i) efficiency of time integration, (ii) high-order grid generation, and (iii) robustness. See, e.g., Wang et al. (2013) for more discussion and details. The authors also identify efficient hp-adaptivity as a major issue, however control of adaptivity and error estimates is a general problem for all numerical schemes, not specific to high-order DG.

Discontinuous Galerkin methods are in general very well suited for explicit time stepping, e.g., low storage Runge-Kutta time integrators. As the mass matrix is local and most often diagonal, no inversion is needed so that the computational complexity of a single time step is very low. Typical for explicit time integration, a CFL-like time step restriction is necessary to keep the simulations stable. For DG, the maximum CFL number not only depends on the mesh size and the fastest wave speed (or equivalent viscous speed), but also on the polynomial degree of the approximation. In industrial applications, such an explicit time step restriction could turn out to be prohibitive, resulting in an inefficient overall framework. Naturally, a remedy to this issue is implicit time integration methods, in particular implicit high-order Runge-Kutta methods. However, as it turns out, high-order DG methods in three spatial dimensions result in huge block dense algebraic equation systems that are notoriously difficult to solve efficiently. And without a proper preconditioner, no benefit from implicit time integration is left over. (In many cases, one can observe a negative speed up compared to explicit Runge-Kutta time integration).

High-order grid generation is a subtle problem not easy to realize at first. Often high-order DG methods are praised for their capabilities of using unstructured meshes. A major issue, however, is that gaining the full benefit of the high-order DG approach requires meshes with high-order approximations at curved boundaries. The automatic construction of curvilinear high quality meshes, with, e.g., boundary layers, is still an open research problem. There are strategies and work arounds available, e.g., an open source software solution that post processes a straight-sided mesh.[3] But for a fully operable process chain, much more research and development is needed.

The third major issue, that of robustness, is more subtle than the time integration efficiency and high-order mesh generation, but at least as important, if not more important—without a stable approximation, there is no need to talk about efficiency at all. Also no high-order mesh is a remedy when the scheme is not robust and blows up during the simulation. DG methods have natural inbuilt upwind-like dissipation from (approximate) Riemann solvers and are often characterized as more stable than, for instance, their continuous counterpart, the standard Galerkin finite element methods.

---

[1] https://cordis.europa.eu/docs/results/30/30719/122807181-6_en.pdf.

[2] https://www.dlr.de/as/desktopdefault.aspx/tabid-7027/11654_read-27492/.

[3] https://www.hopr-project.org.

However, with the high-order of DG methods there comes a whole bag of robustness issues. Most prominently, when simulating problems with discontinuities such as shock waves, Gibbs-type oscillations occur along with possible violations of physical solution bounds, such as positivity of density and pressure, that lead to failure of the algorithm. The inbuilt upwind-like dissipation in the DG methodology is not enough for stability.

Even without discontinuities, other nonlinear solution features may bring high-order DG methods to failure. In realistic applications of turbulent flows, the resolution is orders of magnitude smaller than necessary for grid convergence. While most DG variants work perfectly fine for well-resolved problems, e.g., in a grid convergence study, it turns out to be non-trivial to construct robust DG schemes for under-resolved problems.

Not all of the issues presented above have been solved to a satisfactory level for the DG method; in fact most of these issues are active areas of research. A comprehensive discussion on all aspects and possible solution strategies would clearly go beyond the scope of a single book chapter, especially since we aim to present the details of the theory in the spirit of the three DG books mentioned above, i.e., with maximum detail.

In comparison to the three DG books that essentially cover the prior state of the art, we focus on the issue of robustness for under-resolved flows, e.g., when simulating turbulence, and aim to present the advances made in the last decade. We first introduce the mathematical and algorithmic building blocks of the DG method in Sect. 3.2. In Sect. 3.3, we present the underlying partial differential equations for compressible fluid dynamics, with a focus on their mathematical stability properties. In Sect. 3.4, we provide a detailed description of the spectral element framework on curvilinear unstructured hexahedral grids and its corresponding DG variant in Sect. 3.5. The main strategy to get a provably stable nodal DG scheme is presented in Sect. 3.6 with an outlook and possible extensions discussed in the last Sect. 3.7.

Nomenclature

Notation used throughout this chapter is adapted from Gassner et al. (2018):

$\mathbb{P}^N$	Space of polynomials of degree $\leqslant N$
$\mathbb{I}^N$	Polynomial interpolation operator
(x, y, z)	Physical space coordinates
(ξ, η, ζ)	Reference space coordinates
$\vec{v}$	Vector in three-dimensional space
$\vec{n} = n_1\hat{x} + n_2\hat{y} + n_3\hat{z}$	Physical space normal vector
$\hat{n} = \hat{n}^1\hat{\xi} + \hat{n}^2\hat{\eta} + \hat{n}^3\hat{\zeta}$	Reference space normal vector
$\mathbf{u}$	Continuous quantity
$\mathbf{U}$	Polynomial approximation

$\overset{\leftrightarrow}{\mathbf{f}}, \overset{\leftrightarrow}{\tilde{\mathbf{f}}}$	Block vector of Cartesian flux and contravariant flux
$\mathcal{D}$	$(N+1) \times (N+1)$ Matrix
$\underline{\mathrm{B}}$	5×5 Matrix
$\mathfrak{B}$	15×15 Block matrix.

Spectral Calculus Toolbox

Algorithms developed to numerically model physical problems are typically designed to solve discrete approximations of partial differential equations (PDEs). Often, such PDEs are derived in the framework of differential and integral calculus and can be formulated in terms of first-order differential operators, such as the divergence. These PDEs express fundamental physical laws like the conservation of mass, momentum, and total energy.

The continuum operators are equipped with important differential and integral identities, e.g., the derivative of a constant function is zero, or integration-by-parts. To demonstrate conservation and stability of numerical approximations as well as to accurately capture the physics of a solution it is beneficial for the discretization to mimic as many of these important differential and integral identities as possible.

This section provides the groundwork and discussion of a discrete spectral calculus for nodal spectral element methods. The basic principles of polynomial interpolation and high-order Gauss-type quadrature provide the tools needed to develop high-order, conservative, and stable approximations for PDEs written as conservation laws or balance laws.

Spectral methods owe their roots to the solutions of PDEs by orthogonal polynomial expansions (Gottlieb and Orszag 1977; Kreiss and Oliger 1973). Spectral element methods today approximate solutions of PDEs with piecewise polynomials equivalent to finite series of Legendre polynomials, e.g.,

$$U(x) = \sum_{k=0}^{N} \hat{C}_k L_k(x), \tag{1}$$

which is a polynomial of degree N. As a shorthand, we let $\mathbb{P}^N$ denote the space of polynomials of degree less than or equal to N and write $U \in \mathbb{P}^N$. Whereas finite difference methods approximate the solutions of PDEs only at a finite set of discrete points, U_j, spectral element methods are akin to finite element methods in that the approximation is well-defined at all points.

Four features characterize Discontinuous Galerkin (DG) spectral element methods:

- Approximation of the solutions and fluxes by piecewise high-order polynomials that represent polynomial expansions.
- Approximation of integrals and inner products by high-order Gauss-type quadratures.

- A weak formulation of the original differential equations.
- Coupling of elements through the use of a numerical flux (aka "Riemann Solver").

In this section, we review the background for the first two features, namely, the approximation of functions by Legendre polynomial expansions and Gauss-Lobatto quadratures used by discontinuous Galerkin spectral element methods. Along the way, we develop a discrete calculus that mirrors the continuous one, which will allow us to use familiar operations on discretely defined functions.

Legendre Polynomials and Series

That the Legendre polynomial $L_k(x)$ in (1) is a polynomial of degree k is seen through the three term recurrence relation it satisfies

$$\begin{aligned} L_{k+1}(x) &= \frac{2k+1}{k+1} x L_k(x) - \frac{k}{k+1} L_{k-1}(x), \quad x \in [-1, 1], \\ L_0(x) &= 1, \\ L_1(x) &= x. \end{aligned} \tag{2}$$

Two Legendre polynomials L_k and L_n are orthogonal with respect to the $\mathbb{L}^2(-1, 1)$ inner product

$$\langle L_k, L_n \rangle = \int_{-1}^{1} L_k L_n \,\mathrm{d}x = \frac{2}{2k+1} \delta_{kn}, \tag{3}$$

where $\mathbb{L}^2(-1, 1)$ is the space of square integrable functions on the interval $[-1, 1]$. That is, all functions u for which $\|u\|^2 = \langle u, u \rangle < \infty$. The Kronecker delta is non-zero only when the subscripts match

$$\delta_{kn} = \begin{cases} 1 & k = n \\ 0 & \text{otherwise} \end{cases}, \tag{4}$$

so it follows that the norm of L_k is given by

$$\|L_k\|^2 = \frac{2}{2k+1}. \tag{5}$$

The Legendre polynomials form a basis for the space $\mathbb{L}^2(-1, 1)$, which means that any square integrable function u on $[-1, 1]$ can be represented as an *infinite* series in Legendre polynomials

$$u = \sum_{k=0}^{\infty} \hat{u}_k L_k(x) \quad \text{for all} \quad u \in \mathbb{L}^2(-1, 1)\,. \tag{6}$$

The spectral coefficients $\hat{u}_k$ are found as usual through orthogonal projection

$$\langle u, L_n \rangle = \sum_{k=0}^{\infty} \hat{u}_k \langle L_k, L_n \rangle = \sum_{k=0}^{\infty} \hat{u}_k \|L_k\|^2 \delta_{kn} = \hat{u}_n \|L_n\|^2, \tag{7}$$

so that

$$\hat{u}_n = \frac{\langle u, L_n \rangle}{\|L_n\|^2}, \qquad n = 0, 1, \ldots, \infty. \tag{8}$$

The best approximation of u by the polynomial $U \in \mathbb{P}^N$ defined in (1) is to choose $\hat{C} = \hat{u}$ for $k = 0, 1, 2, \ldots, N$, because then the error is orthogonal to the approximation space,

$$||u - U|| = \left\| \sum_{k=N+1}^{\infty} \hat{u}_k L_k(x) \right\| . \tag{9}$$

Series truncation is the orthogonal projection of $\mathbb{L}^2(-1, 1)$ onto $\mathbb{P}^N(-1, 1)$ with respect to the continuous inner product, and we call that approximation $U = \mathbb{P}^N(u)$, where $\mathbb{P}^N$ is called the truncation operator.

Legendre Polynomial Interpolation

Alternatively, a function u can be approximated by a polynomial interpolant of degree N that passes through $N + 1$ nodal points. Polynomial spectral element methods approximate a function $u(x)$ as a high-order Legendre polynomial interpolant $U(x)$ represented in (1) where the spectral interpolation coefficients $\hat{C}_k$ are determined so that

$$\mathbb{I}^N(u)(x_j) = U(x_j) = u(x_j) \quad j = 0, 1, \ldots, N. \tag{10}$$

We will use the notation $\mathbb{I}^N(u)(x)$ or just $\mathbb{I}^N(u)$ without the argument to denote the polynomial interpolant of order N, with $\mathbb{I}^N$ being the interpolation operator. Upper case functions like $U(x)$ will denote polynomial interpolants, whereas lower case functions can be anything, unless otherwise noted or convention dictates otherwise.

The preferable approach to find the coefficients $\hat{C}_k$ to satisfy (10) is to mimic the orthogonal projection process used to find the $\hat{u}_k$. Suppose that we have a *discrete inner product* $\langle \cdot, \cdot \rangle_N$ with the property

$$\langle L_k, L_n \rangle_N = ||L_n||_N^2 \, \delta_{kn}, \tag{11}$$

where $||L_n||_N^2 = \langle L_n, L_n \rangle_N$. Then the interpolation coefficients $\hat{C}_n$, for $n = 0, 1, 2, \ldots, N$ could be computed without solving a full Vandermonde matrix system as

$$\hat{C}_n = \frac{\langle u, L_n \rangle_N}{||L_n||_N^2} \quad n = 0, 1, \ldots, N, \tag{12}$$

since

$$\langle u, L_n \rangle_N = \sum_{k=0}^{\infty} \hat{C}_k \langle L_k(x), L_n \rangle_N = \sum_{k=0}^{\infty} \hat{C}_k \, ||L_n||_N^2 \, \delta_{kn} = \hat{C}_n ||L_n||_N^2. \tag{13}$$

An alternative—and equivalent—representation of the interpolant U is to use the *Lagrange* or *nodal* form

$$U(x) = \sum_{j=0}^{N} U_j \ell_j(x), \tag{14}$$

where $U_j = U\left(x_j\right)$ and the $\ell_j(x)$ are the Lagrange interpolating polynomials with nodes at the same points x_j

$$\ell_j(x) = \prod_{i=0; i \neq j}^{N} \frac{(x - x_i)}{\left(x_j - x_i\right)} \in \mathbb{P}^N, \tag{15}$$

that clearly possess the Kronecker delta property

$$\ell_j(x_n) = \delta_{jn}. \tag{16}$$

It remains, then to find an appropriate quadrature for the discrete inner product and interpolation nodes.

Legendre Gauss Quadrature and the Discrete Inner Product

We can construct a discrete inner product with the desired properties by approximating the true inner product with a form of Gauss-Legendre quadrature known as the *Gauss-Lobatto-Legendre* (or just Gauss-Lobatto) quadrature rule. The Gauss-Lobatto quadrature approximation of a function f is a weighted sum of nodal values,

$$\int_N f(x)\,\mathrm{d}x \equiv \sum_{j=0}^{N} f(x_j) w_j, \tag{17}$$

where the nodes x_j are

$$x_j = +1, -1, \text{ and the zeros of } L'_N(x), \tag{18}$$

and the quadrature weights are

$$w_j = \frac{2}{N(N+1)} \frac{1}{\left[L_N(x_j)\right]^2}. \tag{19}$$

The Gauss-Lobatto rule is exact when the integrand is a polynomial of degree $2N - 1$ so we can write

$$\int_N f(x)\,\mathrm{d}x = \int_{-1}^{1} f(x)\,\mathrm{d}x \quad \text{for all } f \in \mathbb{P}^{2N-1}. \tag{20}$$

The quadrature rule (17) allows us to define the discrete inner product as

$$\langle u, v\rangle_N \equiv \sum_{j=0}^{N} u(x_j)v(x_j)w_j, \tag{21}$$

which has the desired orthogonality properties. If we replace u and v by the Legendre polynomials L_k and L_n, then provided $k + n \leqslant 2N - 1$,

$$\langle L_k, L_n\rangle_N = \int_{-1}^{1} L_k L_n\,\mathrm{d}x = ||L_k||_N^2\,\delta_{kn}. \tag{22}$$

The quadrature is therefore exact for all the inner products except L_N with itself. That means for all $k < N$, $||L_k||_N = ||L_k||$. The last case needs to be computed directly and separately leading to

$$||L_k||_N^2 = \begin{cases} \dfrac{2}{2k+1} & k < N \\ \dfrac{2}{N} & k = N \end{cases}. \tag{23}$$

We can expose several useful properties from the definition of the discrete inner product. For any function $g(x)$, (21) and the interpolation conditions (10) imply that

$$\left\langle \mathbb{I}^N(g), V\right\rangle_N = \sum_{j=0}^{N} g_j V_j w_j = \langle g, V\rangle_N \quad \text{for all } V \in \mathbb{P}^N. \tag{24}$$

So any time the argument on the left is seen, it can be viewed as the interpolant of that argument. For example, for $U, V, W \in \mathbb{P}^N$,

$$\langle UV, W\rangle_N = \left\langle \mathbb{I}^N(UV), W\right\rangle_N . \tag{25}$$

Also, the exactness of the quadrature (20) implies that if the product $UV \in \mathbb{P}^{2N-1}$,

$$\langle U, V\rangle_N = \langle U, V\rangle . \tag{26}$$

For instance, the discrete inner product of a polynomial approximation with its first derivative is exact

$$\langle U, U_x\rangle_N = \langle U, U_x\rangle . \tag{27}$$

However, it also follows that for $U \in \mathbb{P}^N$

$$||U||_N \neq ||U|| . \tag{28}$$

Nevertheless, the discrete norm is *equivalent* to the continuous norm in that for $U \in \mathbb{P}^N$ the discrete norm is bounded from above and below by the continuous norm (Canuto and Quarteroni 1982)

$$||U|| \leqslant ||U||_N \leq \sqrt{2+\frac{1}{N}}\,||U|| \leqslant \sqrt{3}\,||U|| . \tag{29}$$

The discrete norm is never smaller than the true $\mathbb{L}^2$ norm and is never more than about 73% larger than the true norm. Equivalence says that if an approximation converges in the discrete norm, it converges in the continuous norm, too.

Aliasing Error

Since the Gauss-Lobatto quadrature (17) is exact only for polynomials of degree $2N-1$, the question is raised as to what errors are created by using the discrete inner product (21) compared to the exact integral (3). This is of particular interest when the function being interpolated is not a polynomial or not a polynomial of sufficiently low-order. The answer comes by comparing the exact coefficients $\hat{u}_k$ and the interpolation coefficients $\hat{C}_k$.

To relate the exact and interpolation coefficients, we substitute (6) into (12) and use the discrete orthogonality to see that

$$\hat{C}_k = \frac{\left\langle \sum\limits_{n=0}^{\infty} \hat{u}_n L_n, L_k\right\rangle_N}{||L_k||_N^2} = \hat{u}_k + \frac{\sum\limits_{n=N+1}^{\infty} \hat{u}_n \langle L_n, L_k\rangle_N}{||L_k||_N^2} \equiv \hat{u}_k + \hat{A}_k \quad k = 0, 1, \ldots, N. \tag{30}$$

So (30) shows that the interpolation coefficients computed with quadrature are the exact coefficients plus an *aliasing* error that depends on the discrete projection of L_n

onto L_k for $n > N$. The interpolant can therefore be written as

$$\begin{aligned}
\mathbb{I}^N(u) &= \sum_{k=0}^{N} \hat{C}_k L_k(x) = \sum_{k=0}^{N} \hat{u}_k L_k(x) + \sum_{k=0}^{N} \left\{ \frac{1}{||L_k||_N^2} \sum_{n=N+1}^{\infty} \hat{u}_n \langle L_n, L_k \rangle_N \right\} L_k(x) \\
&= u - \sum_{k=N+1}^{\infty} \hat{u}_k L_k(x) + \sum_{k=0}^{N} \left\{ \frac{1}{||L_k||_N^2} \sum_{n=N+1}^{\infty} \hat{u}_n \langle L_n, L_k \rangle_N \right\} L_k(x) \\
&= u + \left\{ \sum_{k=N+1}^{\infty} \hat{u}_k L_k(x) \right\} + \sum_{k=0}^{N} \hat{A}_k L_k(x).
\end{aligned} \tag{31}$$

So the interpolant is the actual function plus two errors. The first error is the truncation error, $u - \mathbb{P}^N u$, due to the finite number of modes available. The second error is the aliasing error, which arises because discretely the higher order modes have a non-zero contribution to the low-order modes, as represented in the fact that the discrete inner products do not vanish.

The indices on the sums over k in (31) show that the truncation and aliasing errors are orthogonal to each other, since

$$\left\langle \sum_{k=N+1}^{\infty} \hat{u}_k L_k, \sum_{n=0}^{N} A_n L_n \right\rangle = \sum_{k=N+1}^{\infty} \sum_{n=0}^{N} \hat{u}_k A_n \langle L_k, L_n \rangle = 0.$$

Note also that (31) says that if $u \in \mathbb{P}^N$ for which $\hat{u}_k = 0,\ k > N$, then $\mathbb{I}^N(u) = u$, as expected.

The projection result (25) shows that the discrete inner product of a compound argument (function of polynomials) with a polynomial introduces aliasing errors. Compound arguments appear when projecting a flux, like $f(U) = \frac{1}{2}U^2$ for the Burgers equation, onto the basis functions.

For *polynomial* compound functions, where the result is the product of polynomials, the aliasing error created by discrete inner products can be eliminated by *consistent integration*, more commonly referred to as "overintegration," at the cost of extra evaluations of the function. The idea is to evaluate the inner product at $M > N$ points so that the discrete inner product is exact. For the product $UV \in \mathbb{P}^{2N}$, for example, the problem is to find M so that for $W \in \mathbb{P}^N$

$$\langle UV, W \rangle_M = \int_{-1}^{1} UVW \,\mathrm{d}x. \tag{32}$$

The product $UVW \in \mathbb{P}^{3N}$ and the Gauss-Lobatto quadrature is exact for arguments in $\mathbb{P}^{2M-1}$. Therefore, there is no aliasing error if

$$3N = 2M - 1 \quad \text{or} \quad M = \frac{3N+1}{2}. \tag{33}$$

In other words, aliasing due to the discrete inner product can be eliminated by evaluating the inner product at

$$M > \frac{3}{2}N \tag{34}$$

points, 50% more than used in the interpolation. More generally, for a polynomial function $F \in \mathbb{P}^p$, aliasing can be avoided when taking

$$M > \frac{p+1}{2}. \tag{35}$$

It should be clear, however, that if F is not a polynomial function, then (25) implies that aliasing will be present except in the limit as $M \to \infty$, i.e., taking an infinite number of quadrature points and converging the discrete inner product to the continuous one.

Spectral Differentiation

Derivatives of functions are approximated by the derivatives of their polynomial interpolants

$$u' \approx \left(\mathbb{I}^N(u)\right)'. \tag{36}$$

The interpolant can be represented in either the nodal or modal form. The choice can be made solely on efficiency considerations.

In Legendre spectral element methods, derivative approximations are computed by matrix-vector multiplication where the vector holds the nodal values of the approximation and the matrix holds derivatives of the Lagrange interpolating polynomials. The derivative of the Lagrange form interpolant is

$$U' = \sum_{n=0}^{N} U_n \ell_n'(x). \tag{37}$$

When evaluated at the Gauss-Lobatto points $x_j,\ j = 0, 1, \ldots, N$,

$$U_j' = \sum_{n=0}^{N} U_n \ell_n'(x_j) = \sum_{n=0}^{N} U_n \mathcal{D}_{jn} \tag{38}$$

where $\mathcal{D}_{jn} = \ell_n'(x_j)$ are the elements of the *derivative matrix*, $\mathcal{D}$. Thus, derivatives are computed by matrix-vector multiplication

$$\mathcal{U}' = \mathcal{D}\mathcal{U}, \tag{39}$$

where $\mathcal{U} = [U_0\ U_1\ \dots\ U_N]^T$.

One often noted feature of the approximation (38) is that the approximation to the derivative includes only points in the domain, even up to the boundary points, independent of the approximation order. This is a feature not held, for instance, with high-order finite difference methods, where external "ghost points" appear if the stencil is used near the boundaries.

Derivatives of functions of a polynomial, like a flux $f(U)$, a product UV, or other compound quantity are computed nodally. For instance, the product $Q = UV \in \mathbb{P}^{2N}$ is computed by $\mathcal{D}\mathcal{Q}$ where $Q_j = U_j V_j$. Differentiation of the product is therefore equivalent to

$$(UV)' \approx \left(\mathbb{I}^N(UV)\right)'. \tag{40}$$

As a result, there is an aliasing error associated with representing the product as a polynomial of degree N.

One unfortunate aspect of polynomial differentiation is that *differentiation and interpolation do not commute*, i.e.,

$$\left(\mathbb{I}^N(u)\right)' \neq \mathbb{I}^{N-1}\left(u'\right). \tag{41}$$

As a consequence of interpolation and differentiation not commuting, common rules like the product and chain rules do not hold except in special cases. For example, the product rule

$$\left(\mathbb{I}^N(UV)\right)' \neq \mathbb{I}^{N-1}\left(U'V\right) + \mathbb{I}^{N-1}\left(UV'\right) \tag{42}$$

unless $UV \in \mathbb{P}^N$.

Spectral Accuracy

The truncation, $\mathbb{P}^N(u)$, interpolation, $\mathbb{I}^N(u)$, derivative, $\left(\mathbb{I}^N(u)\right)'$, and Gauss-Lobatto quadrature, $\int_N u\,\mathrm{d}x$, approximations all possess what is known as spectral accuracy: The rate of convergence depends only on the smoothness of u. For very smooth functions they converge extremely fast, and with enough smoothness, they converge exponentially fast. In this section, we review these facts and their meaning and leave the technical derivations and more precise forms to Canuto et al. (2007).

Spectral accuracy follows from the fact that the truncation error (c.f. (9))

$$\left|\left|u - \mathbb{P}^N(u)\right|\right| = \left|\left|\sum_{k=N+1}^{\infty} \hat{u}_k L_k(x)\right|\right| = \sum_{k=N+1}^{\infty} \frac{2}{2k+1} \left|\hat{u}_k\right|^2 \tag{43}$$

depends on the size of the modal coefficients, $\hat{u}_k$. From Fourier analysis, we know that the smoother u is, the faster those coefficients decay. With $\hat{u}_{N+1}$ being the dominant mode, the truncation error decays $\sim \left|\hat{u}_{N+1}\right|$. So we get the relationship between smoothness and truncation error through the modal coefficients.

To write the error convergence more precisely, we define the Sobolev norm,

$$||u||^2_{H^m} = \sum_{n=0}^{m} \left|\left|\frac{d^n u}{dx^n}\right|\right|^2, \tag{44}$$

so the smoother the function u is, the larger the index m for which the norm is finite, i.e., $||u||_{H^m} < \infty$. Note that when $m = 0$, the Sobolev norm reduces to the $\mathbb{L}^2(-1, 1)$ energy norm. Also, if $||u||^2_{H^m} < \infty$ for some m, then the energy norm of u and each derivative individually up to order m is also bounded.

In terms of the Sobolev norm, the truncation error satisfies

$$\left|\left|u - \mathbb{P}^N(u)\right|\right| \leqslant CN^{-m}\,||u||_{H^m}, \tag{45}$$

where C is a generic constant. Equation (45) is the statement of spectral accuracy. We see directly that for a fixed smoothness implied by m, the error converges like N^{-m}, which is rapid for large m. For fixed N, the convergence rate increases as m increases. If all derivatives exist, so that we can take $m \to \infty$, the approximation is said to have infinite order convergence.

The interpolation error seen in (31) is the sum of the truncation error, which decays spectrally fast, plus the aliasing error (30), which also depends on the rate of decay of the modal coefficients. As a result, the interpolation error is also spectral, though larger than the truncation error and requires slightly more smoothness (Canuto et al. 2007). Ultimately, the interpolation error is also bounded like in (45) but only if $m > 1/2$.

Since the interpolation error converges spectrally fast, it is not surprising that the error in the derivative of the interpolant is spectrally accurate too, though at a lower rate. For the derivative,

$$\left|\left|u' - \left(\mathbb{I}^N(u)\right)'\right|\right| \leqslant CN^{1-m}\,||u||_{H^m}. \tag{46}$$

Note that since the differentiation error is spectrally accurate, it follows that the product rule error in (42) also converges spectrally fast.

Finally, the discrete inner product, and by extension the quadrature is spectrally accurate. For a function u and polynomial $\phi \in \mathbb{P}^N$,

$$\left|\langle u, \phi\rangle - \langle u, \phi\rangle_N\right| \leqslant CN^{-m}\,||u||_{H^m}\,||\phi||. \tag{47}$$

Spectral convergence becomes exponential convergence if u is so smooth that it is analytic (in the complex variables sense) in some ellipse in the complex plane around the foci $-1, 1$. Exponential convergence is sometimes confused with spectral accuracy, whereas it is instead a special case. Recently Xie et al. (2013) have shown that

$$\max_{|x|\leqslant 1} \left|u - \mathbb{I}^N(u)\right| \leqslant C\,(\rho)\, N^{3/2} e^{-N\ln(\rho)}, \tag{48}$$

and

$$\max_{0\leqslant j\leqslant N} \left|u' - \left(\mathbb{I}^N(u)\right)'\right| \leqslant C\,(\rho)\, N^{7/2} e^{-N\ln(\rho)}, \tag{49}$$

where ρ increases with the size of the ellipse, and hence the region of analyticity. For large enough N, the exponential decay dominates the polynomial growth. Gauss quadrature is also exponentially convergent for analytic functions by virtue of the interpolation convergence.

The Discrete Inner Product and Summation-by-Parts

One of the most important properties of the discrete inner product (21) with regard to the methods we develop here is the summation-by-parts (SBP) property. The summation-by-parts property is the discrete equivalent of the integration-by-parts property

$$\left\langle u, v'\right\rangle = \int_{-1}^{1} u v' \,\mathrm{d}x = u v|_{-1}^{1} - \left\langle u', v\right\rangle, \tag{50}$$

or equivalently

$$\left\langle u, v'\right\rangle + \left\langle u', v\right\rangle = u v|_{-1}^{1}\,. \tag{51}$$

If $U, V \in \mathbb{P}^N$ then $UV' \in \mathbb{P}^{2N-1}$ and $U'V \in \mathbb{P}^{2N-1}$. Using the exactness between the integral and the Gauss-Lobatto quadrature,

$$\left\langle U, V'\right\rangle_N = \left\langle U, V'\right\rangle = UV|_{-1}^{1} - \left\langle U', V\right\rangle = UV|_{-1}^{1} - \left\langle U', V\right\rangle_N\,. \tag{52}$$

Therefore, the *summation-by-parts* formula is

$$\left\langle U, V'\right\rangle_N = UV|_{-1}^{1} - \left\langle U', V\right\rangle_N\,, \tag{53}$$

or equivalently,

$$\left\langle U, V'\right\rangle_N + \left\langle U', V\right\rangle_N = UV|_{-1}^{1}\,, \tag{54}$$

which is the discrete equivalent of the integration-by-parts property (51) held by the continuous integral.

It is interesting to note that if $U \in \mathbb{P}^N$ and $V \in \mathbb{P}^N$, the summation-by-parts formula (54) gives

$$UV|_{-1}^{1} = \int_N \left(\mathbb{I}^N(UV)\right)' \mathrm{d}x = \int_N U'V \,\mathrm{d}x + \int_N UV' \,\mathrm{d}x, \tag{55}$$

which says that the mean value of the error due to the lack of commutativity of interpolation and differentiation is zero.

Integral Quantities in Matrix-Vector Form

Since the nodal degrees of freedom can be represented as a vector, integral quantities like the inner product and summation-by-parts can be written in matrix-vector form. For instance, let us define the diagonal mass matrix, whose entries are the Gauss-Lobatto quadrature weights,

$$\mathcal{M} = \begin{bmatrix} w_0 & & 0 \\ & \ddots & \\ 0 & & w_N \end{bmatrix}, \tag{56}$$

with which we can write the discrete inner product as

$$\langle U, V \rangle_N = \sum_{j=0}^{N} U_j w_j V_j = \mathcal{U}^T \mathcal{M} \mathcal{V}. \tag{57}$$

Similarly, the quadrature of F can be expressed as

$$\int_N F(x)\,\mathrm{d}x = \langle 1, F \rangle_N = 1^T \mathcal{M} \mathcal{F}. \tag{58}$$

Written in matrix form, we can show that that the summation-by-parts formula (54) is solely a property of the derivative and mass matrices and a boundary matrix

$$\mathcal{B} = \begin{bmatrix} -1 & & & & 0 \\ & 0 & & & \\ & & \ddots & & \\ & & & 0 & \\ 0 & & & & 1 \end{bmatrix}. \tag{59}$$

If we write (54) in summation form,

$$\sum_{i=0}^{N} U_i w_i \left(\mathcal{D}\mathcal{V}\right)_i + \sum_{i=0}^{N} w_i \left(\mathcal{D}\mathcal{U}\right)_i V_i = \left\{U_N V_N - U_0 V_0\right\}, \tag{60}$$

we see the equivalent matrix-vector equation

$$\mathcal{U}^T \mathcal{M}\mathcal{D}\mathcal{V} + (\mathcal{M}\mathcal{D}\mathcal{U})^T \mathcal{V} = \mathcal{U}^T \mathcal{B}\mathcal{V}, \tag{61}$$

which can be factored as

$$\mathcal{U}^T \left\{\mathcal{M}\mathcal{D} + (\mathcal{M}\mathcal{D})^T - \mathcal{B}\right\} \mathcal{V} = 0. \tag{62}$$

Since the polynomials from which the nodal values in the vectors $\mathcal{U}$, $\mathcal{V}$ are arbitrary, the components are linearly independent and it follows that

$$\mathcal{M}\mathcal{D} + (\mathcal{M}\mathcal{D})^T = \mathcal{B}. \tag{63}$$

Commonly, the matrix $\mathcal{Q} = \mathcal{M}\mathcal{D}$ is defined leaving

$$\mathcal{Q} + \mathcal{Q}^T = \mathcal{B}. \tag{64}$$

The relation (64) is a matrix expression of the summation-by-parts property, (54). Summation-by-parts with matrix operators was introduced in the finite difference community, e.g., Kreiss and Olliger (1972), Kreiss and Scherer (1977, 1974), Strand (1994) and, e.g., the review article Svärd and Nordström (2014), with the goal to mimic finite element type energy estimates with local stencil based differentiation operators, i.e., finite differences. The matrix expression (64) shows that collocation type spectral elements with Gauss-Lobatto nodes may structurally be interpreted as diagonal norm summation-by-parts finite difference methods.

From (64) it is possible to assess the structure of $\mathcal{Q}$ and determine many of its entries. With the aid of the Lagrange nodal polynomial basis functions and the collocated Gauss-Lobatto quadrature, it is easy to see that the entries of the $\mathcal{Q}$ matrix can be directly computed as

$$\mathcal{Q}_{ij} = \left\langle \ell'_j, \ell_i \right\rangle_N, \quad i, j = 0, ..., N. \tag{65}$$

Due to the consistency of polynomial interpolation, it follows that it is possible to exactly represent a constant function. From this consistency it follows that the derivative of a constant can be computed exactly, which translates into the matrix condition

$$\mathcal{D}\,\mathbf{1} = 0 \Rightarrow \mathcal{Q}\,\mathbf{1} = 0 \Rightarrow \sum_{j=0}^{N} \mathcal{Q}_{ij} = 0, \quad i = 0, ..., N. \tag{66}$$

That is, the sum of rows of the matrix $\mathcal{Q}$ (or $\mathcal{D}$) are equal to zero. Directly multiplying (64) with a vector containing only ones, $\mathbf{1}$, and using (66), we get for the sum of the columns

$$\sum_{i=0}^{N} \mathcal{Q}_{ij} = \begin{cases} -\ 1, & i = 0, \\ \ \ \ 0, & i = 1, ..., N-1, \\ +\ 1, & i = N. \end{cases} \tag{67}$$

If we assess the diagonal entries of (64), we immediately get

$$\mathcal{Q}_{00} = -\frac{1}{2}, \quad \mathcal{Q}_{NN} = \frac{1}{2}, \quad \mathcal{Q}_{ii} = 0,\ i = 1, ..., N-1. \tag{68}$$

Lastly, the $\mathcal{Q}$ matrix is almost skew-symmetric, i.e.,

$$\mathcal{Q}_{ij} = -\mathcal{Q}_{ji}, \quad \forall i, j \text{ (except for } \mathcal{Q}_{00} \text{ and } \mathcal{Q}_{NN}). \tag{69}$$

Extension to Multiple Space Dimensions

In multiple space dimensions, functions are approximated by tensor products of the one-dimensional interpolants on the quadrilateral $E_2 = [-1, 1] \times [-1, 1]$ or hexahedron $E_3 = [-1, 1]^3$. We let

$$\vec{x} = (x, y, z) = (x_1, x_2, x_3) = x\hat{x} + y\hat{y} + z\hat{z} = \sum_{i=0}^{3} x_i \hat{x}_i, \tag{70}$$

where $\hat{x}$, $\hat{y}$, $\hat{z}$ are the unit vectors in the three coordinate directions, with the similar definition in two dimensions. For functions $u(x, y)$ and $u(x, y, z)$ defined in E_2 and E_3, the Lagrange forms of the interpolant are

$$\mathbb{I}^N(u)(x, y) = \sum_{i,j=0}^{N} u_{ij} \ell_i(x) \ell_j(y), \tag{71}$$

and

$$\mathbb{I}^N(u)(x, y, z) = \sum_{i,j,k=0}^{N} u_{ijk} \ell_i(x) \ell_j(y) \ell_k(z), \tag{72}$$

where $u_{ijk} = u\left(x_i, y_j, z_k\right)$, etc. The x_i, y_j and z_k are located at the Gauss-Lobatto nodes. For notational simplicity, we have assumed the same polynomial order in each space direction, though this is not necessary in practice. With this assumption, we will denote the space of polynomials of degree N in each space direction also as $\mathbb{P}^N$.

Tensor Product Spectral Differentiation

The tensor product makes the computation of partial derivatives simple and efficient. For example,

$$\left.\frac{\partial U}{\partial x}\right|_{ijk} = \sum_{n,m,l=0}^{N} U_{nml}\ell_n'(x_i)\ell_m(y_j)\ell_l(y_k) = \sum_{n=0}^{N} U_{njk}\ell_n'(x_i) = \sum_{n=0}^{N} U_{njk}\mathcal{D}_{in}, \tag{73}$$

where the j and k sums drop out due to the Kronecker delta property of the Lagrange basis (16).

Let us assume that the nodal values are stored in an array format, and let us represent an array slice that defines a vector by a colon, ":". Then the $x - derivative$ can be computed plane-by-plane

$$\mathcal{U}_{jk} = \mathcal{D}U_{:jk}, \quad j,k = 0, 1, \ldots, N. \tag{74}$$

Similar relations hold for the other partial derivatives, which allows us to write the spectral gradient as

$$\begin{aligned}\vec{\nabla}_x U_{ijk} &= \sum_{n=0}^{N} U_{njk}\mathcal{D}_{in}\hat{x} + \sum_{n=0}^{N} U_{ink}\mathcal{D}_{jn}\hat{y} + \sum_{n=0}^{N} U_{ijn}\mathcal{D}_{kn}\hat{z} \\ &= \mathcal{D}U_{:jk}\hat{x} + \mathcal{D}U_{i:j}\hat{y} + \mathcal{D}U_{ij:}\hat{z}\,,\end{aligned} \tag{75}$$

and the divergence as

$$\begin{aligned}\vec{\nabla}_x \cdot \vec{F}_{ijk} &= \sum_{n=0}^{N} (F_1)_{njk}\mathcal{D}_{in} + \sum_{n=0}^{N} (F_2)_{ink}\mathcal{D}_{jn} + \sum_{n=0}^{N} (F_3)_{ijn}\mathcal{D}_{kn} \\ &= \mathcal{D}(F_1)_{:jk} + \mathcal{D}(F_2)_{i:k} + \mathcal{D}(F_3)_{ij:}\ .\end{aligned} \tag{76}$$

Discrete Inner Products and Summation-by-parts

The discrete inner product is defined using the Gauss-Lobatto rule in each space direction. In 3D,

$$\langle U, V\rangle_N = \sum_{i,j,k=0}^{N} U_{ijk}V_{ijk}w_i w_j w_k \equiv \sum_{i,j,k=0}^{N} U_{ijk}V_{ijk}w_{ijk}. \tag{77}$$

As in one space dimension, the discrete inner product is exact when the degree of UV is $2N-1$ or less in each direction, i.e.,

$$\langle U, V\rangle_N = \langle U, V\rangle \quad \text{for all } UV \in \mathbb{P}^{2N-1}. \tag{78}$$

Because of the tensor product, it turns out that summation-by-parts still holds in multiple space dimensions. We show how in three space dimensions and for the partial derivatives in x. Let $U,\, V \in \mathbb{P}^N$. Then

$$\begin{aligned} U &= \sum_{n,m,l=0}^{N} U_{nml}\ell_n(x)\ell_m(y)\ell_l(z) \\ V_x &= \sum_{\mu,\nu,\lambda=0}^{N} V_{\mu\nu\lambda}\ell'_\mu(x)\ell_\nu(y)\ell_\lambda(z) \end{aligned} \tag{79}$$

and so

$$\langle U, V_x\rangle_N = \sum_{n,m,l=0}^{N}\sum_{\mu,\nu,\lambda=0}^{N} U_{nml}V_{\mu\nu\lambda}\left\langle \ell_n\ell_m\ell_l, \ell'_\mu\ell_\nu\ell_\lambda\right\rangle_N. \tag{80}$$

The discrete inner products in the sums factorize,

$$\left\langle \ell_n\ell_m\ell_l, \ell'_\mu\ell_\nu\ell_\lambda\right\rangle_N = \left(\int_N \ell_n\ell'_\mu\,\mathrm{d}x\right)\left(\int_N \ell_m\ell_\nu\,\mathrm{d}y\right)\left(\int_N \ell_l\ell_\lambda\,\mathrm{d}z\right). \tag{81}$$

We then use summation-by-parts on the first factor

$$\left\langle \ell_n\ell_m\ell_l, \ell'_\mu\ell_\nu\ell_\lambda\right\rangle_N = \left(\ell_n\ell_\mu\Big|_{x=-1}^{1} - \int_N \ell'_n\ell_\mu\,\mathrm{d}x\right)\left(\int_N \ell_m\ell_\nu\,\mathrm{d}y\right)\left(\int_N \ell_l\ell_\lambda\,\mathrm{d}z\right) \tag{82}$$

and recombine the discrete inner product

$$\left\langle \ell_n\ell_m\ell_l, \ell'_\mu\ell_\nu\ell_\lambda\right\rangle_N = \int_N \ell_n\ell_\mu\Big|_{x=-1}^{1}\ell_m\ell_\nu\ell_l\ell_\lambda\,\mathrm{d}y\mathrm{d}z - \left\langle \ell'_n\ell_m\ell_l, \ell_\mu\ell_\nu\ell_\lambda\right\rangle_N. \tag{83}$$

Substituting the discrete inner product (83) into (80) gives the summation-by-parts formula

$$\langle U, V_x\rangle_N = \int_N UV\Big|_{x=-1}^{1}\,\mathrm{d}y\mathrm{d}z - \langle U_x, V\rangle_N. \tag{84}$$

The surface quadrature is precisely

$$\int_N UV|_{x=-1}^{1}\,\mathrm{d}y\mathrm{d}z \equiv \sum_{j,k=0}^{N} U(1,y_j,z_k)V(1,y_j,z_k)w_j w_k - \sum_{j,k=0}^{N} U(-1,y_j,z_k)V(-1,y_j,z_k)w_j w_k$$
$$= \sum_{j,k=0}^{N} U_{Njk}V_{Njk}w_j w_k - \sum_{j,k=0}^{N} U_{0jk}V_{0jk}w_j w_k. \tag{85}$$

Equivalent results hold for the y and z derivatives and in two space dimensions.

Multidimensional Summation-by-Parts and Divergence Theorem

The summation-by-parts property extends to three dimensions and to the divergence theorem. Let $\vec{F} \in \mathbb{P}^N$ be a vector

$$\vec{F} = \sum_{i=1}^{d} F_i \hat{x}_i, \tag{86}$$

where $d = 2, 3$ is the number of space dimensions. Then by adding the summation-by-parts property (84) for any $V \in \mathbb{P}^N$ and for each component of the vector $\vec{F}$ and its corresponding derivative, we get the multidimensional summation-by-parts theorem

$$\left\langle \vec{\nabla}_x \cdot \vec{F}, V \right\rangle_N = \int_{\partial E,N} \vec{F}\cdot\hat{n} V\,\mathrm{dS} - \left\langle \vec{F}, \vec{\nabla}_x V \right\rangle_N. \tag{87}$$

If we then set $V = 1$, then we get the discrete divergence theorem

$$\int_{E,N} \vec{\nabla}_x \cdot \vec{F}\,\mathrm{d}\vec{x} = \left\langle \vec{\nabla}_x \cdot \vec{F}, 1 \right\rangle_N = \int_{\partial E,N} \vec{F}\cdot\hat{n}\,\mathrm{dS}. \tag{88}$$

The divergence theorem is used to show conservation. In fact, we can say even more. With $V = 1$ the quadrature is exact and so the discrete conservation is actually conservative in the integral sense.

Summary

We summarize the results of this section by showing the continuous and discrete equivalents for the spectral calculus in Table 3.1.

Table 3.1 Summary of calculus computations and rules on $E_3 = [-1, 1]^3$

Continuous setting	Discrete setting
$u, v, \vec{f} \in \mathbb{L}^2(E)$	$U, V, \vec{F} \in \mathbb{P}^N$
$\langle u, v\rangle = \int_E uv\,\mathrm{d}\vec{x}$	$\langle U, V\rangle_N = \sum_{i,j,k=0}^{N} U_{ijk}V_{ijk}w_i w_j w_k$
$\lVert u\rVert^2 = \langle u, u\rangle$	$\lVert U\rVert_N^2 = \langle U, U\rangle_N$
$\vec{\nabla}_x u = u_x\hat{x} + u_y\hat{y} + u_z\hat{z}$	$\vec{\nabla}_x U_{ijk} = \left(\sum_{n=0}^{N} U_{njk}\mathcal{D}_{in}\right)\hat{x} + \left(\sum_{n=0}^{N} U_{ink}\mathcal{D}_{jn}\right)\hat{y} + \left(\sum_{n=0}^{N} U_{ijn}\mathcal{D}_{kn}\right)\hat{z}$
$\vec{\nabla}_x \cdot \vec{f} = (f_1)_x + (f_2)_y + (f_3)_z$	$\vec{\nabla}_x \cdot \vec{F}_{ijk} = \sum_{n=0}^{N} (F_1)_{njk}\mathcal{D}_{in} + \sum_{n=0}^{N} (F_2)_{ink}\mathcal{D}_{jn} + \sum_{n=0}^{N} (F_3)_{ijn}\mathcal{D}_{kn}$
$\left\langle\vec{\nabla}_x \cdot \vec{f}, v\right\rangle = \int_{\partial E} \vec{f}\cdot\hat{n}v\,\mathrm{dS} - \left\langle\vec{f}, \vec{\nabla}_x v\right\rangle$	$\left\langle\vec{\nabla}_x \cdot \vec{F}, V\right\rangle_N = \int_{\partial E,N} \vec{F}\cdot\hat{n}V\,\mathrm{dS} - \left\langle\vec{F}, \vec{\nabla}_x V\right\rangle_N$
$\int_E \vec{\nabla}_x \cdot \vec{f}\,\mathrm{d}\vec{x} = \int_{\partial E} \vec{f}\cdot\hat{n}\,\mathrm{dS}$	$\int_{\partial E,N} \vec{\nabla}_x \cdot \vec{F}\,\mathrm{d}\vec{x} = \int_{\partial E,N} \vec{F}\cdot\hat{n}\,\mathrm{dS}$
$(uv)' = u'v + uv'$	$(\mathbb{I}^N(UV))' \neq \mathbb{I}^{N-1}(U'V) + \mathbb{I}^{N-1}(UV')$

The Compressible Navier–Stokes Equations

Compressible viscous flows are modeled by the Navier–Stokes equations,

$$\mathbf{u}_t + \sum_{i=1}^{3}\frac{\partial \mathbf{f}_i}{\partial x_i} = \frac{1}{\mathrm{Re}}\sum_{i=1}^{3}\frac{\partial \mathbf{f}_i^v\left(\mathbf{u}, \vec{\nabla}_x\mathbf{u}\right)}{\partial x_i}. \tag{89}$$

The state vector contains the conservative variables of the density, ρ, the momenta, $\rho\vec{v} = (\rho v_1\ \rho v_2\ \rho v_3)^T$ and total energy ρE per unit volume,

$$\mathbf{u} = \begin{bmatrix}\rho\\ \rho\vec{v}\\ \rho E\end{bmatrix} = \begin{bmatrix}\rho\\ \rho v_1\\ \rho v_2\\ \rho v_3\\ \rho E\end{bmatrix}. \tag{90}$$

In standard form, the components of the advective flux are

$$\mathbf{f}_1 = \begin{bmatrix}\rho v_1\\ \rho v_1^2 + p\\ \rho v_1 v_2\\ \rho v_1 v_3\\ \rho v_1 H\end{bmatrix}\quad \mathbf{f}_2 = \begin{bmatrix}\rho v_2\\ \rho v_2 v_1\\ \rho v_2 v_2 + p\\ \rho v_2 v_3\\ \rho v_2 H\end{bmatrix}\quad \mathbf{f}_3 = \begin{bmatrix}\rho v_3\\ \rho v_3 v_1\\ \rho v_3 v_2\\ \rho v_3 v_3 + p\\ \rho v_3 H\end{bmatrix}, \tag{91}$$

where p is the pressure and

$$H = E + \frac{p}{\rho}, \quad E = e + \frac{1}{2}|\vec{v}|^2, \quad e = \frac{1}{\gamma - 1}\frac{p}{\rho}. \tag{92}$$

The equations have been scaled with respect to free-stream reference values so that the Reynolds number is

$$\mathrm{Re} = \frac{\rho_\infty V_\infty L}{\mu_\infty}, \tag{93}$$

where L is the length scale and V_∞ is the free-stream velocity. Additionally, the Mach number and Prandtl numbers are

$$\mathrm{M}_\infty = \frac{V_\infty}{\sqrt{\gamma \mathrm{R} T_\infty}}, \quad \mathrm{Pr} = \frac{\mu_\infty C_p}{\lambda_\infty}. \tag{94}$$

Written in terms of the primitive variables, the viscous fluxes are

$$\begin{aligned}
\mathbf{f}_1^v &= \left[0 \quad \tau_{11} \quad \tau_{12} \quad \tau_{13} \quad \left(\left(\sum\nolimits_{j=1}^{3} v_j \tau_{1j}\right) + \lambda \frac{\partial T}{\partial x}\right)\right]^T, \\
\mathbf{f}_2^v &= \left[0 \quad \tau_{21} \quad \tau_{22} \quad \tau_{23} \quad \left(\left(\sum\nolimits_{j=1}^{3} v_j \tau_{2j}\right) + \lambda \frac{\partial T}{\partial y}\right)\right]^T, \\
\mathbf{f}_3^v &= \left[0 \quad \tau_{31} \quad \tau_{32} \quad \tau_{33} \quad \left(\left(\sum\nolimits_{j=1}^{3} v_j \tau_{3j}\right) + \lambda \frac{\partial T}{\partial z}\right)\right]^T,
\end{aligned} \tag{95}$$

where

$$\tau_{ij} = \mu\left(\frac{\partial v_j}{\partial x^i} + \frac{\partial v_i}{\partial x^j}\right) - \frac{2}{3}\mu\left(\vec{\nabla}_x \cdot \vec{v}\right)\delta_{ij}, \quad \lambda = \frac{\mu}{(\gamma - 1)\mathrm{Pr}\mathrm{M}_\infty^2}, \tag{96}$$

and the temperature is

$$T = \gamma \mathrm{M}_\infty^2 \frac{p}{\rho}. \tag{97}$$

To simplify the presentation, we define *block vectors* (with the double arrow), for instance, the block vector flux,

$$\overleftrightarrow{\mathbf{f}} = \begin{bmatrix} \mathbf{f}_1 \\ \mathbf{f}_2 \\ \mathbf{f}_3 \end{bmatrix}. \tag{98}$$

The spatial gradient of a state vector is a block vector,

$$\vec{\nabla}_x \mathbf{u} = \begin{bmatrix} \mathbf{u}_x \\ \mathbf{u}_y \\ \mathbf{u}_z \end{bmatrix}. \quad (99)$$

The dot product of two block vectors is defined by

$$\overleftrightarrow{\mathbf{f}} \cdot \overleftrightarrow{\mathbf{g}} = \sum_{i=1}^{3} \mathbf{f}_i{}^T \mathbf{g}_i, \quad (100)$$

and the dot product of a block vector with a vector is a state vector,

$$\vec{g} \cdot \overleftrightarrow{\mathbf{f}} = \sum_{i=1}^{3} g_i \mathbf{f}_i. \quad (101)$$

With this notation the divergence of a flux is defined as

$$\vec{\nabla}_x \cdot \overleftrightarrow{\mathbf{f}} = \sum_{i=1}^{3} \frac{\partial \mathbf{f}_i}{\partial x_i}, \quad (102)$$

which allows the Navier–Stokes equations to be written compactly as an advection-diffusion like equation

$$\mathbf{u}_t + \vec{\nabla}_x \cdot \overleftrightarrow{\mathbf{f}} = \frac{1}{\mathrm{Re}} \vec{\nabla}_x \cdot \overleftrightarrow{\mathbf{f}}^{\,v} \left(\mathbf{u}, \vec{\nabla}_x \mathbf{u} \right). \quad (103)$$

As part of the approximation procedure, it is customary to represent the solution gradients as a new variable to get a first-order system of equations

$$\begin{aligned} \mathbf{u}_t + \vec{\nabla}_x \cdot \overleftrightarrow{\mathbf{f}} &= \frac{1}{\mathrm{Re}} \vec{\nabla}_x \cdot \overleftrightarrow{\mathbf{f}}^{\,v} (\mathbf{u}, \overleftrightarrow{\mathbf{q}}) \\ \overleftrightarrow{\mathbf{q}} &= \vec{\nabla}_x \mathbf{u}. \end{aligned} \quad (104)$$

To understand the growth of small perturbations in the fluid state, one also studies linearized forms of the Navier–Stokes equations, (103). When linearized about a constant state, the Navier–Stokes equations can be written in the form

$$\mathbf{u}_t + \sum_{j=1}^{3} \frac{\partial \underline{\mathrm{A}}_j \mathbf{u}}{\partial x_j} = \frac{1}{\mathrm{Re}} \sum_{i=1}^{3} \frac{\partial}{\partial x_i} \left(\sum_{j=1}^{3} \underline{\mathrm{B}}_{ij} \frac{\partial \mathbf{u}}{\partial x_j} \right), \quad (105)$$

where $\mathbf{u} = [\delta\rho\ \delta v_1\ \delta v_2\ \delta v_3\ \delta p]^T$ is the perturbation from the constant state reference values. The coefficient matrices $\underline{A}_j$ and $\underline{B}_{ij}$ are constant in the linear approximation of the equations.

To again simplify the notation for use in the analysis, we define a block vector of matrices, e.g.,

$$\overleftrightarrow{\mathbf{A}} = \begin{bmatrix} \underline{A}_1 \\ \underline{A}_2 \\ \underline{A}_3 \end{bmatrix}, \tag{106}$$

and full block matrix

$$\mathfrak{B} = \begin{bmatrix} \underline{B}_{11} & \underline{B}_{12} & \underline{B}_{13} \\ \underline{B}_{21} & \underline{B}_{22} & \underline{B}_{23} \\ \underline{B}_{31} & \underline{B}_{32} & \underline{B}_{33} \end{bmatrix}. \tag{107}$$

Then the product rule applied to the divergence of the flux in (105) can be written as

$$\vec{\nabla}_x \cdot \overleftrightarrow{\mathbf{f}} = \left(\vec{\nabla}_x \cdot \overleftrightarrow{\mathbf{A}}\right)\mathbf{u} + \left(\overleftrightarrow{\mathbf{A}}\right)^T \vec{\nabla}_x \mathbf{u}, \tag{108}$$

where

$$\overleftrightarrow{\mathbf{f}} = \begin{bmatrix} \underline{A}_1\mathbf{u} \\ \underline{A}_2\mathbf{u} \\ \underline{A}_3\mathbf{u} \end{bmatrix}, \quad \left(\overleftrightarrow{\mathbf{A}}\right)^T = \left[\underline{A}_1\ \underline{A}_2\ \underline{A}_3\right]. \tag{109}$$

The nonconservative advective form of the linearized Navier–Stokes equations can therefore be written as an advection-diffusion equation

$$\mathbf{u}_t + \left(\vec{\nabla}_x \cdot \overleftrightarrow{\mathbf{A}}\right)\mathbf{u} + \left(\overleftrightarrow{\mathbf{A}}\right)^T \vec{\nabla}_x \mathbf{u} = \frac{1}{\mathrm{Re}}\vec{\nabla}_x \cdot \left(\mathfrak{B}\vec{\nabla}_x \mathbf{u}\right). \tag{110}$$

Averaging the conservative and nonconservative forms gives a <u>split form</u> of the PDE

$$\mathbf{u}_t + \frac{1}{2}\left\{\vec{\nabla}_x \cdot \overleftrightarrow{\mathbf{f}} + \left(\vec{\nabla}_x \cdot \overleftrightarrow{\mathbf{A}}\right)\mathbf{u} + \left(\overleftrightarrow{\mathbf{A}}\right)^T \vec{\nabla}_x \mathbf{u}\right\} = \frac{1}{\mathrm{Re}}\vec{\nabla}_x \cdot \left(\mathfrak{B}\vec{\nabla}_x \mathbf{u}\right). \tag{111}$$

Note that with constant advection matrices, the divergence $\vec{\nabla}_x \cdot \overleftrightarrow{\mathbf{A}}$ is zero.

Boundedness of Energy and Entropy

With suitable boundary conditions, small perturbations,

$$\mathbf{u} = [\delta\rho\ \delta v_1\ \delta v_2\ \delta v_3\ \delta p]^T, \tag{112}$$

are bounded by the initial data in that

$$||\mathbf{u}(T)|| \leqslant C\,||\mathbf{u}_0||, \tag{113}$$

where $||\mathbf{u}||^2 = \int_\Omega |\mathbf{u}|^2\,\mathrm{dV}$ is the "energy norm" on a domain Ω. Bounding the energy norm guarantees that the individual components of the perturbed state are bounded at any fixed time. They won't blow up.

We show boundedness of the perturbations by multiplying the split form (111) by an arbitrary $\mathbb{L}^2(\Omega)$ test function, φ, and integrating over the domain to get a weak form of the equation. In inner product notation, that weak form is

$$\langle \mathbf{u}_t, \varphi\rangle + \frac{1}{2}\left\{\left\langle \vec{\nabla}_x \cdot \overleftrightarrow{\mathbf{f}}, \varphi\right\rangle + \left\langle \left(\overleftrightarrow{\mathbf{A}}\right)^T \vec{\nabla}_x \mathbf{u}, \varphi\right\rangle\right\} = \frac{1}{\mathrm{Re}}\left\langle \vec{\nabla}_x \cdot \left(\mathfrak{B}\vec{\nabla}_x \mathbf{u}\right), \varphi\right\rangle. \tag{114}$$

As with the nonlinear equations in (104), we introduce the intermediate block vector $\overleftrightarrow{\mathbf{q}} = \vec{\nabla}_x \mathbf{u}$ to produce a first-order system

$$\begin{aligned} \langle \mathbf{u}_t, \varphi\rangle + \frac{1}{2}\left\{\left\langle \vec{\nabla}_x \cdot \overleftrightarrow{\mathbf{f}}, \varphi\right\rangle + \left\langle \left(\overleftrightarrow{\mathbf{A}}\right)^T \vec{\nabla}_x \mathbf{u}, \varphi\right\rangle\right\} &= \frac{1}{\mathrm{Re}}\left\langle \vec{\nabla}_x \cdot \left(\mathfrak{B}\overleftrightarrow{\mathbf{q}}\right), \varphi\right\rangle \\ \left\langle \overleftrightarrow{\mathbf{q}}, \overleftrightarrow{\vartheta}\right\rangle &= \left\langle \vec{\nabla}_x \mathbf{u}, \overleftrightarrow{\vartheta}\right\rangle, \end{aligned} \tag{115}$$

where the auxiliary equation for $\overleftrightarrow{\mathbf{q}}$ is multiplied by the test function $\overleftrightarrow{\vartheta}$ and integrated over the domain.

We then apply the multidimensional integration-by-parts law to the second and fourth terms, which contain flux divergence, to separate surface (physical boundary) and volume contributions. If, further, we define

$$\overleftrightarrow{\mathbf{f}}^{(T)}(\varphi) = \begin{bmatrix} \underline{\mathrm{A}}_1^T \varphi \\ \underline{\mathrm{A}}_2^T \varphi \\ \underline{\mathrm{A}}_3^T \varphi \end{bmatrix}, \tag{116}$$

then we can re-write the first equation of (115) as

$$\begin{aligned}
\langle \mathbf{u}_t, \varphi \rangle &+ \frac{1}{2}\left\{ \left\langle \vec{\nabla}_x \mathbf{u}, \overleftrightarrow{\mathbf{f}}^{(T)}(\varphi) \right\rangle - \left\langle \overleftrightarrow{\mathbf{f}}, \vec{\nabla}_x \varphi \right\rangle \right\} \\
&+ \int_{\partial\Omega} \left(\frac{1}{2}\left(\overleftrightarrow{\mathbf{f}} \cdot \vec{n} \right) - \frac{1}{\mathrm{Re}} \left(\left(\mathfrak{B} \vec{\nabla}_x \mathbf{u} \right) \cdot \vec{n} \right) \right)^T \varphi \, \mathrm{dS} \\
&= -\frac{1}{\mathrm{Re}} \left\langle \mathfrak{B} \overleftrightarrow{\mathbf{q}}, \vec{\nabla}_x \varphi \right\rangle,
\end{aligned} \tag{117}$$

where $\vec{n}$ is the physical space outward normal to the surface.

Looking ahead a few steps, we see that the first term in (117) becomes the time derivative of the energy if we replace the test function φ by the solution $\mathbf{u}$. Furthermore, the second term would vanish with the same substitution if the advection matrices $\underline{A}_i$ were symmetric, leaving only boundary terms, which can be controlled with boundary conditions. This roadmap suggests the need for symmetry in the equations.

The system (117) for the linearized compressible Navier–Stokes equations, although not symmetric, is known to be symmetrizable by a single constant symmetrization matrix, $\underline{S}$, and there are multiple symmetrizers (Abarbanel and Gottlieb 1981) from which to choose. We denote the symmetrized matrices as

$$\underline{A}_j^s = \underline{S}^{-1}\underline{A}_j\underline{S} = \left(\underline{A}_j^s\right)^T \quad \text{and} \quad \underline{B}_{ij}^s = \underline{S}^{-1}\underline{B}_{ij}\underline{S} = \left(\underline{B}_{ij}^s\right)^T \geqslant 0. \tag{118}$$

Explicit representations of the symmetrizer and coefficient matrices are presented in Nordström and Svärd (2005). Furthermore, the symmetrized block matrix $\mathfrak{B}^s = \mathfrak{S}^{-1}\mathfrak{B}\mathfrak{S}$, where

$$\mathfrak{S} = \begin{bmatrix} \underline{S} & 0 & 0 \\ 0 & \underline{S} & 0 \\ 0 & 0 & \underline{S} \end{bmatrix}, \tag{119}$$

is the diagonal block matrix of the symmetrizer, is symmetric and non-negative.

To symmetrize the system and obtain an energy bound at the same time, we let $\varphi = \left(\underline{S}^{-1}\right)^T \underline{S}^{-1}\mathbf{u}$ in (117), which includes symmetrization as part of the test function. Then

$$\begin{aligned}
\left\langle \underline{S}^{-1}\mathbf{u}_t, \underline{S}^{-1}\mathbf{u} \right\rangle &+ \frac{1}{2}\left\{ \left\langle \vec{\nabla}_x \mathbf{u}, \overleftrightarrow{\mathbf{f}}^{(T)}\left(\left(\underline{S}^{-1}\right)^T \underline{S}^{-1}\mathbf{u} \right) \right\rangle - \left\langle \mathfrak{S}^{-1}\overleftrightarrow{\mathbf{f}}, \vec{\nabla}_x \left(\underline{S}^{-1}\mathbf{u}\right) \right\rangle \right\} \\
&+ \int_{\partial\Omega} \left(\frac{1}{2}\underline{S}^{-1}\left(\overleftrightarrow{\mathbf{f}} \cdot \vec{n} \right) - \frac{1}{\mathrm{Re}}\underline{S}^{-1} \left(\left(\mathfrak{B} \vec{\nabla}_x \mathbf{u} \right) \cdot \vec{n} \right) \right)^T \underline{S}^{-1}\mathbf{u} \, \mathrm{dS} \\
&= -\frac{1}{\mathrm{Re}} \left\langle \mathfrak{S}^{-1}\mathfrak{B} \overleftrightarrow{\mathbf{q}}, \vec{\nabla}_x \left(\underline{S}^{-1}\mathbf{u}\right) \right\rangle.
\end{aligned} \tag{120}$$

To simplify the notation, let us define the symmetric state vector as $\mathbf{u}^s = \underline{S}^{-1}\mathbf{u}$ and examine the terms in (120) separately. First,

$$\langle \underline{S}^{-1}\mathbf{u}_t, \underline{S}^{-1}\mathbf{u}\rangle = \frac{1}{2}\frac{d}{dt}\,||\mathbf{u}^s||^2 \tag{121}$$

provides the time derivative of the energy of the symmetrized state. Next, the volume term for the diffusion can be written as

$$\left\langle \mathfrak{S}^{-1}\mathfrak{B}\overleftrightarrow{\mathbf{q}}, \vec{\nabla}_x\left(\underline{S}^{-1}\mathbf{u}\right)\right\rangle = \left\langle \mathfrak{B}^s\overleftrightarrow{\mathbf{q}}^s, \vec{\nabla}_x\mathbf{u}^s\right\rangle. \tag{122}$$

Making the changes on the boundary terms,

$$\begin{aligned}&\int\limits_{\partial\Omega}\left(\frac{1}{2}\underline{S}^{-1}\left(\overleftrightarrow{\mathbf{f}}\cdot\vec{n}\right) - \frac{1}{\mathrm{Re}}\underline{S}^{-1}\left(\left(\mathfrak{B}\vec{\nabla}_x\mathbf{u}\right)\cdot\vec{n}\right)\right)^T \underline{S}^{-1}\mathbf{u}\,\mathrm{dS}\\ &= \int\limits_{\partial\Omega}\left(\frac{1}{2}\left(\overleftrightarrow{\mathbf{f}}^s\cdot\vec{n}\right) - \frac{1}{\mathrm{Re}}\left(\left(\mathfrak{B}^s\vec{\nabla}_x\mathbf{u}^s\right)\cdot\vec{n}\right)\right)^T \mathbf{u}^s\,\mathrm{dS},\end{aligned} \tag{123}$$

where

$$\overleftrightarrow{\mathbf{f}}^s = \begin{bmatrix}\underline{A}_1^s\mathbf{u}^s\\ \underline{A}_2^s\mathbf{u}^s\\ \underline{A}_3^s\mathbf{u}^s\end{bmatrix}. \tag{124}$$

The most interesting terms in (120) are the volume flux terms. The solution flux term is

$$\left\langle \mathfrak{S}^{-1}\overleftrightarrow{\mathbf{f}}, \vec{\nabla}_x\left(\underline{S}^{-1}\mathbf{u}\right)\right\rangle = \left\langle \overleftrightarrow{\mathbf{f}}^s, \vec{\nabla}_x\mathbf{u}^s\right\rangle, \tag{125}$$

and the test function flux term are now the same, for

$$\begin{aligned}\left\langle \vec{\nabla}_x\mathbf{u}, \overleftrightarrow{\mathbf{f}}^{(T)}\left(\left(\underline{S}^{-1}\right)^T\underline{S}^{-1}\mathbf{u}\right)\right\rangle &= \left\langle \mathfrak{S}\vec{\nabla}_x\underline{S}^{-1}\mathbf{u}, \left(\mathfrak{S}^{-1}\overleftrightarrow{\mathbf{f}}\left(\mathbf{u}^s\right)\right)^T\right\rangle\\ &= \left\langle \vec{\nabla}_x\underline{S}^{-1}\mathbf{u}, \left(\mathfrak{S}^{-1}\overleftrightarrow{\mathbf{f}}\left(\mathbf{u}^s\right)\mathfrak{S}\right)^T\right\rangle\\ &= \left\langle \vec{\nabla}_x\mathbf{u}^s, \overleftrightarrow{\mathbf{f}}^s\left(\mathbf{u}^s\right)\right\rangle.\end{aligned} \tag{126}$$

Finally, when we set $\overleftrightarrow{\vartheta} = \left(\mathfrak{S}^{-1}\right)^T\mathfrak{S}^{-1}\mathfrak{B}\overleftrightarrow{\mathbf{q}}$ in the second equation of (115)

$$\left\langle \overleftrightarrow{\mathbf{q}}, \left(\mathfrak{S}^{-1}\right)^T\mathfrak{S}^{-1}\mathfrak{B}\overleftrightarrow{\mathbf{q}}\right\rangle = \left\langle \vec{\nabla}_x\mathbf{u}, \left(\mathfrak{S}^{-1}\right)^T\mathfrak{S}^{-1}\mathfrak{B}\overleftrightarrow{\mathbf{q}}\right\rangle \tag{127}$$

we see that

$$\left\langle \vec{\nabla}_x \mathbf{u}^s, \mathfrak{B}^s \overleftrightarrow{\mathbf{q}}^s \right\rangle = \left\langle \overleftrightarrow{\mathbf{q}}^s, \mathfrak{B}^s \overleftrightarrow{\mathbf{q}}^s \right\rangle \geqslant 0. \tag{128}$$

Gathering all the terms, the flux volume terms cancel due to the equivalence of (125) and (126), leaving

$$\begin{aligned} \frac{1}{2}\frac{d}{dt}\left|\left|\mathbf{u}^s\right|\right|^2 + \int_{\partial\Omega} \left(\frac{1}{2}\left(\overleftrightarrow{\mathbf{f}}^s \cdot \vec{n}\right) - \frac{1}{\mathrm{Re}}\left(\left(\mathfrak{B}^s \vec{\nabla}_x \mathbf{u}^s\right) \cdot \vec{n}\right)\right)^T \mathbf{u}^s \,\mathrm{dS} \\ = -\frac{1}{\mathrm{Re}}\left\langle \overleftrightarrow{\mathbf{q}}^s, \mathfrak{B}^s \overleftrightarrow{\mathbf{q}}^s \right\rangle \leqslant 0. \end{aligned} \tag{129}$$

We see, then, that any growth in the energy, defined as the $\mathbb{L}^2$ norm, is determined by the boundary integral,

$$\frac{1}{2}\frac{d}{dt}\left|\left|\mathbf{u}^s\right|\right|^2 \leqslant -\int_{\partial\Omega} \left(\frac{1}{2}\left(\overleftrightarrow{\mathbf{f}}^s \cdot \vec{n}\right) - \frac{1}{\mathrm{Re}}\left(\left(\mathfrak{B}^s \vec{\nabla}_x \mathbf{u}^s\right) \cdot \vec{n}\right)\right)^T \mathbf{u}^s \,\mathrm{dS}. \tag{130}$$

Integrating in time over the interval $[0, T]$,

$$\left|\left|\mathbf{u}^s(T)\right|\right|^2 \leqslant \left|\left|\mathbf{u}^s(0)\right|\right|^2 - \int_0^T \int_{\partial\Omega} \left(\left(\overleftrightarrow{\mathbf{f}}^s \cdot \vec{n}\right) - \frac{2}{\mathrm{Re}}\left(\left(\mathfrak{B}^s \vec{\nabla}_x \mathbf{u}^s\right) \cdot \vec{n}\right)\right)^T \mathbf{u}^s \,\mathrm{dS}. \tag{131}$$

To properly pose the problem, initial and boundary data must be specified. The value at $t = 0$ is replaced by initial data $\mathbf{u}_0^s$. As for the physical boundary terms, Nordström and Svärd (2005) show that the matrices can be split in characteristic fashion into incoming and outgoing information with boundary data specified along the incoming characteristics

$$\begin{aligned} \mathrm{PBT} &= \int_{\partial\Omega} \left(\left(\overleftrightarrow{\mathbf{f}}^s \cdot \vec{n}\right) - \frac{2}{\mathrm{Re}}\left(\left(\mathfrak{B}^s \vec{\nabla}_x \mathbf{u}^s\right) \cdot \vec{n}\right)\right)^T \mathbf{u}^s \,\mathrm{dS} \\ &= \int_{\partial\Omega} \mathbf{w}^{+T} \Lambda^+ \mathbf{w}^+ \,\mathrm{dS} - \int_{\partial\Omega} \mathbf{g}^T \left|\Lambda^-\right| \mathbf{g} \,\mathrm{dS}, \end{aligned} \tag{132}$$

where $\Lambda^+ > 0$ and $\Lambda^- < 0$. We will assume here that no energy is introduced by the boundary data, and so set $\mathbf{g} = 0$. As a result,

$$\mathrm{PBT} = \int_{\partial\Omega} \left(\left(\overleftrightarrow{\mathbf{f}}^s \cdot \vec{n}\right) - \frac{2}{\mathrm{Re}}\left(\left(\mathfrak{B}^s \vec{\nabla}_x \mathbf{u}^s\right) \cdot \vec{n}\right)\right)^T \mathbf{u}^s \,\mathrm{dS} = \int_{\partial\Omega} \mathbf{w}^{+T} \Lambda^+ \mathbf{w}^+ \,\mathrm{dS} \geqslant 0, \tag{133}$$

so that

$$\left|\left|\mathbf{u}^s(T)\right|\right| \leqslant \left|\left|\mathbf{u}_0^s\right|\right|. \tag{134}$$

Finally, since $\mathbf{u}^s = \underline{S}^{-1}\mathbf{u}$, $\mathbf{u} = \underline{S}\mathbf{u}^s$,

$$\frac{1}{\left\|\underline{S}\right\|_2}\left\|\mathbf{u}\right\| \leqslant \left\|\mathbf{u}^s\right\| \leqslant \left\|\underline{S}^{-1}\right\|_2\left\|\mathbf{u}\right\|, \tag{135}$$

where $\|\cdot\|_2$ is the matrix 2-norm. Therefore, we have the desired result,

$$\left|\left|\mathbf{u}(T)\right|\right| \leqslant C\left|\left|\mathbf{u}_0\right|\right|. \tag{136}$$

The analysis of the linearized compressible Navier–Stokes equations (105) provides an $\mathbb{L}^2(\Omega)$ bound, (136), on the solution "energy". One might think that an analogous statement of the solution "energy" should hold for a nonlinear systems of PDEs. Generally, though, linear stability estimates in the $\mathbb{L}^2(\Omega)$ sense are insufficient to exclude unphysical solution behavior like expansion shocks (Merriam 1987). To eliminate the possibility of such phenomena the notion of the "energy" estimate must be generalized for nonlinear systems.

To motivate this generalized solution estimate strategy, we take a slight detour to examine important concepts from thermodynamics. Thermodynamic laws provide rules to decide how physical systems **cannot** behave and act as guidelines for what solution behavior is physically meaningful and what is not.

The first law of thermodynamics concerns the conservation of the total energy in a closed system (already present as the fifth equation in the Navier–Stokes equations, (89)–(91)). The second law states that the entropy of a closed physical system tends to increase over time and, importantly, that it cannot decrease.

Though somewhat esoteric, the second law of thermodynamics regulates how energies are allowed to transfer within a system. For reversible processes the entropy remains constant over time (isentropic) and the time derivative of the total system entropy is zero. For irreversible processes the entropy increases and that time derivative is positive. Solution dynamics where the total system entropy shrinks in time are never observed and are deemed unphysical. A smooth solution that satisfies the system of nonlinear PDEs, like (103), corresponds to a reversible process. One of the difficulties, either analytically or numerically, of nonlinear PDEs with a dominant hyperbolic character is that the solution may develop discontinuities (e.g., shocks) regardless of the continuity of the initial conditions (Evans 2012). Such a discontinuous solution corresponds to an irreversible process and increases entropy.

So, the laws of thermodynamics play a pivotal role because they intrinsically provide admissibility criteria and select physically relevant solutions (Lax 1954, 1967; Tadmor 1987). As given in the compressible Navier–Stokes equations (103), the total entropy is not part of the state vector of conservative variables $\mathbf{u}$. However, we know that the total entropy *is* a conserved quantity for reversible (isentropic) processes. So where is this conservation law "hiding"? It turns out that there are additional conserved quantities, including the entropy, that are not explicitly built into the nonlinear system, but are still a consequence of the PDE.

To reveal an auxiliary conservation law for the second law of thermodynamics, we define a convex (mathematical) entropy function $s = s(\mathbf{u})$ that is a scalar function and depends nonlinearly on the conserved variables. From this it will be possible to generalize the previous $\mathbb{L}^2(\Omega)$ bound for the solution energy $\|\mathbf{u}\|$ and instead develop a stability bound on the mathematical entropy of the form

$$\int\limits_{\Omega} s(\mathbf{u}(T))\,\mathrm{dV} \leqslant \int\limits_{\Omega} s(\mathbf{u}(0))\,\mathrm{dV} + \mathrm{PBT}, \tag{137}$$

where PBT are the physical boundary terms. This statement of entropy stability (137) provides a bound on the entropy function in terms of the initial condition and appropriate boundary conditions, analogous to the linear bound (131). Further, as the mathematical entropy is a convex function of the solution $\mathbf{u}$, the entropy bound also leads to a bound on an associated norm of $\mathbf{u}$ (Merriam 1987; Dutt 1988).

Entropy stability for a single entropy does not give nonlinear stability, but it does give a stronger estimate than linear stability (Merriam 1989; Tadmor 2003), which is formally only appropriate for small perturbations to the equations. For the compressible Navier–Stokes equations an appropriate entropy pair $(s, \vec{f}^S)$, consists of the scalar entropy function

$$s = s(\mathbf{u}) = -\frac{\rho\varsigma}{\gamma - 1} = -\frac{\rho\,(\ln p - \gamma \ln \rho)}{\gamma - 1}, \tag{138}$$

where $\varsigma = \ln p - \gamma \ln \rho$ is the physical entropy, and the associated entropy flux

$$\vec{f}^S = s\,\vec{v}\,. \tag{139}$$

Note that the mathematical entropy s is taken as the negative of the physical entropy so that the mathematical entropy is bounded in time, just like the energy measured by the $\mathbb{L}^2$ norm. This allows us to write the more mathematically common type of bound (137). From here on we will use the term entropy to refer to the mathematical entropy, not the physical, thermodynamic entropy.

We also introduce entropy variables, the vector $\mathbf{w}$ being the derivative of the entropy with respect to the conservative state variables,

$$\mathbf{w} = \frac{\partial s}{\partial \mathbf{u}} = \left[\frac{\gamma - \varsigma}{\gamma - 1} - \frac{\rho|\vec{v}|^2}{2p},\ \frac{\rho v_1}{p},\ \frac{\rho v_2}{p},\ \frac{\rho v_3}{p},\ \frac{-\rho}{p}\right]^T, \tag{140}$$

with the convexity property

$$\mathbf{k}^T \frac{\partial^2 s}{\partial \mathbf{u}^2}\mathbf{k} > 0, \quad \forall \mathbf{k} \neq 0, \tag{141}$$

if $\rho > 0$ and $p > 0$ (Carpenter et al. 2014; Tadmor 2003; Dutt 1988).

The positivity requirement on the density and the temperature, $T \propto p/\rho$, ensures a one-to-one mapping between conservative and entropy variables. This constraint is unfortunately not a by-product of the entropy stability estimate for the thermodynamic entropy. Hence, entropy stability is not a true nonlinear stability statement and further criteria (up to this point unknown for the three-dimensional compressible Navier–Stokes equations) are necessary. Consequently, entropy stable discretizations can (and do) produce invalid solutions with negative density or temperature and need additional strategies to guarantee positivity.

The entropy variables are introduced because they contract the entropy pair, meaning that they satisfy the relations

$$\mathbf{w}^T\, \mathbf{u}_t = \left(\frac{\partial s}{\partial \mathbf{u}}\right)^T \mathbf{u}_t = s_t(\mathbf{u}), \tag{142}$$

and

$$\mathbf{w}^T\, \vec{\nabla}_x \cdot \overleftrightarrow{\mathbf{f}} = \vec{\nabla}_x \cdot \vec{f}^S. \tag{143}$$

The contraction allows us to convert a system of advection equations (in this instance the compressible Euler equations)

$$\mathbf{u}_t + \vec{\nabla}_x \cdot \overleftrightarrow{\mathbf{f}} = 0 \tag{144}$$

to a scalar advection equation for the entropy simply by multiplying by the entropy variables $\mathbf{w}$,

$$\mathbf{w}^T \left(\mathbf{u}_t + \vec{\nabla}_x \cdot \overleftrightarrow{\mathbf{f}}\right) = s_t + \vec{\nabla}_x \cdot \vec{f}^S = 0. \tag{145}$$

(Cf. how multiplying by the solution state $\mathbf{u}$ in the linear analysis above converts the system to a scalar equation for the mathematical energy.)

Furthermore, the viscous flux can be re-written in terms of the gradient of the entropy variables

$$\overleftrightarrow{\mathbf{f}}^v \left(\mathbf{u}, \vec{\nabla}_x \mathbf{u}\right) = \mathfrak{B}^S\, \vec{\nabla}_x \mathbf{w}, \tag{146}$$

where $\mathfrak{B}^S$ satisfies

$$\underline{\mathbf{B}}_{ij}^S = (\underline{\mathbf{B}}_{ji}^S)^T, \quad \sum_{i=1}^{d}\sum_{j=1}^{d} \left(\frac{\partial \mathbf{w}}{\partial x_i}\right)^T \underline{\mathbf{B}}_{ij}^S \left(\frac{\partial \mathbf{w}}{\partial x_j}\right) \geqslant 0, \quad \forall \mathbf{w}, \tag{147}$$

if $p > 0$ and $\mu > 0$ (Carpenter et al. 2014; Tadmor and Zhong 2006; Dutt 1988). Formally, the entropy variables (140) can be used to re-write the compressible Navier–Stokes equations into a symmetric and non-negative form as shown by Dutt (1988), again analogous to the linear analysis symmetrization procedure.

Using the contraction properties of the entropy variables, we can construct a bound on the mathematical entropy of the form

$$\bar{s}(T) \leqslant \bar{s}(0), \tag{148}$$

where

$$\bar{s} = \langle s(\mathbf{u}), 1\rangle = \int_{\Omega} s(\mathbf{u})\, \mathrm{dV} \tag{149}$$

is the total entropy, provided that suitable boundary conditions are applied.

We find the bound (148) much as we did when we found the energy bound for the linear system. We multiply the first equation by the entropy variables and the second equation by the viscous flux, and integrate over the domain to get the weak forms

$$\begin{aligned} \langle \mathbf{w}(\mathbf{u}), \mathbf{u}_t\rangle + \left\langle \mathbf{w}(\mathbf{u}), \vec{\nabla}_x \cdot \overleftrightarrow{\mathbf{f}}\right\rangle &= \frac{1}{\mathrm{Re}} \left\langle \mathbf{w}(\mathbf{u}), \vec{\nabla}_x \cdot \overleftrightarrow{\mathbf{f}}^{\,v}\right\rangle, \\ \left\langle \overleftrightarrow{\mathbf{q}}, \overleftrightarrow{\mathbf{f}}^{\,v}\right\rangle &= \left\langle \vec{\nabla}_x \mathbf{w}, \overleftrightarrow{\mathbf{f}}^{\,v}\right\rangle. \end{aligned} \tag{150}$$

Next we use the properties of the entropy pair (142) and (143) to contract the left side of the first equation of (150), and use multidimensional integration-by-parts on the right-hand side to get

$$\begin{aligned} \langle s_t(\mathbf{u}), 1\rangle + \left\langle \vec{\nabla}_x \cdot \vec{f}^{S}, 1\right\rangle &= \frac{1}{\mathrm{Re}} \int_{\partial\Omega} \mathbf{w}^T(\mathbf{u}) \left(\overleftrightarrow{\mathbf{f}}^{\,v} \cdot \vec{n}\right) \mathrm{dS} - \frac{1}{\mathrm{Re}} \left\langle \vec{\nabla}_x \mathbf{w}(\mathbf{u}), \overleftrightarrow{\mathbf{f}}^{\,v}\right\rangle, \\ \left\langle \overleftrightarrow{\mathbf{q}}, \overleftrightarrow{\mathbf{f}}^{\,v}\right\rangle &= \left\langle \vec{\nabla}_x \mathbf{w}, \overleftrightarrow{\mathbf{f}}^{\,v}\right\rangle. \end{aligned} \tag{151}$$

Inserting the second equation of (151) into the first and applying the identity (146) gives

$$\langle s_t(\mathbf{u}), 1\rangle + \left\langle \vec{\nabla}_x \cdot \vec{f}^{S}, 1\right\rangle = \frac{1}{\mathrm{Re}} \int_{\partial\Omega} \mathbf{w}^T(\mathbf{u}) \left(\overleftrightarrow{\mathbf{f}}^{\,v} \cdot \vec{n}\right) \mathrm{dS} - \frac{1}{\mathrm{Re}} \left\langle \overleftrightarrow{\mathbf{q}}, \mathfrak{B}^S \overleftrightarrow{\mathbf{q}}\right\rangle. \tag{152}$$

Finally, we use the property (147) as well as multidimensional integration-by-parts to integrate the flux divergence on the left side to get the estimate

$$\frac{d}{dt}\bar{s} \leqslant \int\limits_{\partial\Omega} \left(-\vec{f}^{S} \cdot \vec{n} + \frac{1}{\mathrm{Re}} \mathbf{w}^{T}(\mathbf{u}) \left(\overleftrightarrow{\mathbf{f}}^{v} \cdot \vec{n} \right) \right) \mathrm{dS}. \tag{153}$$

This entropy estimate is precisely that given previously in (137) except the form of the physical boundary terms is now explicitly given for the compressible Navier–Stokes equations. Boundary conditions then need to be specified so that the bound on the entropy depends only on the boundary data. We will assume here that boundary data are given so that the right-hand side is non-positive so that the entropy will not increase in time. For more thorough discussion on boundary conditions for the Navier–Stokes equations, see, e.g., Dutt (1988), Dalcin et al. (2019), Hindenlang et al. (2019). Integrating (153) in time then gives the desired result, (148).

Construction of Curvilinear Spectral Elements

The general goal for the nodal DG method is to use the Lagrange polynomial basis with Gauss-Lobatto interpolation nodes to approximate the solutions as high-order polynomial interpolants, (71) or (72). This appears to restrict the approximation to the simple quadrilateral E_2 or hexahedron E_3. To overcome this severe limitation, we use a process to extend the methods to completely general geometries. That process consists of three steps:

1. The domain Ω is subdivided into quadrilateral or hexahedral elements, e^k, $k = 1, 2, \ldots, K$.
2. A mapping is created from the computational space coordinate $\vec{\xi}$ on the reference element, E_2 or E_3, onto the physical space coordinate $\vec{x}$ for each element e^k.
3. The equations are re-written in terms of the computational space coordinate on the reference element.

It is on the reference element with the mapped equations that the DG approximation is then created using the spectral approximation tools derived in Sect. 3.2. The result will be a DG spectral element method to approximate the solution of conservation laws in three-dimensional geometries, e.g., (103), that has as many of the properties of the continuous equations as possible, e.g., (136) or (148).

A significant advantage of approximating the equations in computational (or reference) space is that they can be derived independently of the element shape and depend only on the transformation defined in Step 2. A specific advantage is that high-order spectral approximations for the reference domain E_3 have been previously described in Sect. 3.2.8, with spectral accuracy coming from the Gauss-Lobatto quadrature and the Lagrange polynomial basis ansatz.

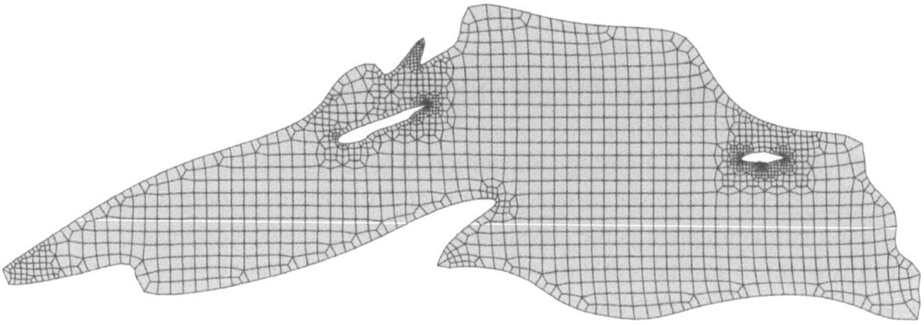

(a) Quadrilateral mesh of Lake Superior.

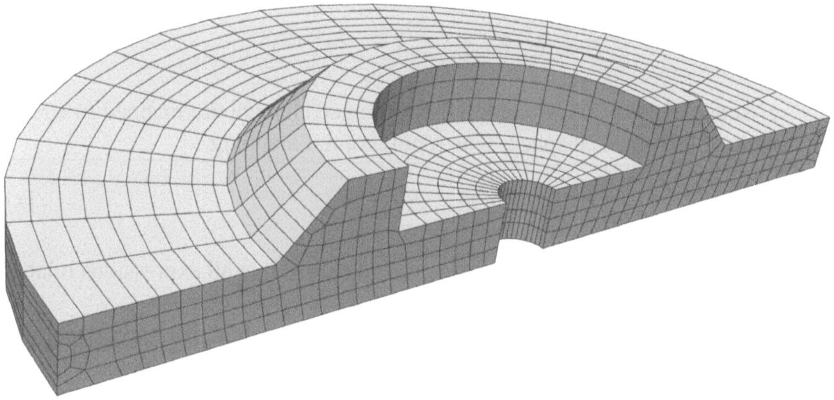

(b) Hexahedral mesh.

Fig. 3.1 Example meshes in **a** two and **b** three spatial dimensions

Subdividing the Domain: Spectral Element Mesh Generation

The first step in the approximation is to subdivide a domain Ω into a mesh of non-overlapping elements e^k, $k = 1, 2, \ldots, K$. We restrict the discussion to quadrilateral or hexahedral elements because the forthcoming DG approximation will be built from a tensor product ansatz as in Sect. 3.2.8. Two examples of such meshes are given in Figure 3.1. The generation of such meshes, especially with curved boundary information, is outside the scope of this chapter. As mentioned in the Prologue, high-order mesh generation is a difficult task with many open issues. For instance, it is necessary that the elements are valid and non-inverted, with non-negative mapping Jacobians. This, however, is non-trivial for boundary layer meshes with high aspect ratios near boundaries with high curvature. We refer the interested reader to Geuzaine and Remacle (2009), Hindenlang (2014) and the references therein for more details.

Mapping Elements from the Reference Element

More complex elements than E_2 or E_3 that have curved boundaries can be accommodated by a transformation from the reference elements onto the physical elements e^k. We adopt the naming convention where the domain E_d is called the reference element or computational domain and the element onto which computational domain is mapped is the physical domain.

From the mesh in the previous section, the physical domain has been divided into a set of (possibly curved) elements $\{e^k\}_{k=1}^K$. As before, the physical domain coordinates will be denoted by $\vec{x} = (x, y, z)^T$ within an element e^k. Analogously, the computational domain coordinates are defined as

$$\vec{\xi} = (\xi, \eta, \zeta)^T = \left(\xi^1, \xi^2, \xi^3\right)^T, \tag{154}$$

in the reference element E_3. Points in the reference element are mapped to each of the element, e^k, in physical space, with a polynomial mapping

$$\vec{x} = \vec{X}^k(\vec{\xi}). \tag{155}$$

In the following we will ignore the index, k, and realize that all expressions relate to any given element, e^k.

To maintain generality, (155) is an algebraic transformation that maps the boundaries of the reference element to the boundaries of the physical element and interior to interior. In this section, we demonstrate how to create a three-dimensional transformation from E_3 to a curved hexahedron. The two-dimensional mappings for quadrilateral elements are described in the book by Kopriva (2009).

The most common approach to generate the mapping $\vec{X}(\vec{\xi})$ is to use transfinite interpolation introduced by Gordon and Hall (1973). The idea is to interpolate between (possibly curved) boundaries with a polynomial to guarantee a smooth transformation between the computational and physical domains. The simplest transfinite interpolation, and the one almost always used in practice, is the linear blending formula, which uses a linear interpolation between boundaries.

In three space dimensions the physical domain is bounded by six curved faces Γ_i, $i = 1, \ldots, 6$, as depicted in Figure 3.2. Although it may be possible to define the boundary curves through analytic functions, we show later that constraints like free-stream preservation require that the curves are polynomials in their arguments. As a result, the faces are approximated by polynomials of degree N, written in the Lagrange basis. For example, the third boundary face Γ_3 is approximated as

$$\Gamma_3 \approx \mathbb{I}^N(\Gamma_3) = \sum_{i,j=0}^{N} (\Gamma_3)_{ij}\, \ell_i(\xi)\ell_j(\eta). \tag{156}$$

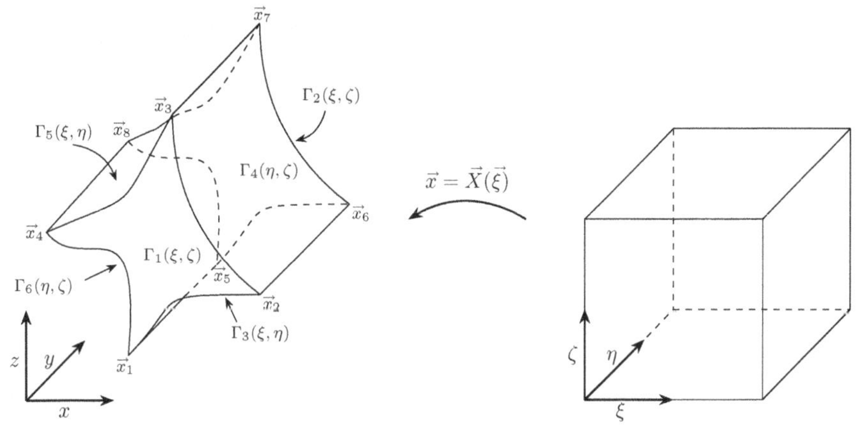

Fig. 3.2 Example mapping from computational to physical coordinates

Approximating the boundary to the same polynomial order, N, as the solution is called isoparametric.

The transformation is derived by linear interpolation between opposing faces. As such, one first creates a linear interpolation between two faces, say Γ_3 and Γ_5

$$\vec{X}_{35}(\xi,\eta) = \frac{1}{2}\{(1-\zeta)\Gamma_3(\xi,\eta) + (1+\zeta)\Gamma_5(\xi,\eta)\}. \tag{157}$$

Similarly, linear interpolation is constructed for the other four faces as

$$\begin{aligned}\vec{X}_{12}(\xi,\zeta) &= \frac{1}{2}\{(1-\eta)\Gamma_1(\xi,\zeta) + (1+\eta)\Gamma_2(\xi,\zeta)\},\\ \vec{X}_{64}(\eta,\zeta) &= \frac{1}{2}\{(1-\xi)\Gamma_6(\eta,\zeta) + (1+\xi)\Gamma_4(\eta,\zeta)\}.\end{aligned} \tag{158}$$

The final mapping will be a combination of the six face interpolants and three linear interpolations between them, starting with the sum

$$\begin{aligned}\vec{\Sigma}(\xi,\eta,\zeta) = \frac{1}{2}\{&(1-\xi)\Gamma_6(\eta,\zeta) + (1+\xi)\Gamma_4(\eta,\zeta) + (1-\eta)\Gamma_1(\xi,\zeta)\\ &+ (1+\eta)\Gamma_2(\xi,\zeta) + (1-\zeta)\Gamma_3(\xi,\eta) + (1+\zeta)\Gamma_5(\xi,\eta)\}.\end{aligned} \tag{159}$$

Unfortunately, the combination (159) no longer always matches at the faces

$$\vec{\Sigma}(-1,\eta,\zeta) = \Gamma_6(\eta,\zeta) + \frac{1}{2}\{(1-\eta)\Gamma_1(-1,\zeta) + (1+\eta)\Gamma_2(-1,\zeta) + (1-\zeta)\Gamma_3(-1,\eta) + (1+\zeta)\Gamma_5(-1,\eta)\}, \quad (160)$$

$$\vec{\Sigma}(1,\eta,\zeta) = \Gamma_4(\eta,\zeta) + \frac{1}{2}\{(1-\eta)\Gamma_1(1,\zeta) + (1+\eta)\Gamma_2(1,\zeta) + (1-\zeta)\Gamma_3(1,\eta) + (1+\zeta)\Gamma_5(1,\eta)\}, \quad (161)$$

$$\vec{\Sigma}(\xi,-1,\zeta) = \Gamma_1(\xi,\zeta) + \frac{1}{2}\{(1-\xi)\Gamma_6(-1,\zeta) + (1+\xi)\Gamma_4(-1,\zeta) + (1-\zeta)\Gamma_3(\xi,-1) + (1+\zeta)\Gamma_5(\xi,-1)\}, \quad (162)$$

$$\vec{\Sigma}(\xi,1,\zeta) = \Gamma_2(\xi,\zeta) + \frac{1}{2}\{(1-\xi)\Gamma_6(1,\zeta) + (1+\xi)\Gamma_4(1,\zeta) + (1-\zeta)\Gamma_3(\xi,1) + (1+\zeta)\Gamma_5(\xi,1)\}, \quad (163)$$

$$\vec{\Sigma}(\xi,\eta,-1) = \Gamma_3(\xi,\eta) + \frac{1}{2}\{(1-\eta)\Gamma_1(\xi,-1) + (1+\eta)\Gamma_2(\xi,-1) + ((1-\xi)\Gamma_6(\eta,-1) + (1+\xi)\Gamma_4(\eta,-1)\}, \quad (164)$$

$$\vec{\Sigma}(\xi,\eta,1) = \Gamma_5(\xi,\eta) + \frac{1}{2}\{(1-\eta)\Gamma_1(\xi,1) + (1+\eta)\Gamma_2(\xi,1) + (1-\xi)\Gamma_6(\eta,1) + (1+\xi)\Gamma_4(\eta,1)\}. \quad (165)$$

To match the faces, correction terms must be subtracted in the ξ, η, and ζ directions to cancel the additional terms that appear in the braces of (160)— (165). These linear corrections are

$$\vec{C}^{\xi} = \left(\frac{1-\xi}{4}\right)\left[(1-\eta)\Gamma_1(-1,\zeta) + (1+\eta)\Gamma_2(-1,\zeta) + (1-\zeta)\Gamma_3(-1,\eta) + (1+\zeta)\Gamma_5(-1,\eta)\right] + \left(\frac{1+\xi}{4}\right)\left[(1-\eta)\Gamma_1(1,\zeta) + (1+\eta)\Gamma_2(1,\zeta) + (1-\zeta)\Gamma_3(1,\eta) + (1+\zeta)\Gamma_5(1,\eta)\right], \quad (166)$$

$$\vec{C}^{\eta} = \left(\frac{1-\eta}{4}\right)\left[(1-\xi)\Gamma_6(-1,\zeta) + (1+\xi)\Gamma_4(-1,\zeta) + (1-\zeta)\Gamma_3(\xi,-1) + (1+\zeta)\Gamma_5(\xi,-1)\right] + \left(\frac{1+\eta}{4}\right)\left[(1-\xi)\Gamma_6(1,\zeta) + (1+\xi)\Gamma_4(1,\zeta) + (1-\zeta)\Gamma_3(\xi,1) + (1+\zeta)\Gamma_5(\xi,1)\right], \quad (167)$$

and

$$\vec{C}^{\zeta} = \left(\frac{1-\zeta}{4}\right)\left[(1-\eta)\Gamma_1(\xi,-1) + (1+\eta)\Gamma_2(\xi,-1) + (1-\xi)\Gamma_6(\eta,-1) + (1+\xi)\Gamma_4(\eta,-1)\right] + \left(\frac{1+\zeta}{4}\right)\left[(1-\eta)\Gamma_1(\xi,1) + (1+\eta)\Gamma_2(\xi,1) + (1-\xi)\Gamma_6(\eta,1) + (1+\xi)\Gamma_4(\eta,1)\right]. \quad (168)$$

However, subtracting the correction terms (166), (167), and (168) from (159) removes the interior contribution twice. Thus, to complete the correction to (159), one adds back the transfinite map of the reference cube to a straight-sided hexahedral element,

$$
\begin{aligned}
\vec{X}_H(\vec{\xi}) =& \frac{1}{8}\{\vec{x}_1(1-\xi)(1-\eta)(1-\zeta)+\vec{x}_2(1+\xi)(1-\eta)(1-\zeta) \\
&+\vec{x}_3(1+\xi)(1+\eta)(1-\zeta)+\vec{x}_4(1-\xi)(1+\eta)(1-\zeta) \\
&+\vec{x}_5(1-\xi)(1-\eta)(1+\zeta)+\vec{x}_6(1+\xi)(1-\eta)(1+\zeta) \\
&+\vec{x}_7(1+\xi)(1+\eta)(1+\zeta)+\vec{x}_8(1-\xi)(1+\eta)(1+\zeta)\},
\end{aligned} \tag{169}
$$

where $\vec{x}_i$, $i = 1, \ldots, 8$ are the locations of the corners of the hexahedron.

The final transfinite interpolation with linear blending for a curved-sided hexahedron is therefore

$$
\vec{X}(\vec{\xi}) = \vec{\Sigma}(\vec{\xi}) - \frac{1}{2}\left[\vec{C}^{\xi}+\vec{C}^{\eta}+\vec{C}^{\zeta}\right]+\vec{X}_H(\vec{\xi}), \tag{170}
$$

where the correction terms (166), (167), and (168) are further divided by two, otherwise they would contribute at each of the twelve edges twice.

Transforming Equations from Physical to Reference Domains

The mapping (170) provides a mechanism to connect differential operators in the computational domain to the physical domain. Under the mapping, the equations themselves, e.g., (103), are transformed as well, essentially being an exercise of the chain rule. Specifically, the differential operators of the divergence, gradient, and curl change form due to the mapping.

Rather than simply apply the chain rule, we summarize a general approach that uses ideas from differential geometry to transform equations between the reference and physical coordinate systems. This approach better exposes properties of the transformations that should be satisfied by the approximation. Full discussions of these general derivations can be found in Farrashkhalvat and Miles (2003), Knupp and Steinberg (1993) and in Hesthaven and Warburton (2008), Kopriva (2009) for the particular context of spectral methods.

The differential transformations are described in terms of two sets of independent coordinate basis vectors. The first is the *covariant basis*

$$
\vec{a}_i = \frac{\partial \vec{X}}{\partial \xi^i} \quad i = 1, 2, 3, \tag{171}
$$

whose components lie tangent to the transformation of a coordinate line in the computational space. Conveniently, the covariant basis vectors can be computed directly from the mapping between the reference element and physical space, $\vec{X}$, (170).

The second basis is the *contravariant basis*, whose components are normal to the transformation of the coordinate lines,

$$
\vec{a}^i = \vec{\nabla}_x \xi^i, \; i = 1, 2, 3. \tag{172}
$$

The contravariant basis vectors, for instance, point in the direction of the normal at a physical boundary. These two bases are not necessarily orthogonal, and will not be, unless the transformation is conformal. At first glance, it appears that the inverse mapping $\vec{\xi} = \vec{X}^{-1}(\vec{x})$ is needed to compute the contravariant basis vectors $\vec{a}^i,\ i = 1, 2, 3$. But this is not the case, once we have a way to represent the gradient in reference space.

A differential surface element can be written terms of the reference space coordinates by way of the cross product

$$\mathrm{d}\mathrm{S}^i = \vec{a}_j \mathrm{d}\xi^j \times \vec{a}_k \mathrm{d}\xi^k = (\vec{a}_j \times \vec{a}_k)\mathrm{d}\xi^j \mathrm{d}\xi^k, \quad (i, j, k) \text{ cyclic} \tag{173}$$

from which a volume element can be generated by extending the the surface element (173) in the normal direction

$$\mathrm{dV} = \vec{a}_i \cdot (\vec{a}_j \times \vec{a}_k)\mathrm{d}\vec{\xi} = J\mathrm{d}\vec{\xi}, \quad (i, j, k) \text{ cyclic.} \tag{174}$$

Writing the volume element this way exposes the Jacobian of the transformation in terms of the covariant basis vectors,

$$J = \vec{a}_1 \cdot (\vec{a}_2 \times \vec{a}_3). \tag{175}$$

Using the usual pillbox approach, the divergence is derived from the surface and volume differentials as

$$\vec{\nabla}_x \cdot \vec{f} = \frac{1}{J} \sum_{i=1}^{3} \frac{\partial}{\partial \xi^i} \left((\vec{a}_j \times \vec{a}_k) \cdot \vec{f} \right). \tag{176}$$

From the divergence it is possible to find an important identity satisfied by the covariant basis vectors. Under the assumption that the flux vector $\vec{f}$ is an arbitrary constant state, i.e., $\vec{f} = \vec{c}$, (176) simplifies to

$$0 = \sum_{i=1}^{3} \frac{\partial}{\partial \xi^i} (\vec{a}_j \times \vec{a}_k). \tag{177}$$

The statement (177) is one form of the *metric identities*. From (177) it is possible to re-write the divergence (176) into an equivalent form

$$\vec{\nabla}_x \cdot \vec{f} = \frac{1}{J} \sum_{i=1}^{3} (\vec{a}_j \times \vec{a}_k) \cdot \frac{\partial \vec{f}}{\partial \xi^i}. \tag{178}$$

From the alternative form of the divergence (178) it is straightforward to see that the *gradient* of some scalar function g in reference coordinates is

$$\vec{\nabla}_x g = \frac{1}{J}\sum_{i=1}^{3}(\vec{a}_j \times \vec{a}_k)\frac{\partial g}{\partial \xi^i}. \tag{179}$$

If we replace g by $g = \xi^i$, $i = 1, 2, 3$ in the gradient, (179), we relate the contravariant vectors (172) to the covariant

$$\vec{\nabla}_x \xi^i = \frac{1}{J}\sum_{m=1}^{3}(\vec{a}_j \times \vec{a}_k)\frac{\partial \xi^i}{\partial \xi^m}. \tag{180}$$

But $\frac{\partial \xi^i}{\partial \xi^m} = \delta_{im}$ so the sum simplifies to a definition of the volume weighted contravariant vectors in terms of the covariant,

$$J\vec{\nabla}_x \xi^i = J\vec{a}^i = \vec{a}_j \times \vec{a}_k, \quad (i, j, k) \text{ cyclic}. \tag{181}$$

Therefore, the contravariant basis can be computed from the covariant basis, which in turn can be computed directly from the transformation of the reference element to a physical element.

Now it is possible to write the metric identities (177) compactly in terms of the contravariant vectors,

$$\sum_{i=1}^{3}\frac{\partial J\vec{a}^i}{\partial \xi^i} = \vec{\nabla}_\xi \cdot \left(J\vec{a}^i\right) = 0. \tag{182}$$

Since a contravariant vector points in the direction of the normal along an element face, it is easy to construct a normal. The (outward) pointing normal vectors in the physical coordinate in terms of the reference coordinates are

$$\vec{n}_i = \frac{J\vec{a}^i}{|J\vec{a}^i|} = \frac{|J|}{J}\frac{\vec{a}_j \times \vec{a}_k}{|\vec{a}_j \times \vec{a}_k|}. \tag{183}$$

The transformation allows the normal in physical space to be related to the normal in reference space. The reference space normal (but not normalized) vectors are directly written in the contravariant basis (181)

$$\hat{n}^i = J\vec{a}^i, \quad i = 1, 2, 3. \tag{184}$$

Going back to (173) and (181), we can write the size of the surface differential elements as $\hat{s}$,

$$\begin{aligned}
\xi = \pm 1 \quad &: \quad \hat{s}(\eta, \zeta) = \left|J\vec{a}^1(\pm 1, \eta, \zeta)\right|, \\
\eta = \pm 1 \quad &: \quad \hat{s}(\xi, \zeta) = \left|J\vec{a}^2(\xi, \pm 1, \zeta)\right|, \\
\zeta = \pm 1 \quad &: \quad \hat{s}(\xi, \eta) = \left|J\vec{a}^3(\xi, \eta, \pm 1)\right|.
\end{aligned} \tag{185}$$

Table 3.2 Differential operators in physical and computational coordinates

Physical element	Reference element
$\vec{\nabla}_x \cdot \vec{f}$	$\frac{1}{J}\sum_{i=1}^{3}\frac{\partial}{\partial \xi^i}\left(J\vec{a}^i \cdot \vec{f}\right)$
$\vec{\nabla}_x g$	$\frac{1}{J}\sum_{i=1}^{3} J\vec{a}^i \frac{\partial g}{\partial \xi^i}$
$\vec{\nabla}_x \times \vec{f}$	$\frac{1}{J}\sum_{i=1}^{3}\frac{\partial}{\partial \xi^i}\left(J\vec{a}^i \times \vec{f}\right)$

The surface elements (185) are continuous across the element interface since the contravariant vectors (181) and the covariant vectors (171) are defined to be tangent to the shared face.

Now we can relate the two normal vector representations, either in physical space (183) or reference space (184), through

$$\hat{n}^i = \frac{J\vec{a}^i}{\hat{s}}\hat{s} = \vec{n}_i\,\hat{s}, \quad i = 1, 2, 3, \tag{186}$$

for the appropriate surface element $\hat{s}$ corresponding to a particular face. Since the $\hat{s}$ are continuous between elements sharing a face, the normal vector only changes sign.

Summary

To extend approximations defined on a square or cube to a general quadrilateral or hexahedron, we re-write differential operators in physical coordinates in terms of reference space coordinates through the contravariant basis vectors (181). We summarize the common differential operators of divergence, gradient, and curl in Table 3.2.

The divergence operator can be written compactly by defining the volume weighted contravariant flux vector $\vec{\tilde{f}}$, whose components are $\tilde{f}^i = J\vec{a}^i \cdot \vec{f}$. In terms of the contravariant flux, the divergence looks similar in both physical and reference spaces,

$$\vec{\nabla}_x \cdot \vec{f} = \frac{1}{J}\sum_{i=1}^{3}\frac{\partial}{\partial \xi^i}\left(J\vec{a}^i \cdot \vec{f}\right) = \frac{1}{J}\sum_{i=1}^{3}\frac{\partial \tilde{f}^i}{\partial \xi^i} = \frac{1}{J}\vec{\nabla}_\xi \cdot \vec{\tilde{f}}. \tag{187}$$

Building a Modern Discontinuous Galerkin Spectral Element Approximation

In this section, we use the derivations of Sect. 3.4.3 and apply a mapping from physical to reference space to the compressible Navier–Stokes equations written in mixed form (104) and derive a DG spectral element approximation for that system.

To extend the transformation of the gradient and divergence operators of Table 3.2 to a system of partial differential equations, we define a matrix of the metric terms,

$$\mathfrak{M} = \begin{bmatrix} Ja_1^1\underline{I} & Ja_1^2\underline{I} & Ja_1^3\underline{I} \\ Ja_2^1\underline{I} & Ja_2^2\underline{I} & Ja_2^3\underline{I} \\ Ja_3^1\underline{I} & Ja_3^2\underline{I} & Ja_3^3\underline{I} \end{bmatrix} \tag{188}$$

with a 5×5 identity matrix $\underline{I}$, to match the size of the Navier–Stokes state variables. With (188), the transformation of the gradient of a state vector is

$$\vec{\nabla}_x \mathbf{u} = \begin{bmatrix} \mathbf{u}_x \\ \mathbf{u}_y \\ \mathbf{u}_z \end{bmatrix} = \frac{1}{J}\mathfrak{M} \begin{bmatrix} \mathbf{u}_\xi \\ \mathbf{u}_\eta \\ \mathbf{u}_\zeta \end{bmatrix} = \frac{1}{J}\mathfrak{M}\vec{\nabla}_\xi \mathbf{u} \tag{189}$$

and the transformation of the divergence is

$$\vec{\nabla}_x \cdot \overleftrightarrow{\mathbf{f}} = \frac{1}{J}\vec{\nabla}_\xi \cdot \left(\mathfrak{M}^T \overleftrightarrow{\mathbf{f}}\right). \tag{190}$$

Moreover, the matrix (188) allows us to define contravariant block vectors

$$\overleftrightarrow{\tilde{\mathbf{f}}} = \mathfrak{M}^T \overleftrightarrow{\mathbf{f}}. \tag{191}$$

Applying the differential operator transformations (189) and (190) as well as the contravariant block vector notation (191), we get the transformed compressible Navier–Stokes equations

$$\begin{aligned} J\mathbf{u}_t + \vec{\nabla}_\xi \cdot \overleftrightarrow{\tilde{\mathbf{f}}}(\mathbf{u}) &= \frac{1}{\text{Re}}\vec{\nabla}_\xi \cdot \overleftrightarrow{\tilde{\mathbf{f}}}^v\left(\mathbf{u}, \overleftrightarrow{\mathbf{q}}\right), \\ J\overleftrightarrow{\mathbf{q}} &= \mathfrak{M}\vec{\nabla}_\xi \mathbf{w}. \end{aligned} \tag{192}$$

Note, to build an approximation that accounts for the entropy, we have taken the auxiliary variable $\overleftrightarrow{\mathbf{q}}$ to be the gradient of the entropy variables (140), to match the continuous equations (150).

We first apply the polynomial ansatz from the DG toolbox described in Sect. 3.2.8 to approximate the solution, fluxes, entropy variables, etc. as interpolants written in the Lagrange basis (72). These quantities are denoted with capital letters, e.g., $\mathbf{u} \approx \mathbf{U}$.

We then generate weak forms of the equations as in Sect. 3.3.1: We multiply the transformed equations (192) by test functions φ and $\overleftrightarrow{\vartheta}$, also polynomials, and integrate over the reference element E_d. Any integrals in the weak formulation are approximated with the Gauss-Lobatto quadrature and the quadrature points are collocated with the interpolation points, as discussed in Sect. 3.2.7. Finally, we apply multidimensional summation-by-parts (87) to move derivatives off the fluxes and onto the test functions, generating boundary terms. The result is the set of two weak forms,

$$
\begin{aligned}
&\langle \mathbb{I}^N(J)\mathbf{U}_t, \varphi \rangle_N + \int\limits_{\partial E, N} \varphi^T \left\{\mathbf{F}_n - \mathbf{F}_n^{v}\right\} \hat{s}\, \mathrm{dS} - \left\langle \overleftrightarrow{\tilde{\mathbf{F}}}, \vec{\nabla}_\xi \varphi \right\rangle_N = -\frac{1}{\mathrm{Re}} \left\langle \overleftrightarrow{\tilde{\mathbf{F}}}^{v}, \vec{\nabla}_\xi \varphi \right\rangle_N , \\
&\left\langle \mathbb{I}^N(J)\overleftrightarrow{\mathbf{Q}}, \overleftrightarrow{\vartheta} \right\rangle_N = \int\limits_{\partial E, N} \mathbf{W}^T \left(\overleftrightarrow{\vartheta} \cdot \vec{n} \right) \hat{s}\, \mathrm{dS} - \left\langle \mathbf{W}, \vec{\nabla}_\xi \cdot \left(\mathfrak{M}^T \overleftrightarrow{\vartheta} \right) \right\rangle_N .
\end{aligned}
\tag{193}
$$

Here, we use a compact notation for the normal fluxes, i.e., the normal flux in physical space $\mathbf{F}_n = \left(\overleftrightarrow{\mathbf{F}} \cdot \vec{n} \right)$.

A result of the DG polynomial ansatz is that solution values at element interfaces are discontinuous, and thus, the surface fluxes are not uniquely defined. This presents a problem to uniquely determine the normal fluxes, $\mathbf{F}_n$. To resolve this, the elements are coupled through the boundaries as in a finite volume scheme with appropriate *numerical flux functions* denoted by $\mathbf{F}_n^*$, $\mathbf{F}_n^{v,*}$ and $\mathbf{W}^*$.

The numerical fluxes are functions of two states, one to the left and one to the right of the interface, e.g., $\mathbf{F}_n^* (\mathbf{U}_L, \mathbf{U}_R)$. They must also be consistent, i.e., $\mathbf{F}_n^* (\mathbf{U}, \mathbf{U}) = \overleftrightarrow{\mathbf{f}} (\mathbf{U}) \cdot \vec{n}$ so that if there is no jump, the exact flux in the normal direction is recovered.

Other conditions, we will see, are still needed to ensure stability of the numerical scheme.

With the discontinuities at element interfaces resolved, we can perform another application of the multidimensional summation-by-parts (87) on the first equation in (193) to move derivatives from the test functions back to the transformed flux vectors

$$\left\langle \mathbb{I}^N(J)\mathbf{U}_t, \varphi \right\rangle_N + \left\langle \vec{\nabla}_\xi \cdot \mathbb{I}^N(\overset{\leftrightarrow}{\tilde{\mathbf{F}}}), \varphi \right\rangle_N + \int_{\partial E, N} \varphi^T \left\{ \mathbf{F}_n^* - \mathbf{F}_n \right\} \hat{s}\, \mathrm{dS}$$

$$= \frac{1}{\mathrm{Re}} \left\langle \vec{\nabla}_\xi \cdot \mathbb{I}^N(\overset{\leftrightarrow}{\tilde{\mathbf{F}}}{}^v), \varphi \right\rangle_N + \frac{1}{\mathrm{Re}} \int_{\partial E, N} \varphi^T \left\{ \mathbf{F}_n^{v,*} - \mathbf{F}_n^v \right\} \hat{s}\, \mathrm{dS}, \tag{194}$$

$$\left\langle \mathbb{I}^N(J)\overset{\leftrightarrow}{\mathbf{Q}}, \overset{\leftrightarrow}{\vartheta} \right\rangle_N = \int_{\partial E, N} \mathbf{W}^{*,T} \left(\overset{\leftrightarrow}{\vartheta} \cdot \vec{n} \right) \hat{s}\, \mathrm{dS} - \left\langle \mathbf{W}, \vec{\nabla}_\xi \cdot \left(\mathfrak{M}^T \overset{\leftrightarrow}{\vartheta} \right) \right\rangle_N .$$

The first equation of (194) is called the strong form of the DG approximation. It will be used later to create an entropy stable method. Note that the surface contributions within each element for the first equation resemble a penalty method in this form, in that it is proportional to the difference between the numerical flux and the flux computed from the interior. Note also that at this point, the multidimensional summation-by-parts, (87), says that (193) and (194) are algebraically equivalent.

Approximations like (193) and (194) have been used in practice for many years, and they "usually" work. Sometimes, however, they are known to be unstable in that the computations blow up with unbounded energy or entropy. Since we have already shown in (130) and (153) that the energy of the linear equations and the entropy of the nonlinear equations is bounded by the boundary contributions, we should expect the numerical schemes to share these properties: i.e., they should be stable.

The problem is that *even for linear fluxes, the approximations* (193) *and* (194) *are not necessarily stable*. Based on what we have seen so far, we should require the numerical schemes to mimic the properties of the continuous equations. To that end, it should not be surprising that we should start with the same split form of the equation, (115), that was used to show boundedness of the continuous solution.

Starting with (115), we construct an alternative, split form DG approximation, by approximating the divergence of the flux with an approximation of the average of the conservative and nonconservative forms,

$$\vec{\nabla}_\xi \cdot \mathbb{I}^N(\overset{\leftrightarrow}{\tilde{\mathbf{F}}}) \approx \frac{1}{2}\mathbb{I}^N \left\{ \vec{\nabla}_\xi \cdot \mathbb{I}^N(\overset{\leftrightarrow}{\tilde{\mathbf{F}}}) + \left(\overset{\leftrightarrow}{\tilde{\mathbf{A}}} \right)^T \vec{\nabla}_\xi \mathbf{U} + \left(\vec{\nabla}_\xi \cdot \overset{\leftrightarrow}{\tilde{\mathbf{A}}} \right) \mathbf{U} \right\}, \tag{195}$$

where $\overset{\leftrightarrow}{\tilde{\mathbf{A}}} = \mathbb{I}^N(\mathfrak{M}^T \mathbb{I}^N(\overset{\leftrightarrow}{\mathbf{A}}))$ and $\overset{\leftrightarrow}{\tilde{\mathbf{F}}} = \mathbb{I}^N(\overset{\leftrightarrow}{\tilde{\mathbf{A}}}\mathbf{U})$. Since we have already assumed that $\vec{\nabla} \cdot \overset{\leftrightarrow}{\mathbf{A}} = 0$ to ensure that any energy growth is the system is due solely to boundary conditions, we will also assume in the following that $\vec{\nabla} \cdot \overset{\leftrightarrow}{\tilde{\mathbf{A}}} = 0$. Alternatively, we can simply drop that term from the approximation, since it will be a spectrally accurate approximation to zero. Doing so will lead to an approximation that is conservative only to within spectral accuracy, but that is less critical for linear systems of equations than for nonlinear.

With the split form approximation to the divergence, the DG approximation of advection terms of (194) becomes

$$\frac{1}{2}\left\langle \vec{\nabla}_\xi \cdot \mathbb{I}^N(\overleftrightarrow{\tilde{\mathbf{F}}}), \varphi \right\rangle_N + \frac{1}{2}\left\langle \left(\overleftrightarrow{\tilde{\mathbf{A}}}\right)^T \vec{\nabla}_\xi \mathbf{U}, \varphi \right\rangle_N + \int_{\partial E,N} \varphi^T \left\{\mathbf{F}_n^* - \mathbf{F}_n\right\} \hat{s}\,\mathrm{dS}. \tag{196}$$

We can write (196) in any one of many algebraically equivalent forms, and then use whichever is convenient for a given purpose. For instance, we can apply the multidimensional summation-by-parts, (87), to the first term and move the coefficient matrix in the second to get the algebraically equivalent form

$$-\frac{1}{2}\left\langle \overleftrightarrow{\tilde{\mathbf{F}}}, \vec{\nabla}_\xi \varphi \right\rangle_N + \frac{1}{2}\left\langle \vec{\nabla}_\xi \mathbf{U}, \overleftrightarrow{\tilde{\mathbf{F}}}^{(T)}(\varphi) \right\rangle_N + \int_{\partial E,N} \varphi^T \left\{\mathbf{F}_n^* - \frac{1}{2}\mathbf{F}_n\right\} \hat{s}\,\mathrm{dS}, \tag{197}$$

where

$$\overleftrightarrow{\tilde{\mathbf{F}}}^{(T)}(\varphi) = \mathbb{I}^N(\overleftrightarrow{\tilde{\mathbf{A}}}^T \varphi) = \mathbb{I}^N \begin{bmatrix} \tilde{\underline{\mathbf{A}}}_1^T \varphi \\ \tilde{\underline{\mathbf{A}}}_2^T \varphi \\ \tilde{\underline{\mathbf{A}}}_3^T \varphi \end{bmatrix}, \tag{198}$$

is the test function flux composed with the transpose of the coefficient matrices.

Continuing on, we can apply the multidimensional summation-by-parts rule to the second term of (197) to get another algebraically equivalent form

$$-\frac{1}{2}\left\langle \overleftrightarrow{\tilde{\mathbf{F}}}, \vec{\nabla}_\xi \varphi \right\rangle_N - \frac{1}{2}\left\langle \mathbf{U}, \vec{\nabla}_\xi \cdot \overleftrightarrow{\tilde{\mathbf{F}}}^{(T)}(\varphi) \right\rangle_N + \int_{\partial E,N} \varphi^T \mathbf{F}_n^* \hat{s}\,\mathrm{dS}. \tag{199}$$

In this form, all derivatives are on the test functions.

Finally, we can add and subtract the second term in (196) and combine the difference to get

$$\begin{aligned} &\left\langle \vec{\nabla}_\xi \cdot \mathbb{I}^N(\overleftrightarrow{\tilde{\mathbf{F}}}), \varphi \right\rangle_N + \int_{\partial E,N} \varphi^T \left\{\mathbf{F}_n^* - \mathbf{F}_n\right\} \hat{s}\,\mathrm{dS} \\ &\quad + \frac{1}{2}\left\langle \mathbb{I}^N(\overleftrightarrow{\tilde{\mathbf{A}}})^T \vec{\nabla}_\xi \mathbf{U} - \vec{\nabla}_\xi \cdot \mathbb{I}^N(\overleftrightarrow{\tilde{\mathbf{F}}}), \varphi \right\rangle_N . \end{aligned} \tag{200}$$

Table 3.3 Equivalent DG approximations to the advective flux divergence

Form	Approximation
Strong [S]	$\frac{1}{2}\left\langle \vec{\nabla}_\xi \cdot \mathbb{I}^N(\overleftrightarrow{\tilde{\mathbf{F}}}), \varphi \right\rangle_N + \frac{1}{2}\left\langle \mathbb{I}^N(\overleftrightarrow{\tilde{\mathbf{A}}})^T \vec{\nabla}_\xi \mathbf{U}, \varphi \right\rangle_N + \int_{\partial E,N} \varphi^T \left\{ \mathbf{F}_n^* - \mathbf{F}_n \right\} \hat{s}\, \mathrm{dS}$
Weak [W]	$-\frac{1}{2}\left\langle \overleftrightarrow{\tilde{\mathbf{F}}}, \vec{\nabla}_\xi \varphi \right\rangle_N - \frac{1}{2}\left\langle \mathbf{U}, \vec{\nabla}_\xi \cdot \overleftrightarrow{\tilde{\mathbf{F}}}^{(T)}(\varphi) \right\rangle_N + \int_{\partial E,N} \varphi^T \mathbf{F}_n^* \hat{s}\, \mathrm{dS}$
Directly Stable [DS]	$-\frac{1}{2}\left\langle \overleftrightarrow{\tilde{\mathbf{F}}}, \vec{\nabla}_\xi \varphi \right\rangle_N + \frac{1}{2}\left\langle \vec{\nabla}_\xi \mathbf{U}, \overleftrightarrow{\tilde{\mathbf{F}}}^{(T)}(\varphi) \right\rangle_N + \int_{\partial E,N} \varphi^T \left\{ \mathbf{F}_n^* - \frac{1}{2}\mathbf{F}_n \right\} \hat{s}\, \mathrm{dS}$
Strong + Correction [SC]	$\left\langle \vec{\nabla}_\xi \cdot \mathbb{I}^N(\overleftrightarrow{\tilde{\mathbf{F}}}), \varphi \right\rangle_N + \int_{\partial E,N} \varphi^T \left\{ \mathbf{F}_n^* - \mathbf{F}_n \right\} \hat{s}\, \mathrm{dS} +$ $\frac{1}{2}\left\langle \mathbb{I}^N(\overleftrightarrow{\tilde{\mathbf{A}}})^T \vec{\nabla}_\xi \mathbf{U} - \vec{\nabla}_\xi \cdot \mathbb{I}^N(\overleftrightarrow{\tilde{\mathbf{F}}}), \varphi \right\rangle_N$

We summarize the equivalent forms for the approximation of the divergence of the flux in Table 3.3.

Since the approximations in Table 3.3 are algebraically equivalent, we can choose which one to use depending on what property of the equations we wish to study. Additionally, their equivalence can be exploited in practice as they can be reduced to the same implementation in code.

For instance, the form of the advective volume quadratures in the [DS] form (197) look like those in the continuous form (117), which was used to show energy boundedness. On the other hand, if we set $\varphi = 1$ for each component in the [W] form (199), the first two terms vanish, leaving only the surface quadrature. The result implies that the approximation is conservative, since the integral (the quadrature is exact when the test function is a polynomial of degree zero) over the volume of the divergence is equal to the integral of the flux over the surface.

Finally, written in the [SC] form, we see that the split form (196) is the original conservative form, (194), plus a correction that is the discrete projection of

$$\mathrm{CN} = \mathbb{I}^N(\overleftrightarrow{\tilde{\mathbf{A}}})^T \vec{\nabla}_\xi \mathbf{U} - \vec{\nabla}_\xi \cdot \mathbb{I}^N(\overleftrightarrow{\tilde{\mathbf{F}}}). \tag{201}$$

The quantity CN is the amount by which the product rule fails to hold due to aliasing when taking the divergence of the linear flux, $\overleftrightarrow{\tilde{\mathbf{F}}} = \mathbb{I}^N(\overleftrightarrow{\tilde{\mathbf{A}}}\mathbf{U})$. In other words, the split form approximation in (195) serves to cancel the product rule (aliasing) error in the divergence approximation.

Finally, since the problem is linear, we use the state rather than the entropy variables in the second equation of (194) for the diffusion approximation

$$\left\langle \mathbb{I}^N(J)\overleftrightarrow{\mathbf{Q}}, \overleftrightarrow{\vartheta} \right\rangle_N = \int_{\partial E,N} \mathbf{U}^{*,T} \left(\overleftrightarrow{\vartheta} \cdot \vec{n} \right) \hat{s}\, \mathrm{dS} - \left\langle \mathbf{U}, \vec{\nabla}_\xi \cdot \left(\mathfrak{M}^T \overleftrightarrow{\vartheta} \right) \right\rangle_N, \tag{202}$$

to match the second equation of (115).

Role of the Split Form Approximation

It is desirable that the approximation match as many properties of the original PDE as possible. Important properties include boundedness of the solution (stability), conservation, free-stream preservation, phase, and dissipation properties. In this section we will show that the split form approximation is stable, and if the metric terms are computed so that they satisfy the metric identities discretely, the approximation is free-stream (or constant state) preserving.

Stability

In this section, we show that the discontinuous Galerkin approximation to the linear system of equations is stable if the split form approximation of the divergence is used. In the process, we will see precisely why the straight forward divergence approximation is not stable, but will often run stably.

Using the [DS] form of the advective terms from Table 3.3, the split form discontinuous Galerkin approximation of the linear Navier–Stokes equations is

$$\begin{aligned}
&\left\langle \mathbb{I}^N(J)\mathbf{U}_t, \varphi \right\rangle_N - \frac{1}{2}\left\langle \overleftrightarrow{\tilde{\mathbf{F}}}, \vec{\nabla}_\xi \varphi \right\rangle_N + \frac{1}{2}\left\langle \vec{\nabla}_\xi \mathbf{U}, \overleftrightarrow{\tilde{\mathbf{F}}}^{(T)}(\varphi) \right\rangle_N + \int_{\partial E,N} \varphi^T \left\{ \mathbf{F}_n^* - \frac{1}{2}\mathbf{F}_n \right\} \hat{s}\, \mathrm{dS} \\
&\qquad = \frac{1}{\mathrm{Re}}\left\langle \vec{\nabla}_\xi \cdot \mathbb{I}^N(\overleftrightarrow{\tilde{\mathbf{F}}}^v), \varphi \right\rangle_N + \frac{1}{\mathrm{Re}} \int_{\partial E,N} \varphi^T \left\{ \mathbf{F}_n^{v,*} - \mathbf{F}_n^v \right\} \hat{s}\, \mathrm{dS}, \\
&\left\langle \mathbb{I}^N(J)\overleftrightarrow{\mathbf{Q}}, \overleftrightarrow{\vartheta} \right\rangle_N = \int_{\partial E,N} \mathbf{U}^{*,T} \left(\overleftrightarrow{\vartheta} \cdot \vec{n} \right) \hat{s}\, \mathrm{dS} - \left\langle \mathbf{U}, \vec{\nabla}_\xi \cdot \left(\mathfrak{M}^T \overleftrightarrow{\vartheta} \right) \right\rangle_N .
\end{aligned} \tag{203}$$

To assess stability, we follow the steps as to show energy boundedness in Sect. 3.3.1. This time we first we set $\overleftrightarrow{\vartheta} = \mathbb{I}^N((\mathfrak{S}^{-1})^T \mathfrak{S}^{-1}\mathfrak{B}\overleftrightarrow{\mathbf{Q}})$ in the second equation of (203). Using the fact that $(\mathfrak{S}^{-1})^T$ commutes with $\mathfrak{M}^T$,

$$\left\langle \mathbb{I}^N(J)\mathfrak{S}^{-1}\overleftrightarrow{\mathbf{Q}}, \mathfrak{S}^{-1}\mathfrak{B}\overleftrightarrow{\mathbf{Q}}\right\rangle_N = \left\langle \mathbb{I}^N(J)\overleftrightarrow{\mathbf{Q}}^s, \mathfrak{B}^s\overleftrightarrow{\mathbf{Q}}^s\right\rangle_N$$
$$= \int_{\partial E,N} \left\{\mathbf{U}^{s,*} - \mathbf{U}^s\right\}^T \mathbf{F}_n^{v,s}\hat{s}\, \mathrm{d}S + \left\langle \vec{\nabla}_\xi \mathbf{U}^s, \overleftrightarrow{\tilde{\mathbf{F}}}^{v,s}\right\rangle_N, \tag{204}$$

where $\mathbf{F}_n^{v,s} = \overleftrightarrow{\mathbf{F}}^{v,s}\cdot\vec{n}$ and $\overleftrightarrow{\tilde{\mathbf{F}}}^{v,s} = \mathbb{I}^N\left(\mathfrak{M}^T\mathfrak{B}^s\overleftrightarrow{\mathbf{Q}}^s\right)$. Therefore,

$$\left\langle \vec{\nabla}_\xi \mathbf{U}^s, \overleftrightarrow{\tilde{\mathbf{F}}}^{v,s}\right\rangle_N = \left\langle \mathbb{I}^N(J)\overleftrightarrow{\mathbf{Q}}^s, \mathfrak{B}^s\overleftrightarrow{\mathbf{Q}}^s\right\rangle_N - \int_{\partial E,N} \left\{\mathbf{U}^{s,*} - \mathbf{U}^s\right\}^T \mathbf{F}_n^{v,s}\hat{s}\, \mathrm{d}S, \tag{205}$$

where

$$\left\langle \mathbb{I}^N(J)\overleftrightarrow{\mathbf{Q}}^s, \mathfrak{B}^s\overleftrightarrow{\mathbf{Q}}^s\right\rangle_N \geqslant 0. \tag{206}$$

Next, we set $\varphi = \left(\mathfrak{S}^{-1}\right)^T\mathfrak{S}^{-1}\mathbf{U} = \left(\mathfrak{S}^{-1}\right)^T\mathbf{U}^s$ in the first equation of (203). The time derivative term becomes

$$\left\langle \mathbb{I}^N(J)\mathbf{U}_t, \left(\mathfrak{S}^{-1}\right)^T\mathfrak{S}^{-1}\mathbf{U}\right\rangle_N = \left\langle \mathbb{I}^N(J)\mathbf{U}_t^s, \mathbf{U}^s\right\rangle_N$$
$$= \frac{1}{2}\frac{d}{dt}\sum_{ijk=0}^{N} \mathbb{I}^N(J)_{ijk}\left|\mathbf{U}^s\right|^2 w_{ijk}. \tag{207}$$

For (207) to represent a norm, equivalent to the continuous energy norm (121), and for (206) to hold, it is necessary that $\mathbb{I}^N(J)_{ijk} > 0$ for all N. This fact should be remembered in the grid generation process to ensure that the energy is always positive. If this is true, then we can write

$$\left\langle \mathbb{I}^N(J)\mathbf{U}_t^s, \mathbf{U}^s\right\rangle_N \equiv \frac{1}{2}\frac{d}{dt}\left|\left|\mathbf{U}^s\right|\right|_{J,N}^2. \tag{208}$$

The advective volume terms in (203) cancel when we substitute for the test function, as they did for the continuous terms, (126), leaving us with

$$\frac{1}{2}\frac{d}{dt}\left|\left|\mathbf{U}^s\right|\right|_{J,N}^2 = -\int\limits_{\partial E,N} \left(\mathbf{U}^s\right)^T \left(\left\{\mathbf{F}_n^{s,*} - \frac{1}{2}\mathbf{F}_n^s\right\} - \frac{1}{\mathrm{Re}}\mathbf{F}_n^{v,s,*}\right)\hat{s}\,\mathrm{dS}$$
$$+\frac{1}{\mathrm{Re}}\int\limits_{\partial E,N} \left\{\mathbf{U}^{s,*} - \mathbf{U}^s\right\}^T \mathbf{F}_n^{v,s}\hat{s}\,\mathrm{dS} \tag{209}$$
$$-\frac{1}{\mathrm{Re}}\left\langle J\overset{\leftrightarrow}{\mathbf{Q}}^s, \mathfrak{B}^s\overset{\leftrightarrow}{\mathbf{Q}}^s\right\rangle_N .$$

Separating the advective and viscous boundary terms, the elemental contribution to the total energy is

$$\frac{1}{2}\frac{d}{dt}\left|\left|\mathbf{U}^s\right|\right|_{J,N}^2 = -\int\limits_{\partial E,N} \left(\mathbf{U}^s\right)^T \left\{\mathbf{F}_n^{s,*} - \frac{1}{2}\mathbf{F}_n^s\right\}\hat{s}\,\mathrm{dS}$$
$$+\frac{1}{\mathrm{Re}}\int\limits_{\partial E,N} \left\{\left(\mathbf{U}^s\right)^T \mathbf{F}_n^{v,s,*} + \left(\mathbf{U}^{s,*}\right)^T \mathbf{F}_n^{v,s} - \left(\mathbf{U}^s\right)^T \mathbf{F}_n^{v,s}\right\}\hat{s}\,\mathrm{dS}$$
$$-\frac{1}{\mathrm{Re}}\left\langle J\overset{\leftrightarrow}{\mathbf{Q}}^s, \mathfrak{B}^s\overset{\leftrightarrow}{\mathbf{Q}}^s\right\rangle_N . \tag{210}$$

The total energy is found by summing over all of the elements. At the element faces, there will be jumps in the solution states and fluxes. To represent those jumps, we introduce the jump operator: For a quantity V defined on the left, L, and right, R, side of an interface with respect to the outward normal,

$$[\![V]\!] \equiv V_R - V_L \tag{211}$$

is the jump operator.

Summing over all elements,

$$\frac{1}{2}\frac{d}{dt}\sum_{k=1}^{K}\left|\left|\mathbf{U}^{s,k}\right|\right|_{J,N}^2 = \sum_{\substack{\text{interior}\\\text{faces}}}\int\limits_{N} \left\{[\![\mathbf{U}^s]\!]^T\mathbf{F}_n^{s,*} - \frac{1}{2}[\![\left(\mathbf{U}^s\right)^T \overset{\leftrightarrow}{\mathbf{F}}^s]\!]\cdot\vec{n}\right\}\hat{s}\,\mathrm{dS}$$
$$-\frac{1}{\mathrm{Re}}\sum_{\substack{\text{interior}\\\text{faces}}}\int\limits_{N} \left\{[\![\mathbf{U}^s]\!]^T\mathbf{F}_n^{v,s,*} + \left(\mathbf{U}^{s,*}\right)^T [\![\overset{\leftrightarrow}{\mathbf{F}}^{v,s}]\!]\cdot\vec{n} - [\![\left(\mathbf{U}^s\right)^T \overset{\leftrightarrow}{\mathbf{F}}^{v,s}]\!]\cdot\vec{n}\right\}\hat{s}\,\mathrm{dS}$$
$$-\frac{1}{\mathrm{Re}}\sum_{k=1}^{K}\left\langle J\overset{\leftrightarrow}{\mathbf{Q}}^{s,k}, \mathfrak{B}^s\overset{\leftrightarrow}{\mathbf{Q}}^{s,k}\right\rangle_N$$
$$+\ \mathrm{PBT}\,, \tag{212}$$

where $\mathbf{U}^{s,k}$ is the (symmetric) solution vector on element k. The quantity PBT represents the physical boundary terms, which we assume are dissipative, i.e., $\mathrm{PBT} \leqslant 0$.

Sufficient conditions for stability are those for which the right-hand side of (212) is always non-positive. Since the third term is always non-positive because $\mathfrak{B}^s > 0$, sufficent conditions are that at each node on the interior element faces, the numerical values $\mathbf{F}_n^{v,s,*}$, $\mathbf{F}_n^{s,*}$ and $\mathbf{U}^{s,*}$ are chosen so that

$$[\![\mathbf{U}^s]\!]^T\mathbf{F}_n^{s,*} - \frac{1}{2}[\![(\mathbf{U}^s)^T \overleftrightarrow{\mathbf{F}}^s]\!] \cdot \vec{n} \leqslant 0, \tag{213}$$

and

$$[\![\mathbf{U}^s]\!]^T\mathbf{F}_n^{v,s,*} + \left(\mathbf{U}^{s,*}\right)^T [\![\overleftrightarrow{\mathbf{F}}^{v,s}]\!] \cdot \vec{n} - [\![(\mathbf{U}^s)^T \overleftrightarrow{\mathbf{F}}^{v,s}]\!] \cdot \vec{n} \geqslant 0. \tag{214}$$

With such conditions satisfied, the norm of the approximate solution is bounded by the initial conditions,

$$\frac{d}{dt}\left|\left|\mathbf{U}^s\right|\right|_{J,N}^2 \equiv \frac{d}{dt}\sum_{k=1}^{K}\left|\left|\mathbf{U}^{s,k}\right|\right|_{J,N}^2 \leqslant 0 \quad \Rightarrow \quad \left|\left|\mathbf{U}^s(T)\right|\right|_{J,N} \leqslant \left|\left|\mathbf{U}_0^s\right|\right|_{J,N}. \tag{215}$$

So what remains is to find suitable numerical fluxes such that (213) and (214) hold. Since the advective part of the equations is hyperbolic, the advective flux can be split according the wave directions relative to the normal direction

$$\overleftrightarrow{\mathbf{F}}^s \cdot \vec{n} = \left(\left(\mathfrak{M}^T\overleftrightarrow{\mathbf{A}}^s\right)\cdot\hat{n}\right)\mathbf{U}^s = \left(\tilde{\overleftrightarrow{\mathbf{A}}}^s \cdot \hat{n}\right)\mathbf{U}^s \equiv \tilde{\underline{\mathbf{A}}}_n^s\mathbf{U}^s = \left(\tilde{\underline{\mathbf{A}}}_n^{s,+} + \tilde{\underline{\mathbf{A}}}_n^{s,-}\right)\mathbf{U}^s, \tag{216}$$

where

$$\tilde{\underline{\mathbf{A}}}_n^{s,\pm} = \frac{1}{2}\left(\tilde{\underline{\mathbf{A}}}_n^s \pm \left|\tilde{\underline{\mathbf{A}}}_n^s\right|\right). \tag{217}$$

From that splitting, we can write the numerical advective flux choosing the left and right states according to the wave direction given by the sign of the eigenvalues as

$$\mathbf{F}_n^{s,*}\left(\mathbf{U}_L^s, \mathbf{U}_R^s\right) = \tilde{\underline{\mathbf{A}}}_n^{s,+}\mathbf{U}_L^s + \tilde{\underline{\mathbf{A}}}_n^{s,-}\mathbf{U}_R^s. \tag{218}$$

We substitute (217) into (218) and rearrange to get a numerical flux

$$\begin{aligned}\mathbf{F}_n^{s,*}\left(\mathbf{U}_L^s, \mathbf{U}_R^s\right) &= \frac{\tilde{\underline{\mathbf{A}}}_n^s\mathbf{U}_L^s + \tilde{\underline{\mathbf{A}}}_n^s\mathbf{U}_R^s}{2} + \frac{\sigma}{2}\left|\tilde{\underline{\mathbf{A}}}_n^s\right|\left(\mathbf{U}_L^s - \mathbf{U}_R^s\right)\\ &= \tilde{\underline{\mathbf{A}}}_n^s\left\{\!\!\left\{\mathbf{U}^s\right\}\!\!\right\} - \frac{\sigma}{2}\left|\tilde{\underline{\mathbf{A}}}_n^s\right|[\![\mathbf{U}^s]\!],\end{aligned} \tag{219}$$

where $\{\!\{\mathbf{U}\}\!\} = \frac{1}{2}\left(\mathbf{U}_L + \mathbf{U}_R\right)$ is the arithmetic mean. For additional flexibility, we have added the parameter σ so that the fully upwind numerical flux corresponds to $\sigma = 1$, whereas $\sigma = 0$ gives the central flux.

With either the upwind or central numerical flux (219), the contribution of the advective fluxes at the faces is dissipative. For any two state vectors,

$$
\begin{aligned}
[\![\mathbf{a}^T\mathbf{b}]\!] &= [\![\sum_{m=1}^{5} a_m b_m]\!] = \sum_{m=1}^{5}[\![a_m b_m]\!] \\
&= \sum_{m=1}^{5}\left(\{a_m\}[\![b_m]\!] + [\![a_m]\!]\{b_m\}\right) = \{\mathbf{a}\}^T[\![\mathbf{b}]\!] + [\![\mathbf{a}]\!]^T\{\mathbf{b}\}\,.
\end{aligned}
\tag{220}
$$

Therefore

$$
\begin{aligned}
[\![(\mathbf{U}^s)^T\overleftrightarrow{\mathbf{F}}^s]\!]\cdot\vec{n} &= \{\!\{\mathbf{U}^s\}\!\}^T[\![\overleftrightarrow{\mathbf{F}}^s]\!]\cdot\vec{n} + [\![\mathbf{U}^s]\!]^T\left\{\!\!\left\{\overleftrightarrow{\mathbf{F}}^s\right\}\!\!\right\}\cdot\vec{n} \\
&= \{\!\{\mathbf{U}^s\}\!\}^T\left(\overleftrightarrow{\mathbf{A}}^s\cdot\vec{n}\right)[\![\mathbf{U}^s]\!] + [\![\mathbf{U}^s]\!]^T\left(\overleftrightarrow{\mathbf{A}}^s\cdot\vec{n}\right)\{\!\{\mathbf{U}^s\}\!\} \\
&= \{\!\{\mathbf{U}^s\}\!\}^T\tilde{\underline{\mathbf{A}}}_n^s[\![\mathbf{U}^s]\!] + [\![\mathbf{U}^s]\!]^T\tilde{\underline{\mathbf{A}}}_n^s\{\!\{\mathbf{U}^s\}\!\} \\
&= 2[\![\mathbf{U}^s]\!]^T\tilde{\underline{\mathbf{A}}}_n^s\{\!\{\mathbf{U}^s\}\!\}\,,
\end{aligned}
\tag{221}
$$

so

$$
[\![\mathbf{U}^s]\!]^T\mathbf{F}_n^{s,*} - \frac{1}{2}[\![(\mathbf{U}^s)^T\overleftrightarrow{\mathbf{F}}^s]\!]\cdot\vec{n} = -\frac{\sigma}{2}[\![\mathbf{U}^s]\!]^T\left|\tilde{\underline{\mathbf{A}}}_n^s\right|[\![\mathbf{U}^s]\!] \leqslant 0\,,
\tag{222}
$$

which satisfies condition (213) for either the central numerical flux or an upwind flux, and the contribution of the advective interface terms to the energy in (212) is non-positive.

We are now left to satisfy (214) for the viscous terms. The simplest choice is to match the equality, which can be done with the Bassi-Rebay-1 (or BR1 for short) numerical flux from Bassi and Rebay (1997), which computes the interface values as simple arithmetic means

$$
\begin{aligned}
\mathbf{U}^{s,*} &= \frac{\mathbf{U}_L^s + \mathbf{U}_R^s}{2} = \{\!\{\mathbf{U}^s\}\!\} \\
\mathbf{F}_n^{v,s,*} &= \left(\frac{\overleftrightarrow{\mathbf{F}}_L^{v,s} + \overleftrightarrow{\mathbf{F}}_R^{v,s}}{2}\right)\cdot\vec{n} = \left\{\!\!\left\{\overleftrightarrow{\mathbf{F}}^{v,s}\right\}\!\!\right\}\cdot\vec{n}\,.
\end{aligned}
\tag{223}
$$

When we make the substitution of the BR1 fluxes into the left side of (214) it becomes (factoring out the normal direction)

$$
\left([\![\mathbf{U}^s]\!]^T\left\{\!\!\left\{\overleftrightarrow{\mathbf{F}}^{v,s}\right\}\!\!\right\} + \{\!\{\mathbf{U}^s\}\!\}^T[\![\overleftrightarrow{\mathbf{F}}^{v,s}]\!] - [\![(\mathbf{U}^s)^T\overleftrightarrow{\mathbf{F}}^{v,s}]\!]\right)\cdot\vec{n}.
\tag{224}
$$

Then replacing the jump in the product using the identity (220), the approximation satisfies (214) because

$$[\![\mathbf{U}^s]\!]^T \left\{\!\!\left\{\overleftrightarrow{\mathbf{F}}^{v,s}\right\}\!\!\right\} + \{\!\{\mathbf{U}^s\}\!\}^T [\![\overleftrightarrow{\mathbf{F}}^{v,s}]\!] - \{\!\{\mathbf{U}^s\}\!\}^T [\![\overleftrightarrow{\mathbf{F}}^{v,s}]\!] - [\![\mathbf{U}^s]\!]^T \left\{\!\!\left\{\overleftrightarrow{\mathbf{F}}^{v,s}\right\}\!\!\right\} = 0, \quad (225)$$

so (224) vanishes exactly. Therefore, the split form approximation (203) is stable in the sense of (215).

We are now in the position to also see why the standard scheme, using only the divergence of the flux polynomial, can work, but is not guaranteed to be stable. In Table 3.3 [SC] shows that the split form approximation of the advective terms is the standard scheme plus a correction term. Alternatively, the standard approximation is the split form minus that correction.

If we subtract the correction term from the split form approximation and insert the results (222), (225) and (206), then the standard approximation satisfies

$$\begin{aligned}\frac{1}{2}\frac{d}{dt}\left|\left|\mathbf{U}^s\right|\right|^2_{J,N} = &-\frac{1}{\mathrm{Re}}\left|\left|\overleftrightarrow{\mathbf{Q}}^s\right|\right|^2_{\mathfrak{B}^s,N} - \frac{\sigma}{2}\sum_{\substack{\text{interior}\\ \text{faces}}}\int_N [\![\mathbf{U}^s]\!]^T \left|\tilde{\mathbf{A}}^s_n\right| [\![\mathbf{U}^s]\!]\, \mathrm{dS}\\ &+\frac{1}{2}\sum_{k=1}^{K}\left|\left\langle\left\{\mathbb{I}^N(\tilde{\mathbf{A}})^T\,\vec{\nabla}_\xi\mathbf{U}^{s,k} - \vec{\nabla}_\xi\cdot\overleftrightarrow{\tilde{\mathbf{F}}}^s(\mathbf{U}^{s,k})\right\},\mathbf{U}^{s,k}\right\rangle_N\right| \\ &+ \mathrm{PBT},\end{aligned} \quad (226)$$

which is (212) plus the correction term contribution. The additional volume term is due to the failure of the product rule to hold for polynomial interpolants due to aliasing.

Equation (226) shows that the physical diffusion and/or the dissipation associated with the numerical flux could counterbalance the product rule error and make the right-hand side non-positive. For well-resolved solutions, the product rule error will be spectrally small, making it likely that the physical and interface dissipations are sufficiently large for stabilization. For under resolved solutions, the aliasing errors may be too large for the approximate solution to stay bounded. For large Reynolds numbers, the physical dissipation may be too small. The artificial dissipation due to the numerical fluxes might be sufficiently large, depending on flux solver is chosen. (For example, a Lax-Friedrichs numerical flux will be more dissipative than the exact upwind one.) Finally, changing from the BR1 to another viscous coupling procedure, coupled with a more dissipative numerical flux might be enough to counteract the aliasing term. But, ultimately, the key to a stable discontinuous Galerkin spectral element method (DGSEM) is the stable approximation of the advective terms, as given in the split form approximation.

The Importance of the Metric Identities

One simple property of fluid flows and the solutions of the associated linearized equations with constant coefficient matrices is that a constant solution stays constant. This property is usually known as free-stream preservation for fluid flows and we will use that term here. It is desirable that free-stream preservation holds for the approximate solution for if it doesn't, waves can spontaneously appear and propagate in an initially constant state even without applied external forces, see, e.g., Kopriva (2006).

We now show that the split form spatial approximation of the constant coefficient linearized Euler equations is free-stream preserving provided that the approximations of metric terms satisfy a form of the metric identities, (182). Using the form [W] in Table 3.3 for the advection terms, the DGSEM approximation of the advection equation is

$$\begin{aligned}\left\langle \mathbb{I}^N(J)\mathbf{U}_t,\varphi\right\rangle_N + \frac{1}{2}\left\langle \vec{\nabla}_\xi\cdot\mathbb{I}^N(\overset{\leftrightarrow}{\tilde{\mathbf{F}}}),\varphi\right\rangle_N + \frac{1}{2}\left\langle \left(\overset{\leftrightarrow}{\tilde{\mathbf{A}}}\right)^T\vec{\nabla}_\xi\mathbf{U},\varphi\right\rangle_N \\ + \int_{\partial E,N}\varphi^T\left\{\mathbf{F}_n^* - \mathbf{F}_n\right\}\hat{s}\,\mathrm{d}S = 0\end{aligned} \tag{227}$$

on each element. We can ignore the contribution of the diffusion terms since they are automatically zero when the gradients are zero.

When $\mathbf{U} = \mathbf{C}$ is constant over all elements, its gradient vanishes and the surface term in (227) vanishes by consistency of the numerical flux. Therefore, if we write out the contravariant flux,

$$\left\langle \mathbb{I}^N(J)\mathbf{U}_t,\varphi\right\rangle_N = -\frac{1}{2}\left\langle \vec{\nabla}_\xi\cdot\mathbb{I}^N(\mathfrak{M}\overset{\leftrightarrow}{\mathbf{A}}\mathbf{C}),\varphi\right\rangle_N \tag{228}$$

Since $\overset{\leftrightarrow}{\mathbf{A}}\mathbf{C}$ is a constant, and φ is arbitrary, the right-hand side of (227) vanishes if and only if for each block of $\mathfrak{M}$,

$$\vec{\nabla}_\xi\cdot\mathbb{I}^N(J\vec{a}^{\,i}) = 0 \quad i = 1,2,3 \tag{229}$$

If we compare (229) with the metric identities (182), we see that the interpolant of the volume weighted contravariant basis vectors must vanish for the approximation to be free-stream preserving. Since differentiation and interpolation do not commute, it is not immediately true that if the metric terms analytically satisfy the metric identities then their interpolants do as well.

It is relatively straightforward to satisfy the metric identities in two spatial dimensions if the boundaries of the elements are polynomials. For such domains,

$$
\begin{aligned}
J\vec{a}^1 &= Y_\xi \hat{x} - X_\eta \hat{y}, \\
J\vec{a}^2 &= -Y_\xi \hat{x} + X_\xi \hat{y}.
\end{aligned} \tag{230}
$$

Therefore, if the mapping $\vec{X} \in \mathbb{P}^N$, which it is if the boundary curves are isoparametric (polynomials of degree N) or less, $\mathbb{I}^N(J\vec{a}^i) = J\vec{a}^i$, and so (229) holds.

It is more complicated to satisfy the metric identities for general hexahedral elements in three spatial dimensions. Direct approximation of the cross product form of the metric terms (181) will not satisfy the discrete metric identities except in special cases because

$$
\sum_{i=1}^{3} \frac{\partial}{\partial \xi^i} \left(\mathbb{I}^N(J\vec{a}^i)\right) = \sum_{i=1}^{3} \frac{\partial}{\partial \xi^i} \left(\mathbb{I}^N\left(\frac{\partial \vec{X}}{\partial \xi^j} \times \frac{\partial \vec{X}}{\partial \xi^k}\right)\right). \tag{231}
$$

Even if $\vec{X} \in \mathbb{P}^N$, the cross product is a polynomial of degree $2N$. Thus, aliasing errors will not allow the outer differentiation to commute with the interpolation operator to allow the terms to cancel.

Special cases for which the cross product form (181) can be used, then, are those where $\mathbb{I}^N(\frac{\partial \vec{X}}{\partial \xi^j} \times \frac{\partial \vec{X}}{\partial \xi^k}) = \frac{\partial \vec{X}}{\partial \xi^j} \times \frac{\partial \vec{X}}{\partial \xi^k}$. Such special cases include

- $\vec{X} \in \mathbb{P}^{N/2}$. If the faces of the hexahedral elements are approximated by half the order of the solution, then the product is a polynomial of degree N and the interpolation is exact.
- The element faces are planar and $N \geqslant 2$. A special case of item 1, the cross product form can be used if the faces are flat.

To avoid such limitations, a general formulation of the metric terms is necessary that satisfies the metric identities. This is achieved by writing the contravariant vector components in a curl form Kopriva (2006), for instance,

$$
Ja_n^i = -\hat{x}_i \cdot \vec{\nabla}_\xi \times \left(\mathbb{I}^N(X_l \vec{\nabla}_\xi X_m)\right), \quad i = 1,2,3, \ n = 1,2,3, \quad (n,m,l) \text{ cyclic.} \tag{232}
$$

Computed this way, $\mathbb{I}^N(J\vec{a}^i) = J\vec{a}^i$ and the divergence of the curl is explicitly zero without the need to commute interpolation and differentiation. Written out in full (232) reads

$$
\begin{aligned}
J\vec{a}^1 &= \left[\left(\mathbb{I}^N(Y_\eta Z)\right)_\zeta - \left(\mathbb{I}^N(Y_\zeta Z)\right)_\eta\right]\hat{x} + \left[\left(\mathbb{I}^N(Z_\eta X)\right)_\zeta - \left(\mathbb{I}^N(Z_\zeta X)\right)_\eta\right]\hat{y} + \left[\left(\mathbb{I}^N(X_\eta Y)\right)_\zeta - \left(\mathbb{I}^N(X_\zeta Y)\right)_\eta\right]\hat{z}, \\
J\vec{a}^2 &= \left[\left(\mathbb{I}^N(Y_\zeta Z)\right)_\xi - \left(\mathbb{I}^N(Y_\xi Z)\right)_\zeta\right]\hat{x} + \left[\left(\mathbb{I}^N(Z_\zeta X)\right)_\xi - \left(\mathbb{I}^N(Z_\xi X)\right)_\zeta\right]\hat{y} + \left[\left(\mathbb{I}^N(X_\zeta Y)\right)_\xi - \left(\mathbb{I}^N(X_\xi Y)\right)_\zeta\right]\hat{z}, \\
J\vec{a}^3 &= \left[\left(\mathbb{I}^N(Y_\xi Z)\right)_\eta - \left(\mathbb{I}^N(Y_\eta Z)\right)_\xi\right]\hat{x} + \left[\left(\mathbb{I}^N(Z_\xi X)\right)_\eta - \left(\mathbb{I}^N(Z_\eta X)\right)_\xi\right]\hat{y} + \left[\left(\mathbb{I}^N(X_\xi Y)\right)_\eta - \left(\mathbb{I}^N(X_\eta Y)\right)_\xi\right]\hat{z}.
\end{aligned} \tag{233}
$$

The Concept of Flux Differencing and Two-Point Fluxes

Although it may not appear so, one feature of the split form approximation is that it can be implemented by a simple modification of the volume integral of a standard DGSEM approximation, thereby taking a code that usually works and transforming it into a code that is provably stable.

To get the implementation form, we use the form $[S]$ and take a tensor product of the Lagrangian basis functions to be the test functions, i.e., $\varphi = \ell_i \ell_j \ell_k \, \mathbf{e}_p$, where p indicates a component of the state vector and $\mathbf{e}_p$ the corresponding unit vector. Since for any state vector polynomial $\mathbf{V} \in \mathbb{P}^N$,

$$\left\langle \mathbf{V}, \ell_i \ell_j \ell_k \, \mathbf{e}_p \right\rangle_N = \sum_{n,m,l=0}^{N} V^p_{nml} \ell_i(\xi_n) \ell_j(\eta_m) \ell_k(\zeta_l) w_{nml} = V^p_{ijk} w_{ijk}. \tag{234}$$

Choosing the test functions in this way for all state components gives for the the volume term with the temporal derivative in $[S]$

$$\langle J \mathbf{U}_t, \varphi \rangle_N \rightarrow J_{ijk} \dot{\mathbf{U}}_{ijk} w_{ijk}. \tag{235}$$

Similarly,

$$\left\langle \vec{\nabla}_\xi \cdot \overleftrightarrow{\tilde{\mathbf{F}}}(\mathbf{U}), \varphi \right\rangle_N \rightarrow w_{ijk} \left\{ \sum_{n=0}^{N} \tilde{\mathbf{F}}^1_{njk} \mathcal{D}_{in} + \sum_{n=0}^{N} \tilde{\mathbf{F}}^2_{ink} \mathcal{D}_{jn} + \sum_{n=0}^{N} \tilde{\mathbf{F}}^3_{ijn} \mathcal{D}_{kn} \right\}, \tag{236}$$

and

$$\left\langle \mathbb{I}^N (\overleftrightarrow{\tilde{\mathbf{A}}})^T \vec{\nabla}_\xi \mathbf{U}, \varphi \right\rangle_N \rightarrow w_{ijk} \left\{ \tilde{\underline{\mathbf{A}}}^1_{ijk} \sum_{n=0}^{N} \mathbf{U}_{njk} \mathcal{D}_{in} + \tilde{\underline{\mathbf{A}}}^2_{ijk} \sum_{n=0}^{N} \mathbf{U}_{ink} \mathcal{D}_{jn} + \tilde{\underline{\mathbf{A}}}^3_{ijk} \sum_{n=0}^{N} \mathbf{U}_{ijn} \mathcal{D}_{kn} \right\}, \tag{237}$$

where the $\tilde{\underline{\mathbf{A}}}^i = J \vec{a}^i \cdot \vec{\underline{\mathbf{A}}}$ are the contravariant coefficient matrix components.

If we add the vanishing terms of the divergence of the coefficient matrices,

$$\left(\vec{\nabla}_\xi \cdot \mathbb{I}^N (\overleftrightarrow{\tilde{\mathbf{A}}}) \mathbf{U}, \varphi \right)_N \rightarrow w_{ijk} \left\{ \sum_{n=0}^{N} \tilde{\underline{\mathbf{A}}}^1_{njk} \mathcal{D}_{in} + \sum_{n=0}^{N} \tilde{\underline{\mathbf{A}}}^2_{ink} \mathcal{D}_{jn} + \sum_{n=0}^{N} \tilde{\underline{\mathbf{A}}}^3_{ijn} \mathcal{D}_{kn} \right\} \mathbf{U}_{ijk}, \tag{238}$$

we can gather the three terms, (236), (237) and (238), to see that

$$\left\langle \vec{\nabla}_\xi \cdot \tilde{\mathbf{F}}(\mathbf{U}), \varphi \right\rangle_N + \left\langle \mathbb{I}^N(\overleftrightarrow{\tilde{\mathbf{A}}}) \cdot \vec{\nabla}_\xi \mathbf{U}, \varphi \right\rangle_N + \left\langle \vec{\nabla}_\xi \cdot \mathbb{I}^N(\overleftrightarrow{\tilde{\mathbf{A}}})\mathbf{U}, \varphi) \right\rangle_N \to$$
$$w_{ijk} \sum_{n=0}^{N} \left\{ \tilde{\mathbf{F}}^1_{njk} + \underline{\tilde{\mathbf{A}}}^1_{ijk}\mathbf{U}_{njk} + \underline{\tilde{\mathbf{A}}}^1_{njk}\mathbf{U}_{ijk} \right\} \mathcal{D}_{in}$$
$$+ w_{ijk} \sum_{n=0}^{N} \left\{ \tilde{\mathbf{F}}^2_{ink} + \underline{\tilde{\mathbf{A}}}^2_{ijk}\mathbf{U}_{ink} + \underline{\tilde{\mathbf{A}}}^2_{ink}\mathbf{U}_{ijk} \right\} \mathcal{D}_{jn} \tag{239}$$
$$+ w_{ijk} \sum_{n=0}^{N} \left\{ \tilde{\mathbf{F}}^3_{ijn} + \underline{\tilde{\mathbf{A}}}^3_{ijk}\mathbf{U}_{ijn} + \underline{\tilde{\mathbf{A}}}^3_{ijn}\mathbf{U}_{ijk} \right\} \mathcal{D}_{kn}.$$

The quantities in the braces in (239) can be interpreted as two-point fluxes. For example, the quantity in braces in the first sum depends on the points ijk and $njk,\ n = 0, \ldots, N$. So let us define the two-point fluxes

$$\begin{aligned}
\overline{\mathbf{F}}^1_{(n,i)jk} &= \tilde{\mathbf{F}}^1_{njk} + \underline{\tilde{\mathbf{A}}}^1_{ijk}\mathbf{U}_{njk} + \underline{\tilde{\mathbf{A}}}^1_{njk}\mathbf{U}_{ijk},\\
\overline{\mathbf{F}}^2_{i(n,j)k} &= \tilde{\mathbf{F}}^2_{ink} + \underline{\tilde{\mathbf{A}}}^2_{ijk}\mathbf{U}_{ink} + \underline{\tilde{\mathbf{A}}}^2_{ink}\mathbf{U}_{ijk},\\
\overline{\mathbf{F}}^3_{ij(n,k)} &= \tilde{\mathbf{F}}^3_{ijn} + \underline{\tilde{\mathbf{A}}}^3_{ijk}\mathbf{U}_{ijn} + \underline{\tilde{\mathbf{A}}}^3_{ijn}\mathbf{U}_{ijk}.
\end{aligned} \tag{240}$$

With two-point fluxes the advective part of the split form approximation (203) looks like the DSGEM implementation presented by Kopriva (2009) except that the fluxes in the derivative sums have been replaced

$$\begin{aligned}
\dot{\mathbf{U}}_{ijk} + \frac{1}{J_{ijk}} \Bigg\{ &\left[\left\{ \tilde{\mathbf{F}}^*_{Njk} - \tilde{\mathbf{F}}_{Njk} \cdot \hat{\xi} \right\} \frac{\delta_{iN}}{w_i} - \left\{ \tilde{\mathbf{F}}^*_{0jk} - \tilde{\mathbf{F}}_{0jk} \cdot \hat{\xi} \right\} \frac{\delta_{i0}}{w_i} + \frac{1}{2} \sum_{n=0}^{N} \overline{\mathbf{F}}^1_{(n,i)jk} \mathcal{D}_{in} \right]\\
&+ \left[\left\{ \tilde{\mathbf{F}}^*_{iNk} - \tilde{\mathbf{F}}_{iNk} \cdot \hat{\eta} \right\} \frac{\delta_{jN}}{w_j} - \left\{ \tilde{\mathbf{F}}^*_{i0k} - \tilde{\mathbf{F}}_{i0k} \cdot \hat{\eta} \right\} \frac{\delta_{j0}}{w_j} + \frac{1}{2} \sum_{n=0}^{N} \overline{\mathbf{F}}^2_{i(n,j)k} \mathcal{D}_{jn} \right]\\
&+ \left[\left\{ \tilde{\mathbf{F}}^*_{ijN} - \tilde{\mathbf{F}}_{ijN} \cdot \hat{\zeta} \right\} \frac{\delta_{kN}}{w_k} - \left\{ \tilde{\mathbf{F}}^*_{ij0} - \tilde{\mathbf{F}}_{ij0} \cdot \hat{\zeta} \right\} \frac{\delta_{0k}}{w_k} + \frac{1}{2} \sum_{n=0}^{N} \overline{\mathbf{F}}^3_{ij(n,k)} \mathcal{D}_{kn} \right] \Bigg\}\\
&= 0.
\end{aligned} \tag{241}$$

We can go further and re-write each of these two-point fluxes in terms of two-point averages. To do so, we add the derivative of a constant, which following (66) is zero,

$$0 = \underline{\tilde{\mathbf{A}}}^1_{ijk}\mathbf{U}_{ijk} \sum_{n=0}^{N} \mathcal{D}_{in} = \sum_{n=0}^{N} \underline{\tilde{\mathbf{A}}}^1_{ijk}\mathbf{U}_{ijk}\mathcal{D}_{in}. \tag{242}$$

Adding (242) to (240), we get a new two-point flux, the first component of which we define by

$$4\tilde{\mathbf{F}}^{\#,1}_{(n,i)jk} = \tilde{\underline{\mathbf{A}}}^{1}_{njk}\mathbf{U}_{njk} + \tilde{\underline{\mathbf{A}}}^{1}_{ijk}\mathbf{U}_{ijk} + \tilde{\underline{\mathbf{A}}}^{1}_{ijk}\mathbf{U}_{njk} + \tilde{\underline{\mathbf{A}}}^{1}_{njk}\mathbf{U}_{ijk} \tag{243}$$

and similarly for the other contravariant fluxes. The right-hand side of (243) can be split into a product of two factors,

$$\tilde{\mathbf{F}}^{\#,1}_{(n,i)jk} = \frac{\left(\tilde{\underline{\mathbf{A}}}^{1}_{njk} + \tilde{\underline{\mathbf{A}}}^{1}_{ijk}\right)}{2}\frac{\left(\mathbf{U}_{ijk} + \mathbf{U}_{njk}\right)}{2}, \tag{244}$$

so that

$$\tilde{\mathbf{F}}^{\#,1}_{(n,i)jk} = \left\{\!\!\left\{\tilde{\underline{\mathbf{A}}}^{1}\right\}\!\!\right\}_{(n,i)jk}\{\mathbf{U}\}_{(n,i)jk}. \tag{245}$$

Since the linear flux $\overset{\leftrightarrow}{\tilde{\mathbf{f}}} = \overset{\leftrightarrow}{\tilde{\mathbf{A}}}\mathbf{U}$, we see that the two-point flux whose divergence is equal to the split form approximation to the divergence of the flux can be expressed as the product of two averages.

With the definition of the two-point flux, we can re-write the sums in (241) with $\overset{\leftrightarrow}{\tilde{\mathbf{F}}}{}^{\#}$, for example,

$$\frac{1}{2}\sum_{n=0}^{N}\bar{\mathbf{F}}^{1}_{(n,i)jk}\mathcal{D}_{in} = \sum_{n=0}^{N} 2\tilde{\mathbf{F}}^{\#,1}_{(n,i)jk}\mathcal{D}_{in}, \tag{246}$$

to give the approximation at each point (and what one would code)

$$\begin{aligned}
\dot{\mathbf{U}}_{ijk} + \frac{1}{J_{ijk}}\Bigg\{&\Bigg[\left\{\tilde{\mathbf{F}}^{*}_{Njk} - \tilde{\mathbf{F}}_{Njk}\cdot\hat{\xi}\right\}\frac{\delta_{iN}}{w_i} - \left\{\tilde{\mathbf{F}}^{*}_{0jk} - \tilde{\mathbf{F}}_{0jk}\cdot\hat{\xi}\right\}\frac{\delta_{i0}}{w_i} + \sum_{n=0}^{N} 2\tilde{\mathbf{F}}^{\#,1}_{(n,i)jk}\mathcal{D}_{in}\Bigg]\\
&+\Bigg[\left\{\tilde{\mathbf{F}}^{*}_{iNk} - \tilde{\mathbf{F}}_{iNk}\cdot\hat{\eta}\right\}\frac{\delta_{jN}}{w_j} - \left\{\tilde{\mathbf{F}}^{*}_{i0k} - \tilde{\mathbf{F}}_{i0k}\cdot\hat{\eta}\right\}\frac{\delta_{j0}}{w_j} + \sum_{n=0}^{N} 2\tilde{\mathbf{F}}^{\#,2}_{i(n,j)k}\mathcal{D}_{jn}\Bigg]\\
&+\Bigg[\left\{\tilde{\mathbf{F}}^{*}_{ijN} - \tilde{\mathbf{F}}_{ijN}\cdot\hat{\zeta}\right\}\frac{\delta_{kN}}{w_k} - \left\{\tilde{\mathbf{F}}^{*}_{ij0} - \tilde{\mathbf{F}}_{ij0}\cdot\hat{\zeta}\right\}\frac{\delta_{0k}}{w_k} + \sum_{n=0}^{N} 2\tilde{\mathbf{F}}^{\#,3}_{ij(n,k)}\mathcal{D}_{kn}\Bigg]\Bigg\}\\
&= 0.
\end{aligned} \tag{247}$$

As a shorthand, we write the divergence operator implied by the three summations in (247) as

$$\begin{aligned}
\vec{\mathbb{D}}\cdot(\overset{\leftrightarrow}{\tilde{\mathbf{F}}})^{\#}(\xi,\eta,\zeta) \equiv 2\sum_{n=0}^{N}\ &\ell_n'(\xi)\tilde{\mathbf{F}}^{\#,1}(\xi,\eta,\zeta;\xi_n,\eta,\zeta)\\
&+\ell_n'(\eta)\tilde{\mathbf{F}}^{\#,2}(\xi,\eta,\zeta;\xi,\eta_n,\zeta)\\
&+\ell_n'(\zeta)\tilde{\mathbf{F}}^{\#,3}(\xi,\eta,\zeta;\xi,\eta,\zeta_n),
\end{aligned} \tag{248}$$

which allows us to add one more equivalent form for the divergence approximation to Table 3.3, namely,

$$\text{Two-Point [T]}: \quad \left\langle \vec{\mathbb{D}} \cdot (\overset{\leftrightarrow}{\tilde{\mathbf{F}}})^{\#}, \varphi \right\rangle_N + \int_{\partial E, N} \varphi^T \left\{ \mathbf{F}_n^* - \mathbf{F}_n \right\} \hat{s}\, \mathrm{dS}. \tag{249}$$

In summary, the split form approximation to the divergence, which includes three terms for $\vec{\nabla}_\xi \cdot \overset{\leftrightarrow}{\tilde{\mathbf{F}}}$, $\left(\vec{\nabla}_\xi \cdot \overset{\leftrightarrow}{\tilde{\mathbf{A}}} \right) \mathbf{U}$ and $\overset{\leftrightarrow}{\tilde{\mathbf{A}}}^T \vec{\nabla}_\xi \mathbf{U}$, can be re-written, represented, and coded in terms of a single two-point flux, $\overset{\leftrightarrow}{\tilde{\mathbf{F}}}^{\#}$. Since it is algebraically equivalent, the approximation written this way is stable, and free-stream preserving if the metric terms are divergence free.

The Final Assembly: A Robust DGSEM

This section serves as the culmination of this book chapter, where we present the details needed to construct an entropy stable DG approximation for the compressible Navier–Stokes equations. In the end, the solution of the numerical method will possess a discrete entropy bound which directly mimics that from the continuous analysis (137). Great care was taken in the previous sections to discuss, contextualize and analyze the components of the DG approximation piecemeal, e.g., high-order accuracy, aliasing, free-stream preservation, split forms, etc., so that we are now fully equipped with a powerful spectral DG toolbox to address this final task.

Systematically, the split form DG approximation for nonlinear PDEs is outlined as follows:

1. The formulation for the linear problem in Sect. 3.5.3 is generalized. Herein it is highlighted that many components are similar, but the approximation of the advective terms undergo a fundamental change in structure.
2. Advective terms are given a primary focus because their proper treatment is critical to demonstrate entropy stability.
3. The stage is then set to present the discrete stability statement.

We create the split form DG approximation from the strong form DG formulation (194), where the auxiliary variable for the approximation to the solution gradient, $\overset{\leftrightarrow}{\mathbf{Q}}$, is written in terms of the entropy variables $\mathbf{W}$. Just as in the previous section, we must address how to treat the contributions of the nonlinear advective and viscous fluxes in the volume of an element (quantities containing a divergence operator) as well as along its surface (quantities denoted with a star).

We approximate the viscous flux contributions first, because their treatment in the high-order split DG approximation for nonlinear problems is straightforward and utilizes well-developed, standard components of the DG toolbox, e.g., Hindenlang et al. (2012), shown in Table 3.1. The divergence of the viscous fluxes in the volume is approximated by

$$\vec{\nabla}_\xi \cdot \mathbb{I}^N(\overset{\leftrightarrow}{\tilde{\mathbf{F}}^v}) \approx \vec{\mathbb{D}}^s \cdot \overset{\leftrightarrow}{\tilde{\mathbf{F}}^v} = \sum_{m=0}^{N} \mathcal{D}_{im}\left(\tilde{\mathbf{F}}_1^v\right)_{mjk} + \sum_{m=0}^{N} \mathcal{D}_{jm}\left(\tilde{\mathbf{F}}_2^v\right)_{imk} + \sum_{m=0}^{N} \mathcal{D}_{km}\left(\tilde{\mathbf{F}}_3^v\right)_{ijm}, \tag{250}$$

where the metric terms are included in the transformed viscous fluxes $\tilde{\mathbf{F}}_l^v, l = 1, 2, 3$.

Analogous to the linear approximation, we approximate the surface contribution of the viscous fluxes with the BR1 numerical flux

$$\mathbf{F}_n^{v,*} = \left\{\!\!\left\{ \overset{\leftrightarrow}{\mathbf{F}^v} \right\}\!\!\right\} \cdot \vec{n}, \quad \mathbf{W}^* = \{\mathbf{W}\}, \tag{251}$$

where, again, the compact notation is used for the arithmetic mean. The only difference in this treatment of the viscous fluxes is the use of the discrete entropy variables and gradients in the auxiliary variable, in contrast to (223) where the solution quantity $\mathbf{U}$ was used. The BR1 terms are neutrally stable for the split form DG approximation of the nonlinear problem, as we show later in Sect. 3.6.2.

Our focus now turns to the advective components in the approximation, which require greater care to produce an entropy stable, split form DG method. From a physical perspective, it makes sense that the advective terms tend to be more troublesome compared to the "nice" viscous terms. The split formulation fundamentally changes the structure of the flux divergence of the advective flux components.

Proper treatment of the volume contribution of the advective flux divergence needed to produce an entropy stable approximation is built from works in the finite difference community (Fisher 2012; Fisher and Carpenter 2013; Fisher et al. 2013; LeFloch and Rohde 2000) and the DG community (Carpenter et al. 2014; Gassner et al. 2016b, 2018). With notation introduced by Gassner et al. (2018), we define the split form DG divergence approximation as

$$\begin{aligned}\vec{\nabla}_\xi \cdot \mathbb{I}^N(\overset{\leftrightarrow}{\tilde{\mathbf{F}}}) \approx \vec{\mathbb{D}} \cdot \overset{\leftrightarrow}{\tilde{\mathbf{F}}}{}^{\#} = \; & 2\sum_{m=0}^{N} \mathcal{D}_{im}\left(\overset{\leftrightarrow}{\mathbf{F}}{}^{\#}(\mathbf{U}_{ijk}, \mathbf{U}_{mjk}) \cdot \left\{\!\!\left\{ J\vec{a}^{\,1} \right\}\!\!\right\}_{(i,m)jk}\right) \\ & + 2\sum_{m=0}^{N} \mathcal{D}_{jm}\left(\overset{\leftrightarrow}{\mathbf{F}}{}^{\#}(\mathbf{U}_{ijk}, \mathbf{U}_{imk}) \cdot \left\{\!\!\left\{ J\vec{a}^{\,2} \right\}\!\!\right\}_{i(j,m)k}\right) \\ & + 2\sum_{m=0}^{N} \mathcal{D}_{km}\left(\overset{\leftrightarrow}{\mathbf{F}}{}^{\#}(\mathbf{U}_{ijk}, \mathbf{U}_{ijm}) \cdot \left\{\!\!\left\{ J\vec{a}^{\,3} \right\}\!\!\right\}_{ij(k,m)}\right)\end{aligned} \tag{252}$$

for each Gauss-Lobatto node i, j, k of an element. As with the approximation of the linear equations, (248), we have introduced $\overleftrightarrow{\mathbf{F}}^{\#}$, an additional two-point volume flux that is symmetric with respect to its arguments and consistent. We write the arithmetic mean in each spatial direction, defined as in (245) compactly, e.g., in the $\xi-$direction as

$$\{\cdot\}_{(i,m)jk} = \frac{1}{2}\left((\cdot)_{ijk} + (\cdot)_{mjk}\right). \tag{253}$$

Note, that the split formulation (252) is analogous to that from the linear analysis (248); however, the treatment of the metric terms differs. The mapping terms have been "peeled off" from the physical flux components. Separating the arithmetic mean of the metric terms from the physical fluxes corresponds to a *dealiasing* of the metric terms, c.f, e.g., Kopriva and Gassner (2014), Kopriva et al. (2019), (which are variable functions themselves when elements have curved sides) and affects stability. Furthermore, it remains crucial for the discrete entropy estimate that the approximation retains free-stream preservation and that the discrete divergence of the metric terms vanish, as discussed in Sect. 3.5.2.

Substituting the divergence discretizations (250) and (252) as well as the BR1 coupling of the viscous fluxes (251) into the strong form DG approximation (194) gives the split form DGSEM:

$$\begin{aligned}
\langle \mathbb{I}^N(J)\mathbf{U}_t, \varphi\rangle_N + \left\langle \vec{\mathbb{D}} \cdot \overleftrightarrow{\tilde{\mathbf{F}}}^{\#}, \varphi \right\rangle_N + \int_{\partial E,N} \varphi^T \left\{\mathbf{F}_n^* - \mathbf{F}_n\right\} \hat{s}\, \mathrm{dS} \\
= \frac{1}{\mathrm{Re}} \left\langle \vec{\mathbb{D}}^S \cdot \overleftrightarrow{\tilde{\mathbf{F}}}^{v}, \varphi \right\rangle_N + \frac{1}{\mathrm{Re}} \int_{\partial E,N} \varphi^T \left\{\mathbf{F}_n^{v,*} - \mathbf{F}_n^{v}\right\} \hat{s}\, \mathrm{dS}, \\
\left\langle \mathbb{I}^N(J)\overleftrightarrow{\mathbf{Q}}, \overleftrightarrow{\vartheta} \right\rangle_N = \int_{\partial E,N} \mathbf{W}^{*,T} \left(\overleftrightarrow{\vartheta} \cdot \vec{n}\right) \hat{s}\, \mathrm{dS} - \left\langle \mathbf{W}, \vec{\nabla}_\xi \cdot \left(\mathfrak{M}^T \overleftrightarrow{\vartheta}\right) \right\rangle_N .
\end{aligned} \tag{254}$$

As presented, the split formulation (254) is incomplete because the two-point volume fluxes, $\overleftrightarrow{\mathbf{F}}^{\#}$, are not yet defined, and the surface coupling of the advective fluxes through $\mathbf{F}_n^*$ remains open. To partially close the question of the surface contributions we connect the choice of the volume flux to the choice of the surface numerical flux through

$$\mathbf{F}_n^* = \overleftrightarrow{\mathbf{F}}^{\#}(\mathbf{U}_L, \mathbf{U}_R) \cdot \vec{n} - \frac{\lambda_{\max}}{2} [\![\mathbf{W}]\!], \tag{255}$$

where $\lambda_{\max}$ is an estimate of the fastest wave speed at the point in question. We use the local Lax-Friedrichs (LLF) numerical flux function as a blueprint in (255) to add numerical dissipation at the element surfaces. This is motivated by the simplicity of the LLF flux and the fact that LLF leads to an entropy stable formulation, as will be shown in Sect. 3.6.2. There are, however, more complex and *selective* dissipation terms (analogous to a Roe 1997 flux) available in the literature, e.g., in Barth (1999), Winters et al. (2017).

So, the final form of the split form DG approximation now completely hinges on the selection of the two-point numerical volume fluxes $\overset{\leftrightarrow}{\mathbf{F}}{}^{\#}$. From Sect. 3.5.3, we know that for linear problems the split formulation is algebraically equivalent to a DG approximation of the advective terms. This equivalence remains true for (252), as well as being a high-order accurate approximation in the volume (Fisher and Carpenter 2013; Gassner et al. 2016b; Ranocha 2018). But what is the "action" of a particular choice of the numerical volume fluxes?

The Choice of the Two-Point Flux

In essence, the split form DG divergence (252) is an abstraction or extension of the standard DGSEM approximation. It encompasses the "classical" DG divergence operator, but offers an impressive ability to recover discrete approximations of alternative forms of the governing equations. To describe how such high-order discretizations of alternative forms of the advective terms are achieved requires an examination of how components of the numerical volume fluxes are constructed. By assumption, the two-point volume fluxes $\overset{\leftrightarrow}{\mathbf{F}}{}^{\#}$ are symmetric with respect to their arguments. So, it is natural that the components of the numerical volume fluxes will be built from some average state (arithmetic or otherwise).

The abstraction of the split form DGSEM (254) provides a powerful framework that offers a convenient construct to generalize, through particular selections of the two-point volume fluxes, a well-trodden technique from the finite difference community for developing split forms of the original governing equations, e.g., Ducros et al. (2000), Pirozzoli (2010), Kennedy and Gruber (2008), Sjögreen et al. (2017), which was put it into the nodal DG context by Gassner (2013), Gassner et al. (2016b). We showed in (245) and (246) that selecting the product of two arithmetic averages is equivalent to a discrete approximation of the split form of a quadratic product, i.e., the average of the conservative form and the advective form of the equations.

The use of two-point fluxes is even more general because it offers a direct translation of various split forms from the continuous level onto the discrete level depending on what is averaged. For example, one can approximate the *cubic split form* for the x-momentum flux divergence in the $x-$direction proposed by Kennedy and Gruber (2008) as

$$\frac{1}{4}\left[(\rho v_1 v_2)_x + \rho(v_1 v_2)_x + v_1(\rho v_2)_x + v_2(\rho v_1)_x + v_1 v_2(\rho)_x + \rho v_2(v_1)_x + \rho v_1(v_2)_x\right]$$
$$\approx 2\sum_{m=0}^{N} \mathcal{D}_{im}\,\{\rho\}_{(i,m)jk}\,\{v_1\}_{(i,m)jk}\,\{v_2\}_{(i,m)jk}\,. \tag{256}$$

Gassner et al. (2016b) provides a dictionary that one can use in order to immediately construct a discrete split formulation from a proposed continuous splitting. The nodal split form DGSEM *inherits* the underlying split form of the equations with a high-order spatial accuracy and remains conservative. This is a somewhat surprising result because the split form is created by averaging particular combinations of the conservative and nonconservative forms of the PDEs. See Gassner (2013), Gassner et al. (2016b) for details.

It appears, though, that translation of a given split form to a two-point flux form is only possible if the splitting on the continuous level is explicitly known. This is a problem when we want to get an entropy conserving (or decreasing) approximation. Tadmor (1984) showed that there always exists a split form (also referred to as a skew-symmetric form) that preserves the mathematical entropy of a PDE system for smooth solutions. Unfortunately, the explicit form of such an entropy conservative splitting is *unknown* for many physically relevant and interesting systems of conservation laws like the compressible Euler equations. Therefore, we need an alternative approach to develop numerical approximations that conserve (or dissipate) the mathematical entropy.

Working with finite volume methods, Tadmor (1987) developed a condition to guarantee that the numerical flux function is entropy conservative. He eschews any knowledge of the split formulation and focuses, instead, on the contraction of the flux derivative into entropy space (143), which we restate here in one space dimension for convenience:

$$\mathbf{w}^T\mathbf{f}_x = f^S_x. \tag{257}$$

The contraction (257) relies on the chain rule, whose discrete recovery is extraordinarily difficult, or often impossible, in practice (Tadmor 2016). To circumvent this obstacle, one applies the product rule to the entropy contraction (257) to re-write it as

$$\mathbf{w}_x^T\mathbf{f} = (\mathbf{w}^T\mathbf{f})_x - f^S_x = (\mathbf{w}^T\mathbf{f} - f^S)_x. \tag{258}$$

Tadmor analyzed this equivalent compatibility condition (258) to determine a numerical surface flux function for finite volume schemes that is discretely entropy conservative. The finite volume method takes the unknowns in each element to be mean values that are naturally discontinuous across element interfaces, see, e.g., LeVeque (2002) for complete details.

The numerical flux Tadmor derived for finite volume approximations carries over to DG approximations, since, as mentioned in Sect. 3.5, the idea to resolve discontinuities with numerical surface fluxes is also used in their construction. We describe the result by considering the contraction (258) at an arbitrary surface point. The flux depends on the discrete values in the current element, denoted with L, and the direct neighbor of that element, denoted with R. Approximating the derivatives in (258) with first-order differences gives Tadmor's *entropy conservation* condition on the numerical surface flux

$$\left(\frac{\mathbf{w}_R - \mathbf{w}_L}{\Delta x}\right)^T \mathbf{f}^{\mathrm{EC}}(\mathbf{u}_L, \mathbf{u}_R) = \frac{\left(\mathbf{w}_R^T \mathbf{f}_R - f_R^S\right) - \left(\mathbf{w}_L^T \mathbf{f}_L - f_L^S\right)}{\Delta x} \tag{259}$$

where Δx is the size of a grid cell. Multiplying through by Δx and utilizing the jump notation (211) the entropy conservation condition on the numerical surface flux is written compactly

$$[\![\mathbf{w}]\!]^T \mathbf{f}^{\mathrm{EC}}(\mathbf{u}_L, \mathbf{u}_R) = [\![\mathbf{w}^T \mathbf{f} - f^S]\!]. \tag{260}$$

It is important to reiterate that the entropy conservative flux, $\mathbf{f}^{\mathrm{EC}}$, is symmetric in its arguments and consistent to the physical flux in the sense that for identical arguments one recovers the physical flux, i.e., $\mathbf{f}^{\mathrm{EC}}(\mathbf{u}, \mathbf{u}) = \mathbf{f}(\mathbf{u})$.

Two interesting aspects of Tadmor's work are: (1) That constructing an entropy conservative surface flux from (260) produces a consistent, low-order finite volume approximation *without* the need to solve a Riemann problem and (2) For systems of nonlinear hyperbolic conservation laws (260) is a single algebraic condition for a vector of unknown flux quantities. Therefore, there exist many "solutions" for $\mathbf{f}^{\mathrm{EC}}$ that yield a numerical surface flux that is entropy conservative by satisfying (260). Care must be taken so that entropy conservative numerical flux function remains physically consistent. One such numerical flux, originally proposed by Tadmor (1987), is defined as a phase integral. Though theoretically useful, this phase integral form is computationally impractical. Thus, over the past 20 years affordable versions of the entropy conservative finite volume surface flux have been developed for a variety of nonlinear hyperbolic systems, c.f, e.g., Chandrashekar (2013), Fjordholm et al. (2011), Winters and Gassner (2016).

We provide here a brief summary of one particular affordable numerical surface flux for the x−direction of the Euler equations with the ideal gas law, since it is relevant to the development of an entropy stable DG approximation for compressible flows. Complete details are provided by Chandrashekar (2013). The crucial idea behind finding a numerically tractable version of an entropy conservative surface flux is the evaluation of its components at various mean states between $\mathbf{u}_L$ and $\mathbf{u}_R$. These mean state expressions can take on incredibly complex forms that depend on the arithmetic mean, the product of arithmetic means or more uncommon quantities like the logarithmic mean (Carlson 1972). We have already introduced notation for the arithmetic mean, e.g., (223). The logarithmic mean of two quantities a_L and a_R takes the form

$$a^{\ln} = \frac{a_L - a_R}{\ln(a_L) - \ln(a_R)}. \tag{261}$$

Note that care must be taken for the logarithmic mean to remain numerically stable when the states are close, $a_L \approx a_R$, as discussed by Ismail and Roe (2009). Also, we introduce an auxiliary variable proportional to the inverse temperature

$$\beta = \frac{p}{2\rho}, \tag{262}$$

which simplifies the form of the entropy variables (140) to

$$\mathbf{w} = \left[\frac{\gamma - \varsigma}{\gamma - 1} - \beta v^2 \,,\, 2\beta v_1 \,,\, 2\beta v_2 \,,\, 2\beta v_3 \,,\, -2\beta\right]^T . \tag{263}$$

Then, Tadmor's entropy conservation condition (260) and many algebraic manipulations determine an analytical expression of an entropy conservative numerical flux for the compressible Euler equations

$$\mathbf{f}^{\text{EC}}(\mathbf{u}_L, \mathbf{u}_R) = \begin{bmatrix} \rho^{\ln}\{v_1\} \\ \rho^{\ln}\{v_1\}^2 + \widehat{p} \\ \rho^{\ln}\{v_1\}\{v_2\} \\ \rho^{\ln}\{v_1\}\{v_3\} \\ \rho^{\ln}\{v_1\}\widehat{H} \end{bmatrix}, \tag{264}$$

where particular average states for the pressure and enthalpy are needed

$$\begin{aligned} \widehat{p} &= \frac{\{\rho\}}{2\{\beta\}}, \\ \widehat{H} &= \frac{1}{2\beta^{\ln}(\gamma-1)} + \frac{\widehat{p}}{\rho^{\ln}} + \{v_1\}^2 + \{v_2\}^2 + \{v_3\}^2 - \frac{1}{2}\left(\{\!\{v_1^2\}\!\} + \{\!\{v_2^2\}\!\} + \{\!\{v_3^2\}\!\}\right). \end{aligned} \tag{265}$$

The numerical surface flux (264) is obviously symmetric with respect to its arguments and it is consistent to the physical flux given in (91).

This slight detour of the discussion to numerical fluxes for low-order finite volume methods actually serves as the backbone for the construction of an entropy conservative (or stable) split form DG approximation, for entropy conservative finite volume flux functions are precisely those to be used as the two-point volume flux functions $\overleftrightarrow{\mathbf{F}}^{\#}$ in the DG flux divergence approximation, (252). In this way, the two-point divergence approximation can recover the action of the entropy conservative split form without an explicit expression of the original equations! The remarkable property of (252) is that it extends the entropy conservative flux form from low-order finite volume approximations to high-order accuracy, as was first demonstrated by Fisher and Carpenter (2013), **provided** the derivative matrix of the high-order method is a diagonal norm SBP operator, as introduced in Sect. 3.2.7. This result unlocks the true power of the two-point DG approximation because, in a sense, the entropy analysis of the high-order numerical scheme reduces to the (somewhat simpler) finite volume problem, (260).

We can finally state a complete version of the split form DG approximation that is entropy conservative (or stable) for nonlinear problems. To do so, we take the volume flux functions to be the entropy conservative surface flux functions from the finite volume approximation, e.g., (264), in each Cartesian direction,

$$\overleftrightarrow{\mathbf{F}}^{\#} = \overleftrightarrow{\mathbf{F}}^{\mathrm{EC}} = \left(\mathbf{F}_1^{\mathrm{EC}}, \mathbf{F}_2^{\mathrm{EC}}, \mathbf{F}_3^{\mathrm{EC}}\right)^T. \tag{266}$$

Note that the method automatically operates on curvilinear geometries because the mapping terms have been separated from the physical flux components in (252). The final split form DG approximation takes the form

$$\begin{aligned}
\left\langle \mathbb{I}^N(J)\mathbf{U}_t, \boldsymbol{\varphi}\right\rangle_N + \left\langle \vec{\mathbb{D}} \cdot \overleftrightarrow{\tilde{\mathbf{F}}}^{\mathrm{EC}}, \boldsymbol{\varphi}\right\rangle_N + \int\limits_{\partial E,N} \boldsymbol{\varphi}^T \left\{\mathbf{F}_n^* - \mathbf{F}_n\right\} \hat{s}\,\mathrm{dS} \\
= \frac{1}{\mathrm{Re}}\left\langle \vec{\mathbb{D}}^s \cdot \overleftrightarrow{\tilde{\mathbf{F}}}^{v}, \boldsymbol{\varphi}\right\rangle_N + \frac{1}{\mathrm{Re}}\int\limits_{\partial E,N} \boldsymbol{\varphi}^T \left\{\mathbf{F}_n^{v,*} - \mathbf{F}_n^{v}\right\} \hat{s}\,\mathrm{dS} \\
\left\langle \mathbb{I}^N(J)\overleftrightarrow{\mathbf{Q}}, \overleftrightarrow{\boldsymbol{\vartheta}}\right\rangle_N = \int\limits_{\partial E,N} \mathbf{W}^{*,T}\left(\overleftrightarrow{\boldsymbol{\vartheta}} \cdot \vec{n}\right)\hat{s}\,\mathrm{dS} - \left\langle \mathbf{W}, \vec{\nabla}_\xi \cdot \left(\mathfrak{M}^T \overleftrightarrow{\boldsymbol{\vartheta}}\right)\right\rangle_N,
\end{aligned} \tag{267}$$

where the surface contributions of the advective fluxes have the form (255).

The Boundedness of the Discrete Entropy

The stage is now set to demonstrate semi-discrete entropy stability of the split form DG method (267) for the compressible Navier–Stokes equations.

Fundamentally, the goal of the discrete entropy analysis is to mimic the continuous analysis performed in Sect. 3.3.1. The key to the continuous entropy analysis was two-fold and required: (i) Integration-by-parts (51) and (ii) Proper contraction of the physical fluxes to become the entropy fluxes (143) (essentially the chain rule). We have on hand discrete, high-order equivalents of both necessary components: (i) A derivative matrix $\mathcal{D}$ with the SBP property (54) and (ii) Two-point flux functions that satisfy Tadmor's entropy conservation condition (260), which are lifted to high-order with the split form DG divergence (252).

We begin from the final split form DG approximation for nonlinear problems (267) and mimic the continuous entropy analysis as closely as possible to get a discrete bound on the entropy. The test function in the first equation is replaced with the polynomial interpolant of the entropy variables $\boldsymbol{\varphi} \leftarrow \mathbf{W}$ and the test function in the second equation is replaced with the viscous fluxes, $\overleftrightarrow{\boldsymbol{\vartheta}} \leftarrow \overleftrightarrow{\mathbf{F}}^{v}$ to obtain

$$\begin{aligned}
\left\langle \mathbb{I}^N(J)\mathbf{U}_t, \mathbf{W}\right\rangle_N + \left\langle \vec{\mathbb{D}} \cdot \overset{\leftrightarrow}{\tilde{\mathbf{F}}}{}^{\mathrm{EC}}, \mathbf{W}\right\rangle_N + \int\limits_{\partial E,N} \mathbf{W}^T \left\{\mathbf{F}_n^* - \mathbf{F}_n\right\} \hat{s}\, \mathrm{dS} \\
= \frac{1}{\mathrm{Re}}\left\langle \vec{\mathbb{D}}^s \cdot \overset{\leftrightarrow}{\tilde{\mathbf{F}}}{}^{v}, \mathbf{W}\right\rangle_N + \frac{1}{\mathrm{Re}}\int\limits_{\partial E,N} \mathbf{W}^T \left\{\mathbf{F}_n^{v,*} - \mathbf{F}_n^{v}\right\} \hat{s}\, \mathrm{dS}, \qquad (268)\\
\left\langle \mathbb{I}^N(J)\overset{\leftrightarrow}{\mathbf{Q}}, \overset{\leftrightarrow}{\mathbf{F}}{}^{v}\right\rangle_N = \int\limits_{\partial E,N} \mathbf{W}^{*,T}\left(\overset{\leftrightarrow}{\mathbf{F}}{}^{v} \cdot \vec{n}\right) \hat{s}\, \mathrm{dS} - \left\langle \mathbf{W}, \vec{\nabla}_\xi \cdot \left(\mathfrak{M}^T \overset{\leftrightarrow}{\mathbf{F}}{}^{v}\right)\right\rangle_N .
\end{aligned}$$

It is possible to condense the expressions in the second equation of (268) from previously introduced notation. That is, $\overset{\leftrightarrow}{\tilde{\mathbf{F}}}{}^{v} = \mathbb{I}^N(\mathfrak{M}^T \overset{\leftrightarrow}{\mathbf{F}}{}^{v})$, $\overset{\leftrightarrow}{\mathbf{F}}{}^{v} \cdot \vec{n} = \mathbf{F}_n^v$ and the standard DG divergence are applied to the viscous flux terms (250) so that

$$\begin{aligned}
\left\langle \mathbb{I}^N(J)\mathbf{U}_t, \mathbf{W}\right\rangle_N + \left\langle \vec{\mathbb{D}} \cdot \overset{\leftrightarrow}{\tilde{\mathbf{F}}}{}^{\mathrm{EC}}, \mathbf{W}\right\rangle_N + \int\limits_{\partial E,N} \mathbf{W}^T \left\{\mathbf{F}_n^* - \mathbf{F}_n\right\} \hat{s}\, \mathrm{dS} \\
= \frac{1}{\mathrm{Re}}\left\langle \vec{\mathbb{D}}^s \cdot \overset{\leftrightarrow}{\tilde{\mathbf{F}}}{}^{v}, \mathbf{W}\right\rangle_N + \frac{1}{\mathrm{Re}}\int\limits_{\partial E,N} \mathbf{W}^T \left\{\mathbf{F}_n^{v,*} - \mathbf{F}_n^{v}\right\} \hat{s}\, \mathrm{dS}, \qquad (269)\\
\left\langle \mathbb{I}^N(J)\overset{\leftrightarrow}{\mathbf{Q}}, \overset{\leftrightarrow}{\mathbf{F}}{}^{v}\right\rangle_N = \int\limits_{\partial E,N} \mathbf{W}^{*,T}\, \mathbf{F}_n^{v}\, \hat{s}\, \mathrm{dS} - \left\langle \mathbf{W}, \vec{\mathbb{D}}^s \cdot \overset{\leftrightarrow}{\tilde{\mathbf{F}}}{}^{v}\right\rangle_N .
\end{aligned}$$

In the continuous entropy analysis, we showed in Sect. 3.3 that the volume has no contribution to the entropy estimate because the contraction of the physical flux divergence into entropy space becomes the entropy flux on the boundary. The split form DG flux divergence (252) that uses the entropy conservative finite volume fluxes precisely mimics this structure discretely. Therefore, it is possible to replace the volume integral (quadrature) of the advective flux in (269) by a surface integral (quadrature) as demonstrated by Gassner et al. (2018)

$$\left\langle \vec{\mathbb{D}} \cdot \overset{\leftrightarrow}{\tilde{\mathbf{F}}}{}^{\mathrm{EC}}, \mathbf{W}\right\rangle_N = \int\limits_{\partial E,N} \left(\vec{F}^S \cdot \vec{n}\right) \hat{s}\, \mathrm{dS} = \int\limits_{\partial E,N} F_n^S\, \hat{s}\, \mathrm{dS}. \qquad (270)$$

At its core, the proof of this property relies on the SBP property and discrete metric identities (229) as well as Tadmor's discrete entropy conservation condition. From (270), the split form DG approximation (269) becomes

$$
\begin{aligned}
&\left\langle \mathbb{I}^N(J)\mathbf{U}_t, \mathbf{W}\right\rangle_N + \int\limits_{\partial E,N} \left(F_n^S + \mathbf{W}^T\left\{\mathbf{F}_n^* - \mathbf{F}_n\right\}\right)\hat{s}\,\mathrm{dS} \\
&\qquad = \frac{1}{\mathrm{Re}}\left\langle \vec{\mathbb{D}}^s \cdot \tilde{\overleftrightarrow{\mathbf{F}}}^v, \mathbf{W}\right\rangle_N + \frac{1}{\mathrm{Re}}\int\limits_{\partial E,N} \mathbf{W}^T\left\{\mathbf{F}_n^{v,*} - \mathbf{F}_n^v\right\}\hat{s}\,\mathrm{dS}, \\
&\left\langle \mathbb{I}^N(J)\overleftrightarrow{\mathbf{Q}}, \overleftrightarrow{\mathbf{F}}^v\right\rangle_N = \int\limits_{\partial E,N} \mathbf{W}^{*,T}\,\mathbf{F}_n^v\,\hat{s}\,\mathrm{dS} - \left\langle \mathbf{W}, \vec{\mathbb{D}}^s \cdot \tilde{\overleftrightarrow{\mathbf{F}}}^v\right\rangle_N .
\end{aligned}
\tag{271}
$$

For the compressible Euler equations there are infinitely many convex entropy functions, $s(\mathbf{u})$, that symmetrize the equations, as shown by Harten (1983); however, Dutt (1988) demonstrated that only the entropy function (138) simultaneously symmetrizes the advective and viscous components of the compressible Navier–Stokes equations.

With this built-in symmetrization in mind, we next examine the first term of the second equation of (271). It is possible to cast the viscous fluxes into an alternative form (146) as the gradients of the entropy variables

$$
\overleftrightarrow{\mathbf{F}}^v = \mathfrak{B}^S \mathfrak{M}^T \vec{\nabla}_\xi \mathbf{W} = \mathfrak{B}^S \overleftrightarrow{\mathbf{Q}}. \tag{272}
$$

The viscous flux matrices $\mathfrak{B}^S$ are symmetric positive definite, (147), and leads to the manipulation

$$
\left\langle \mathbb{I}^N(J)\overleftrightarrow{\mathbf{Q}}, \overleftrightarrow{\mathbf{F}}^v\right\rangle_N = \left\langle \mathbb{I}^N(J)\overleftrightarrow{\mathbf{Q}}, \mathfrak{B}^S\overleftrightarrow{\mathbf{Q}}\right\rangle_N \geqslant \min_{E,N}(\mathbb{I}^N(J))\left\langle \overleftrightarrow{\mathbf{Q}}, \mathfrak{B}^S\overleftrightarrow{\mathbf{Q}}\right\rangle_N \geqslant 0, \tag{273}
$$

provided the interpolant of the element Jacobian is non-negative at the Gauss-Lobatto nodes. Again, see Gassner et al. (2018) for details.

Finally, we substitute the second equation of (271) into the first and apply the estimate (273). This yields an inequality where the volume contribution of the time derivative term is dictated only through the surface contributions of an element

$$
\begin{aligned}
&\left\langle \mathbb{I}^N(J)\mathbf{U}_t, \mathbf{W}\right\rangle_N + \int\limits_{\partial E,N} \left(F_n^S + \mathbf{W}^T\left\{\mathbf{F}_n^* - \mathbf{F}_n\right\}\right)\hat{s}\,\mathrm{dS} \\
&\qquad \leqslant \frac{1}{\mathrm{Re}}\int\limits_{\partial E,N} \left(\mathbf{W}^{*,T}\,\mathbf{F}_n^v + \mathbf{W}^T\left\{\mathbf{F}_n^{v,*} - \mathbf{F}_n^v\right\}\right)\hat{s}\,\mathrm{dS}.
\end{aligned}
\tag{274}
$$

We take a moment to interpret the crucial steps that have just occurred to arrive at the expression (274). The combination of the discrete entropy analysis and the SBP property allowed us to move the advective and viscous flux contributions out of the volume, where we have no control on its behavior, and onto the element boundary, where we do have control through the influence of element neighbors and/or boundary conditions by way of the numerical fluxes.

This movement of all flux influences onto each element's boundaries is a critical intermediate step to mimic the continuous entropy analysis. Now that we have shown how each element contributes to its *local* entropy we are prepared to examine how the discrete entropy will evolve in time globally over the entire domain.

Under the assumption that the chain rule with respect to differentiation in time holds (semi-discrete analysis), the remaining volume term in (274) is the time rate of change of the entropy in an element. From the contraction property of the entropy variables (142) we see that

$$\begin{aligned}\left\langle \mathbb{I}^N(J)\mathbf{U}_t, \mathbf{W}\right\rangle_N &= \sum_{i,j,k}^{N} \omega_{ijk} J_{ijk} \mathbf{W}_{ijk}^T \frac{d\mathbf{U}_{ijk}}{dt} \\ &= \sum_{i,j,k}^{N} \omega_{ijk} J_{ijk} \frac{dS_{ijk}}{dt} = \left\langle \mathbb{I}^N(J)S_t, 1\right\rangle_N .\end{aligned} \tag{275}$$

Moreover, we get the total discrete entropy by summing over all elements in the mesh

$$\frac{d}{dt}\overline{S} = \sum_{k=1}^{K} \left\langle \mathbb{I}^N(J^k)S_t^k, 1\right\rangle_N . \tag{276}$$

Just as in the linear analysis, summing over all elements generates jump terms in the fluxes (advective, viscous and entropy) as well as the entropy variables, whereas the numerical surface flux functions are unique. The physical normal vector $\vec{n}$ is defined uniquely at surfaces to point outward from the current element and into its neighbor so that $\vec{n} = \vec{n}_L = -\vec{n}_R$. With all this in mind, we find that the total discrete entropy satisfies the inequality

$$\begin{aligned}\frac{d}{dt}\overline{S} \leqslant &\sum_{\substack{\text{interior}\\\text{faces}}} \int_N \left\{ [\![\vec{F}^S]\!] \cdot \vec{n} + [\![\mathbf{W}]\!]^T \mathbf{F}_n^* - [\![\mathbf{W}^T \overleftrightarrow{\mathbf{F}}]\!] \cdot \vec{n} \right\} \hat{s}\, \mathrm{dS} \\ &- \frac{1}{\mathrm{Re}} \sum_{\substack{\text{interior}\\\text{faces}}} \int_N \left\{ \mathbf{W}^{*,T} [\![\overleftrightarrow{\mathbf{F}}^v]\!] \cdot \vec{n} + [\![\mathbf{W}]\!]^T \mathbf{F}_n^{v,*} - [\![\mathbf{W}^T \overleftrightarrow{\mathbf{F}}^v]\!] \cdot \vec{n} \right\} \hat{s}\, \mathrm{dS} \\ &+ \mathrm{PBT},\end{aligned} \tag{277}$$

where PBT are the physical boundary terms with proper outward pointing normal orientation

$$
\begin{aligned}
\text{PBT} = & \sum_{\substack{\text{boundary}\\ \text{faces}}} \int_N -\vec{F}^S \cdot \vec{n} + \frac{1}{\text{Re}} \mathbf{W}^T \left(\overleftrightarrow{\mathbf{F}}^v \cdot \vec{n} \right) \, \mathrm{dS} \\
& + \sum_{\substack{\text{boundary}\\ \text{faces}}} \int_N \mathbf{W}^T \mathbf{F}_n^* \, \mathrm{dS} + \frac{1}{\text{Re}} \sum_{\substack{\text{boundary}\\ \text{faces}}} \int_N \mathbf{W}^{*,T} \mathbf{F}_{v,n} + \mathbf{W}^T \mathbf{F}_n^{v,*} \, \mathrm{dS}.
\end{aligned} \tag{278}
$$

Notice that the discrete physical boundary contributions precisely mimic those present in the continuous estimate (153) except for additional dissipation due to the surface fluxes $\mathbf{F}_n^*$, $\mathbf{W}^*$ and $\mathbf{F}_n^{v,*}$ evaluated at the boundaries.

We first investigate the contribution from the advective flux terms at each quadrature point on the interior element faces. The advective numerical surface flux was selected to take the form (255). So, the first part of the total discrete entropy estimate (277) will be

$$
\begin{aligned}
[\![\vec{F}^S]\!] \cdot \vec{n} + [\![\mathbf{W}]\!]^T \mathbf{F}_n^* - [\![\mathbf{W}^T \overleftrightarrow{\mathbf{F}}]\!] \cdot \vec{n} &= [\![\vec{F}^S]\!] \cdot \vec{n} + [\![\mathbf{W}]\!]^T \left(\overleftrightarrow{\mathbf{F}}^{\mathrm{EC}} \cdot \vec{n} - \frac{\lambda_{\max}}{2} [\![\mathbf{W}]\!] \right) - [\![\mathbf{W}^T \overleftrightarrow{\mathbf{F}}]\!] \cdot \vec{n} \\
&= \left([\![\vec{F}^S]\!] + [\![\mathbf{W}]\!]^T \overleftrightarrow{\mathbf{F}}^{\mathrm{EC}} - [\![\mathbf{W}^T \overleftrightarrow{\mathbf{F}}]\!] \right) \cdot \vec{n} - \frac{\lambda_{\max}}{2} [\![\mathbf{W}]\!]^T [\![\mathbf{W}]\!] \\
&= 0 - \frac{\lambda_{\max}}{2} [\![\mathbf{W}]\!]^T [\![\mathbf{W}]\!] \leqslant 0,
\end{aligned} \tag{279}
$$

where the terms involving $\overleftrightarrow{\mathbf{F}}^{\mathrm{EC}}$ vanish by construction from the entropy conservation condition (260) in each Cartesian direction. Also, we note that dissipation must be introduced in an appropriate fashion to ensure the correct sign. In this instance, we took the LLF-type dissipation in terms of the jump in the entropy variables that lead to a guaranteed negative contribution.

Next, we address how the viscous flux terms contribute at the interior element faces. The BR1 discretization (251) was selected for the numerical surface viscous fluxes so that the second part on the right-hand side of (277) becomes

$$
\mathbf{W}^{*,T} [\![\overleftrightarrow{\mathbf{F}}^v]\!] \cdot \vec{n} + [\![\mathbf{W}]\!]^T \mathbf{F}_n^{v,*} - [\![\mathbf{W}^T \overleftrightarrow{\mathbf{F}}^v]\!] \cdot \vec{n} = \left(\{\mathbf{W}\}^T [\![\overleftrightarrow{\mathbf{F}}^v]\!] + [\![\mathbf{W}]\!]^T \left\{\!\!\left\{ \overleftrightarrow{\mathbf{F}}^v \right\}\!\!\right\} - [\![\mathbf{W}^T \overleftrightarrow{\mathbf{F}}^v]\!] \right) \cdot \vec{n}. \tag{280}
$$

From identity (220),

$$
[\![\mathbf{W}^T \overleftrightarrow{\mathbf{F}}^v]\!] = \{\mathbf{W}\}^T [\![\overleftrightarrow{\mathbf{F}}^v]\!] + [\![\mathbf{W}]\!]^T \left\{\!\!\left\{ \overleftrightarrow{\mathbf{F}}^v \right\}\!\!\right\}, \tag{281}
$$

we see that the viscous numerical fluxes (280) at the interior faces vanish exactly

$$
\mathbf{W}^{*,T} [\![\overleftrightarrow{\mathbf{F}}^v]\!] \cdot \vec{n} + [\![\mathbf{W}]\!]^T \mathbf{F}_n^{v,*} - [\![\mathbf{W}^T \overleftrightarrow{\mathbf{F}}^v]\!] \cdot \vec{n} = 0, \tag{282}
$$

as they did for the linear approximation. In this sense, the BR1 treatment of the viscous terms is neutrally stable for the nonlinear compressible flow problem.

From (279), (282) and (278), the final discrete entropy evolution statement is

$$\begin{aligned}\frac{d}{dt}\overline{S} \leqslant & \sum_{\substack{\text{boundary}\\ \text{faces}}} \int_N -\vec{F}^S\cdot\vec{n} + \frac{1}{\text{Re}}\mathbf{W}^T\left(\overleftrightarrow{\mathbf{F}}^v\cdot\vec{n}\right)\,\mathrm{dS} \\ & - \sum_{\substack{\text{boundary}\\ \text{faces}}} \int_N \mathbf{W}^T\overleftrightarrow{\mathbf{F}}^{\text{EC}}\cdot\vec{n} - \sum_{\substack{\text{all}\\ \text{faces}}} \int_N \frac{\lambda_{\max}}{2}[\![\mathbf{W}]\!]^T[\![\mathbf{W}]\!]\,\mathrm{dS} \\ & + \frac{1}{\text{Re}}\sum_{\substack{\text{boundary}\\ \text{faces}}} \int_N \mathbf{W}^{*,T}\mathbf{F}_{v,n} + \mathbf{W}^T\mathbf{F}_n^{v,*}\,\mathrm{dS}.\end{aligned} \tag{283}$$

Notice that the dissipation in the advective fluxes has an influence on the entropy estimate at every surface (physical and interior). Furthermore, the choice of these auxiliary physical boundary terms must ensure that their effect is dissipative to guarantee entropy stability. From another point of view, the additional term gives constraints on the boundary fluxes from which to derive stable boundary conditions, as explored by Dalcin et al. (2019), Hindenlang et al. (2019). If we assume that boundary data is given so that the entropy will not increase in time, e.g., *periodic boundary conditions*, then

$$\frac{d}{dt}\overline{S} \leqslant 0. \tag{284}$$

Integrating over the time interval $t \in [0, T]$, we see that

$$\overline{S}(T) \leqslant \overline{S}(0), \tag{285}$$

which is a discrete equivalent to the entropy bound given in the continuous analysis (148).

Summary

The analysis in this section served to culminate this chapter and describe the components of the entropy stable DGSEM for the compressible Navier–Stokes equations. The approach was systematic to highlight the similarities and differences of the split form DG approximation compared to the "classical" DGSEM. The most crucial change was the abstraction of the volume contributions in the discrete DG divergence operator via the use of a *two-point volume numerical flux*. Furthermore, it served to clarify fundamental changes to the numerical approximation when studying solution estimates for linear and nonlinear problems. Most notably were the use of (i) The gradient of the entropy variables as an auxiliary quantity in the viscous components and (ii) Entropy conservative finite volume flux functions in the two-point volume split formulation.

At its heart, the strengths of the discontinuous Galerkin family of methods are its high-order solution accuracy and low dissipation/dispersion errors, e.g., Winters et al. (2018). To retain these beneficial properties and expand the DGSEM to be provably entropy stable for nonlinear problems, one borrows some of the strongest features of other numerical methods:

- Geometric flexibility from finite element methods.
- Integration-by-parts (summation-by-parts) from spectral methods.
- Split formulations of nonlinear terms from finite difference methods.
- Entropy analysis tools from finite volume methods.

All these components were merged to create the nodal split form DG framework (267) that can approximate the solution of general, nonlinear advection-diffusion equations.

Epilogue

We have surveyed the core components of the split form DG framework, a modern nodal DG variant, herein for the linearized and nonlinear compressible Navier–Stokes equations. A key feature of the split form framework is that it provides a demonstrable improvement to the robustness of the high-order numerical approximation, e.g., Gassner et al. (2016b). A further exploration and (partial) explanation of this beneficial property is provided by Gassner and Winters (2019) or Winters et al. (2018).

The response of the broader high-order numerics community, DG or otherwise, to the split form framework has been immense. As such, it remains an active area of research as the framework is developed and expanded upon in different contexts. For the compressible Navier–Stokes equations, this includes examinations into the development of provably stable boundary conditions (Parsani et al. 2015; Dalcin et al. 2019; Hindenlang et al. 2019) as well as explorations using the split form framework as a "baseline" to which turbulence modeling capabilities are added (Flad and Gassner 2017; Manzanero et al. 2020a; Flad et al. 2020).

A principal aspect of the split form DG framework is its generality. In essence, the stability estimates developed in Secs. 3.5.1 and 3.6.2 rely only on:

1. The SBP property of the derivative matrix $\mathcal{D}$.
2. The formulation of a two-point symmetric flux function to be used in the volume and at the surface.

Because of this, the framework is readily extended to other high-order numerical methods that feature the SBP property, e.g., multi-block finite difference methods (Hicken et al. 2016; Crean et al. 2018) or alternative DG approaches (Pazner and Persson 2019). Additionally, the split forms have been extended to many other systems of PDEs including:

- The shallow water equations (Gassner et al. 2016a; Wintermeyer et al. 2017).
- Euler equations with alternative equations of state (Winters et al. 2019).
- Ideal (Liu et al. 2018) and resistive (Bohm et al. 2018) magnetohydrodynamic (MHD) equations.
- Relativistic Euler and MHD equations (Biswas and Kumar 2019; Wu and Shu 2019).
- Two-phase flows (Renac 2019).
- The Cahn-Hilliard equations (Manzanero et al. 2020d).
- Incompressible Navier–Stokes (INS) (Manzanero et al. 2020b).
- Coupled Cahn-Hilliard and INS (Manzanero et al. 2020c).

The split form technique described herein was designed for curvilinear unstructured hexahedral meshes. Recent extensions have increased the generality and flexibility of the framework to include meshes that contain simplex elements (Chen and Shu 2017; Chan 2018), are non-conforming (Friedrich et al. 2018), or move (Krais et al. 2020; Schnücke et al. 2020; Kopriva et al. 2016). Furthermore, it is possible to create similar entropy stability estimates for interpolation/quadrature node sets that do not include the boundary points (Chan et al. 2019).

The split form technique, through the introduction of the two-point volume fluxes, increases the computational cost of the DG method *locally* on each element. The coupling between DG elements remains weak and the split form DG framework retains the attractive, highly parallelizable nature of the DGSEM (Wintermeyer et al. 2018).

As noted in Sect. 3.3.1, an assumption made within the entropy stability estimate is positivity of particular solution quantities, e.g., the density. For practical simulations, additional shock capturing measure must be incorporated to maintain positivity. However, this must be done carefully to maintain both high-order accuracy and provable entropy stability (Hennemann and Gassner 2020). Moreover, the analysis in Sect. 3.6.2 was done in a semi-discrete sense. Special considerations must be made to develop a fully discrete estimate on the entropy (Friedrich et al. 2019; Ranocha et al. 2020).

Overall, exciting developments continue in the realm of high-order DG methods, and split formulations in general, where numerical methods are designed to mimic important continuous stability estimates of PDE systems. Interestingly, the numerical approximations described in this chapter are rapidly approaching the bleeding edge of the current mathematical knowledge we have for the physical models themselves. As it is quite difficult, and perhaps unwise, to discretely mimic physical properties that we do not understand, the further development of modern high-order methods should be done in close collaboration with researchers from physics, computer science, and mathematics.

In closing, we took writing this book chapter as an opportunity to provide the interested reader with many details on the mathematical derivations but also of the numerical algorithms with the aim to provide a starting point for an actual implementation. In addition to this book chapter, we refer to the open source code FLUXO[4]

[4]https://github.com/project-fluxo.

that implements the 3D curvilinear split form DG methodology with different two-point fluxes. It is our hope that the self-contained derivations and discussions in the previous sections have clarified the motivation and construction of the split form DG method.

References

Abarbanel, S., & Gottlieb, D. (1981). Optimal time splitting for two-and three-dimensional Navier-Stokes equations with mixed derivatives. *Journal of Computational Physics*, *41*(1), 1–33.

Altmann, C., Beck, A. D., Hindenlang, F., Staudenmaier, M., Gassner, G. J., & Munz, C.-D. (2013). An efficient high performance parallelization of a discontinuous Galerkin spectral element method. In Keller, R., Kramer, D., & Weiss, J-P. (eds.), *Facing the Multicore-Challenge III, Lecture Notes in Computer Science*, (pp. 37–47, vo. 7686). Berlin: Springer. ISBN 978-3-642-35892-0.

Baggag, A., Atkins, H., & Keyes, D. (2000). Parallel implementation of the discontinuous Galerkin method. In *Parallel computational fluid dynamics: Towards teraflops, optimization, and novel formulations*, (pp. 115–122).

Barth, T. J. (1999). Numerical methods for gasdynamic systems on unstructured meshes. In Dietmar Kröner, Mario Ohlberger, & Christian Rohde (Eds.), *An Introduction to Recent Developments in Theory and Numerics for Conservation Laws* (Vol. 5, pp. 195–285)., Lecture Notes in Computational Science and Engineering Berlin Heidelberg: Springer.

Bassi, F., & Rebay, S. (1997). A high order accurate discontinuous finite element method for the numerical solution of the compressible Navier-Stokes equations. *Journal of Computational Physics*, *131*, 267–279.

B. Biswas and H. Kumar. Entropy stable discontinuous Galerkin approximation for the relativistic hydrodynamic equations. *arXiv preprint* arXiv:1911.07488, 2019.

Black, K. (1999). A conservative spectral element method for the approximation of compressible fluid flow. *Kybernetika*, *35*(1), 133–146.

Black, K. (2000). Spectral element approximation of convection-diffusion type problems. *Applied Numerical Mathematics*, *33*(1–4), 373–379.

Bohm, M., Winters, A. R., Gassner, G. J., Derigs, D., Hindenlang, F., & Saur, J. (2018). An entropy stable nodal discontinuous Galerkin method for the resistive MHD equations. Part I: Theory and numerical verification. *Journal of Computational Physics*. https://doi.org/10.1016/j.jcp.2018.06.027.

Bonev, B., Hesthaven, J. S., Giraldo, F. X., & Kopera, M. A. (2018). Discontinuous Galerkin scheme for the spherical shallow water equations with applications to tsunami modeling and prediction. *Journal of Computational Physics*, *362*, 425–448.

Canuto, C., & Quarteroni, A. (1982). Approximation results for orthogonal polynomials in Sobolev spaces. *Mathematics of Computation*, *38*(157), 67–86.

Canuto, C., Hussaini, M., Quarteroni, A., & Zang, T. (2007). *Spectral Methods: Evolution to Complex Geometries and Applications to Fluid Dynamics*. Berlin: Springer.

Carlson, B. C. (1972). The logarithmic mean. *The American Mathematical Monthly*, *79*(6), 615–618.

Carpenter, M., Fisher, T., Nielsen, E., & Frankel, S. (2014). Entropy stable spectral collocation schemes for the Navier-Stokes equations: Discontinuous interfaces. *SIAM Journal on Scientific Computing*, *36*(5), B835–B867.

Chan, J. (2018). On discretely entropy conservative and entropy stable discontinuous Galerkin methods. *Journal of Computational Physics*, *362*, 346–374.

Chan, J., Hewett, R. J., & Warburton, T. (2017). Weight-adjusted discontinuous Galerkin methods: wave propogation in heterogeneous media. *SIAM Journal on Scientific Computing, 39*(6), A2935–A2961.

Chan, J., Del Rey Fernández, D. C., & Carpenter, M. H. (2019). Efficient entropy stable Gauss collocation methods. *SIAM Journal on Scientific Computing, 41*(5), A2938–A2966.

Chandrashekar, P. (2013). Kinetic energy preserving and entropy stable finite volume schemes for compressible Euler and Navier-Stokes equations. *Communications in Computational Physics, 14*(1252–1286), 11.

Chen, T., & Shu, C.-W. (2017). Entropy stable high order discontinuous Galerkin methods with suitable quadrature rules for hyperbolic conservation laws. *Journal of Computational Physics, 345*, 427–461.

Cockburn, B., & Shu, C. W. (1991). The Runge-Kutta local projection P^1 discontinuous Galerkin method for scalar conservation laws. *Rairo-Matehmatical Modelling and Numerical Analysis-Modelisation Mathematique et Analyse Numerique, 25*, 337–361.

Cockburn, B., & Shu, C.-W. (1998a). The Runge-Kutta discontinuous Galerkin method for conservation laws V: Multidimensional systems. *Journal of Computational Physics, 141*(2), 199–224.

Cockburn, B., & Shu, C.-W. (1998b). The local discontinuous Galerkin method for time-dependent convection diffusion systems. *SIAM Journal on Numerical Analysis, 35*, 2440–2463.

Cockburn, B., Hou, S., & Shu, C.-W. (1990). The Runge-Kutta local projection discontinuous Galerkin finite element method for conservation laws. IV: The multidimensional case. *Mathematics of Computation, 54*(190), 545–581.

Cockburn, B., Karniadakis, G. E., Shu, C-W. (2000). The development of discontinuous Galerkin methods. In Cockburn, B., Karniadakis, G., & Shu, C.-W. (eds.), *Proceedings of the international symposium on discontinuous galerkin methods*, (pp. 3–50), New York: Springer.

Crean, J., Hicken, J. E., Del Rey Fernández, D. C., Zingg, D. W., & Carpenter, M. H. (2018). Entropy-stable summation-by-parts discretization of the Euler equations on general curved elements. *Journal of Computational Physics, 356*, 410–438.

Dalcin, L., Rojas, D., Zampini, S., Del Rey Fernández, D. C., Carpenter, M. H., & Parsani, M. (2019). Conservative and entropy stable solid wall boundary conditions for the compressible Navier-Stokes equations: Adiabatic wall and heat entropy transfer. *Journal of Computational Physics, 397*, 108775.

Deng, S. (2007). Numerical simulation of optical coupling and light propagation in coupled optical resonators with size disorder. *Applied Numerical Mathematics, 57*(5–7), 475–485. ISSN 0168-9274.

Deng, S.Z., Cai, W., & Astratov, V.N. (2004) Numerical study of light propagation via whispering gallery modes in microcylinder coupled resonator optical waveguides. *Optics Express, 12*(26), 6468–6480, DEC 27 2004. ISSN 1094-4087.

Ducros, F., Laporte, F., Soulères, T., Guinot, V., Moinat, P., & Caruelle, B. (2000). High-order fluxes for conservative skew-symmetric-like schemes in structured meshes: Application to compressible flows. *Journal of Computational Physics, 161*, 114–139.

Dutt, P. (1988). Stable boundary conditions and difference schemes for Navier-Stokes equations. *SIAM Journal on Numerical Analysis, 25*, 245–267.

Evans, L. C. (2012). *Partial differential equations*. American Mathematical Society.

Fagherazzi, S., Furbish, D. J., Rasetarinera, P., & Hussaini, M. Y. (2004a). Application of the discontinuous spectral Galerkin method to groundwater flow. *Advances in Water Resourses, 27*, 129–140.

Fagherazzi, S., Rasetarinera, P., Hussaini, M. Y., & Furbish, D. J. (2004b). Numerical solution of the dam-break problem with a discontinuous Galerkin method. *Journal of Hydraulic Engineering, 130*(6), 532–539. June.

Farrashkhalvat, M., & Miles, J. P. (2003). *Basic Structured Grid Generation: With an introduction to unstructured grid generation*. Butterworth-Heinemann.

Fisher, T., Carpenter, M. H., Nordström, J., Yamaleev, N. K., & Swanson, C. (2013). Discretely conservative finite-difference formulations for nonlinear conservation laws in split form: Theory and boundary conditions. *Journal of Computational Physics*, *234*, 353–375.
Fisher, T. C. (2012). *High-order L_2 stable multi-domain finite difference method for compressible flows*. Ph.D. thesis, Purdue University.
Fisher, T. C., & Carpenter, M. H. (2013). High-order entropy stable finite difference schemes for nonlinear conservation laws: Finite domains. *Journal of Computational Physics*, *252*, 518–557.
Fjordholm, U. S., Mishra, S., & Tadmor, E. (2011). Well-balanced and energy stable schemes for the shallow water equations with discontiuous topography. *Journal of Computational Physics*, *230*(14), 5587–5609.
Flad, D., & Gassner, G. (2017). On the use of kinetic energy preservaing DG-scheme for large eddy simulation. *Journal of Computational Physics*, *350*, 782–795.
Flad, D., Beck, A., & Guthke, P. (2020). A large eddy simulation method for DGSEM using non-linearly optimized relaxation filters. *Journal of Computational Physics*, *408*, 109303.
Friedrich, L., Winters, A. R., Del Rey Fernández, G. J., Gassner, D. C., Parsani, M., & Carpenter, M. H. (2018). An entropy stable h/p non-conforming discontinuous Galerkin method with the summation-by-parts property. *Journal of Scientific Computing*, *77*(2), 689–725.
Friedrich, L., Schnücke, G., Winters, A. R., Del Rey Fernández, D. C., Gassner, G. J., & Carpenter, M. H. (2019). Entropy stable space-time discontinuous Galerkin schemes with summation-by-parts property for hyperbolic conservation laws. *Journal of Scientific Computing*, *80*(1), 175–222.
Gassner, G. (2013). A skew-symmetric discontinuous Galerkin spectral element discretization and its relation to SBP-SAT finite difference methods. *SIAM Journal on Scientific Computing*, *35*(3), A1233–A1253.
Gassner, G., & Kopriva, D. A. (2010). A comparison of the dispersion and dissipation errors of Gauss and Gauss-Lobatto discontinuous Galerkin spectral element methods. *SIAM Journal on Scientific Computing*, *33*(5), 2560–2579.
Gassner, G. J., & Winters, A. R. (2019). A novel robust strategy for discontinuous Galerkin methods in computational physics: Why? When? What? Where? *Submitted to Frontiers in Physics*.
Gassner, G. J., Winters, A. R., & Kopriva, D. A. (2016a). A well balanced and entropy conservative discontinuous Galerkin spectral element method for the shallow water equations. Applied Mathematics and Computation, 272. *Part*, *2*, 291–308.
Gassner, G. J., Winters, A. R., & Kopriva, D. A. (2016b). Split form nodal discontinuous Galerkin schemes with summation-by-parts property for the compressible Euler equations. *Journal of Computational Physics*, *327*, 39–66.
Gassner, G. J., Winters, A. R., Hindenlang, F. J., & Kopriva, D. A. (2018). The BR1 scheme is stable for the compressible Navier-Stokes equations. *Journal of Scientific Computing*, *77*(1), 154–200.
Geuzaine, C., & Remacle, J.-F. (2009). Gmsh: A 3-D finite element mesh generator with built-in pre-and post-processing facilities. *International Journal for Numerical Methods in Engineering*, *79*(11), 1309–1331.
Giraldo, F. X., & Restelli, M. (2008). A study of spectral element and discontinuous Galerkin methods for the Navier-Stokes equations in nonhydrostatic mesoscale atmospheric modeling: Equation sets and test cases. *Journal of Computational Physics*, *227*(8), 3849–3877.
Giraldo, F. X., Hesthaven, J. S., & Warburton, T. (2002). Nodal high-order discontinuous Galerkin methods for the spherical shallow water equations. *Journal of Computational Physics*, *181*(2), 499–525.
Gordon, W. J., & Hall, C. A. (1973). Construction of curvilinear co-ordinate systems and their applications to mesh generation. *International Journal for Numerical Methods in Engineering Engineering*, *7*, 461–477.
Gottlieb, D., & Orszag, S.A. (1977). *Numerical Analysis of Spectral Methods: Theory and Applications*. SIAM-CMBS.
Harten, A. (1983). On the symmetric form of systems of conservation laws with entropy. *Journal of Computational Physics*, *49*, 151–164.

Hennemann, S., & Gassner, G. J. (2020). A provably entropy stable subcell shock capturing approach for high order split form DG. *Submitted to Journal of Computational Physics*.

Hesthaven, J. S., & Warburton, T. (2008). *Nodal discontinuous galerkin methods*. Springer.

Hicken, J. E., Fernández, D. C. D. R., & Zingg, D. W. (2016). Multidimensional summation-by-parts operators: General theory and application to simplex elements. *SIAM Journal on Scientific Computing*, *38*(4), A1935–A1958.

Hindenlang, F. (2014). *Mesh curving techniques for high order parallel simulations on unstructured meshes*. Ph.D. thesis, University of Stuttgart.

Hindenlang, F., Gassner, G. J., Altmann, C., Beck, A., Staudenmaier, M., & Munz, C.-D. (2012). Explicit discontinuous Galerkin methods for unsteady problems. *Computers and Fluids*, *61*, 86–93.

Hindenlang, F. J., Gassner, G. J., Kopriva, D. A. (2019). Stability of wall boundary condition procedures for discontinuous Galerkin spectral element approximations of the compressible Euler equations. arXiv:1901.04924.

Hu, F. Q., Hussaini, M. Y., & Rasetarinera, P. (1999). An analysis of the discontinuous Galerkin method for wave propagation problems. *Journal of Computational Physics*, *151*(2), 921–946.

Ismail, F., & Roe, P. L. (2009). Affordable, entropy-consistent Euler flux functions II: Entropy production at shocks. *Journal of Computational Physics*, *228*(15), 5410–5436.

Karniadakis, G. E., & Sherwin, S. J. (2005). *Spectral/hp element methods for computational fluid dynamics*. Oxford University Press.

Kennedy, C. A., & Gruber, A. (2008). Reduced aliasing formulations of the convective terms within the Navier-Stokes equations for a compressible fluid. *Journal of Computational Physics*, *227*, 1676–1700.

Knupp, P. M., Steinberg, S. (1993). *Fundamentals of grid generation*. CRC-Press.

Kopriva, D. A. (2006). Metric identities and the discontinuous spectral element method on curvilinear meshes. *Journal of Scientific Computing*, *26*(3), 301–327.

Kopriva, D. A. (2009). *Implementing spectral methods for partial differential equations*. Scientific computation.

Kopriva, D. A., & Gassner, G. J. (2014). An energy stable discontinuous Galerkin spectral element discretization for variable coefficient advection problems. *SIAM Journal on Scientific Computing*, *34*(4), A2076–A2099.

Kopriva, D. A., Woodruff, S. L., & Hussaini, M. Y. (2000). Discontinuous spectral element approximation of Maxwell's Equations. In Cockburn, B., Karniadakis, G., Shu, C.-W. (eds.), *Proceedings of the international symposium on discontinuous Galerkin methods* (pp. 355–361), New York: Springer.

Kopriva, D. A., Woodruff, S. L., & Hussaini, M. Y. (2002). Computation of electromagnetic scattering with a non-conforming discontinuous spectral element method. *International Journal for Numerical Methods in Engineering*, *53*(1), 105–122.

Kopriva, D. A., Winters, A. R., Bohm, M., & Gassner, G. J. (2016). A provably stable discontinuous Galerkin spectral element approximation for moving hexahedral meshes. *Computers & Fluids*, *139*, 148–160.

Kopriva, D. A., Hindenlang, F. J., Bolemann, T., & Gassner, G. J. (2019). Free-stream preservation for curved geometrically non-conforming discontinuous Galerkin spectral elements. *Journal of Scientific Computing*, *79*(3), 1389–1408.

Krais, N. Schnücke, G., Bolemann, T., & Gassner, G. (2020). Split form ALE discontinuous Galerkin methods with applications to under-resolved turbulent low-Mach number flows. arXiv:2003.02296.

Kreiss, H.-O., Oliger, J. (1973). *Methods for the approximate solution of time-dependent problems*. World Meteorological Organization, Geneva, 1973. GARP Rept. No.10.

Kreiss, H.-O., & Olliger, J. (1972). Comparison of accurate methods for the integration of hyperbolic equations. *Tellus*, *24*, 199–215.

Kreiss, H.-O., Scherer, G. (1974). Finite element and finite difference methods for hyperbolic partial differential equations. In *Mathematical aspects of finite elements in partial differential equations* (pp. 195–212). Elsevier.

Kreiss, H.-O., Scherer, G. (1977). On the existence of energy estimates for difference approximations for hyperbolic systems. Technical report, Deptpartment of Scientific Computing, Uppsala University.

Lax, P. D. (1954). Weak solutions of nonlinear hyperbolic conservation equations and their numerical computation. *Communications on Pure and Applied Mathematics*, *7*(1), 159–193.

Lax, P. D. (1967). Hyperbolic difference equations: A review of the Courant-Friedrichs-Lewy paper in the light of recent developments. *IBM Journal of Reseach and Development*, *11*(2), 235–238.

LeFloch, P., & Rohde, C. (2000). High-order schemes, entropy inequalities, and nonclassical shocks. *SIAM Journal on Numerical Analysis*, *37*(6), 2023–2060.

LeVeque, R. J. (2020) *Finite Volume Methods for Hyperbolic Problems*. Cambridge University Press.

Liu, Y., Shu, C.-W., & Zhang, M. (2018). Entropy stable high order discontinuous Galerkin methods for ideal compressible MHD on structured meshes. *Journal of Computational Physics*, *354*, 163–178.

Manzanero, J., Ferrer, E., Rubio, G., & Valero, E. (2020a) Design of a Smagorinsky spectral vanishing viscosity turbulence model for discontinuous Galerkin methods. *Computers & Fluids*, 104440.

Manzanero, J., Rubio, G., Kopriva, D. A., Ferrer, E., & Valero, E. (2020b). An entropy-stable discontinuous Galerkin approximation for the incompressible Navier-Stokes equations with variable density and artificial compressibility. *Journal of Computational Physics*, *408*, 109241.

Manzanero, J., Rubio, G., Kopriva, D. A., Ferrer, E., & Valero, E. (2020c) Entropy–stable discontinuous Galerkin approximation with summation–by–parts property for the incompressible Navier–Stokes/Cahn–Hilliard system. *Journal of Computational Physics*, 109363.

Manzanero, J., Rubio, G., Kopriva, D. A., Ferrer, E., & Valero, E. (2020d). A free-energy stable nodal discontinuous Galerkin approximation with summation-by-parts property for the Cahn-Hilliard equation. *Journal of Computational Physics*, *403*, 109072.

Merriam, M. L. (1987). Smoothing and the second law. *Computer Methods in Applied Mechanics and Engineering*, *64*(1–3), 177–193.

Merriam, M. L. (1989). An entropy-based approach to nonlinear stability. *NASA Technical Memorandum*, *101086*(64), 1–154.

Nitsche, J. A. (1971). Über ein Variationsprinzip zur Lösung von Dirichlet-Problemen bei Verwendung von Teilräumen, die keinen Randbedingungen unterworfen sind. *Abh. Math. Sem. Univ. Hamburg*, *36*, 9–15.

Nordström, J., & Svärd, M. (2005). Well-posed boundary conditions for the Navier-Stokes equations. *SIAM Journal on Numerical Analysis*, *43*(3), 1231–1255.

Parsani, M., Carpenter, M. H., & Nielsen, E. J. (2015). Entropy stable wall boundary conditions for the three-dimensional compressible Navier-Stokes equations. *Journal of Computational Physics*, *292*, 88–113.

Pazner, W., & Persson, P.-O. (2019). Analysis and entropy stability of the line-based discontinuous Galerkin method. *Journal of Scientific Computing*, *80*(1), 376–402.

Pirozzoli, S. (2010). Generalized conservative approximations of split convective derivative operators. *Journal of Computational Physics*, *229*(19), 7180–7190.

Ranocha, H. (2018). Comparison of some entropy conservative numerical fluxes for the Euler equations. *Journal of Scientific Computing*, *76*(1), 216–242.

Ranocha, H., Sayyari, M., Dalcin, L., Parsani, M., & Ketcheson, D. I. (2020). Relaxation Runge-Kutta methods: Fully discrete explicit entropy-stable schemes for the compressible Euler and Navier-Stokes equations. *SIAM Journal on Scientific Computing*, *42*(2), A612–A638.

Rasetarinera, P., & Hussaini, M. Y. (2001). An efficient implicit discontinuous spectral Galerkin method. *Journal of Computational Physics*, *172*, 718–738.

Rasetarinera, P., Kopriva, D. A., & Hussaini, M. Y. (2001). Discontinuous spectral element solution of acoustic radiation from thin airfoils. *AIAA Journal*, *39*(11), 2070–2075.

Reed, W. H., & Hill, T. R. (1973) Triangular mesh methods for the neutron transport equation. Technical Report LA-UR-73-479, Los Alamos National Laboratory.

Renac, F. (2019). Entropy stable DGSEM for nonlinear hyperbolic systems in nonconservative form with application to two-phase flows. *Journal of Computational Physics*, *382*, 1–26.

Restelli, M., & Giraldo, F. X. (2009). A conservative discontinuous Galerkin semi-implicit formulation for the Navier-Stokes equations in nonhydrostatic mesoscale modeling. *SIAM Journal on Scientific Computing*, *31*(3), 2231–2257.

Roe, P. L. (1997). Approximate Riemann solvers, parameter vectors, and difference schemes. *Journal of Computational Physics*, *135*(2), 250–258.

Schnücke, G., Krais, N., Bolemann, T., & Gassner, G. J. (2020). Entropy stable discontinuous Galerkin schemes on moving meshes for hyperbolic conservation laws. *Journal of Scientific Computing*, *82*(3), 1–42.

Sjögreen, B., Yee, H. C., & Kotov, D. (2017). Skew-symmetric splitting and stability of high order central schemes. *Journal of Physics: Conference Series*, *837*(1), 012019.

Stanescu, D., Farassat, F., & Hussaini, M. Y. (2002a) Aircraft engine noise scattering - parallel discontinuous Galerkin spectral element method. Paper 2002-0800, AIAA.

Stanescu, D., Xu, J., Farassat, F., & Hussaini, M. Y. (2002b). Computation of engine noise propagation and scattering off an aircraft. *Aeroacoustics*, *1*(4), 403–420.

Strand, B. (1994). Summation by parts for finite difference approximations for d/dx. *Journal of Computational Physics*, *110*,

Svärd, M., & Nordström, J. (2014). Review of summation-by-parts schemes for initial-boundary-value problems. *Journal of Computational Physics*, *268*, 17–38.

Tadmor, E. (1984). Skew-selfadjoint form for systems of conservation laws. *Journal of Mathematical Analysis and Applications*, *103*(2), 428–442.

Tadmor, E. (1987). Entropy functions for symmetric systems of conservation laws. *Journal of Mathematical Analysis and Applications*, *122*(2), 355–359.

Tadmor, E. (2003) Entropy stability theory for difference approximations of nonlinear conservation laws and related time-dependent problems. *Acta Numerica*, *12*, 451–512, 5 (2003).

Tadmor, E. (2016). Perfect derivatives, conservative differences and entropy stable computation of hyperbolic conservation laws. *Discrete and Continuous Dynamical Systems-A*, *36*(8), 4579–4598.

Tadmor, E., & Zhong, W. (2006). Entropy stable approximations of Navier-Stokes equations with no artificial numerical viscosity. *Journal of Hyperbolic Differential Equations*, *3*(3), 529–559.

Wang, Z. J., Fidkowski, K., Abgrall, R., Bassi, F., Caraeni, D., Cary, A., et al. (2013). High-order CFD methods: current status and perspective. *International Journal for Numerical Methods in Fluids*, *72*(8), 811–845.

Wilcox, L. C., Stadler, G., Burstedde, C., & Ghattas, O. (2010). A high-order discontinuous Galerkin method for wave propagation through coupled elastic-acoustic media. *Journal of Computational Physics*, *229*(24), 9373–9396.

Wintermeyer, N., Winters, A. R., Gassner, G. J., & Kopriva, D. A. (2017). An entropy stable nodal discontinuous Galerkin method for the two dimensional shallow water equations on unstructured curvilinear meshes with discontinuous bathymetry. *Journal of Computational Physics*, *340*, 200–242.

Wintermeyer, N., Winters, A. R., Gassner, G. J., & Warburton, T. (2018). An entropy stable discontinuous Galerkin method for the shallow water equations on curvilinear meshes with wet/dry fronts accelerated by GPUs. *Journal of Computational Physics*, *375*, 447–480.

Winters, A. R., & Gassner, G. J. (2016). Affordable, entropy conserving and entropy stable flux functions for the ideal MHD equations. *Journal of Computational Physics*, *301*, 72–108.

Winters, A. R., Derigs, D., Gassner, G. J., & Walch, S. (2017). A uniquely defined entropy stable matrix dissipation operator for high Mach number ideal MHD and compressible Euler simulations. *Journal of Computational Physics*, *332*, 274–289.

Winters, A. R., Moura, R. C., Mengaldo, G., Gassner, G. J., Walch, S., Peiro, J., et al. (2018). A comparative study on polynomial dealiasing and split form discontinuous Galerkin schemes for under-resolved turbulence computations. *Journal of Computational Physics*, *372*, 1–21.

Winters, A. R., Czernik, C., Schily, M. B., & Gassner, G. J. (2019). Entropy stable numerical approximations for the isothermal and polytropic Euler equations. *BIT Numerical Mathematics*, 1–34.

Wu, K., & Shu, C.-W. (2019) Entropy symmetrization and high-order accurate entropy stable numerical schemes for relativistic MHD equations. arXiv:1907.07467.

Xie, Z., Wang, L.-L., & Zhao, X. (2013). On exponential convergence of gegenbauer interpolation and spectral differentiation. *Mathematics of Computation*, *82*, 1017–1036.

Chapter 4
p-Multigrid High-Order Discontinuous Galerkin Solution of Compressible Flows

A. Colombo, A. Ghidoni, G. Noventa, and S. Rebay

Abstract Discontinuous finite element methods are finding widespread use in a wide range of scientific and technical applications since they are among the few available methods for the approximation of partial differential problems that combines high-order accuracy, geometric flexibility, and robustness. The price to pay for the robustness, accuracy, and flexibility of these methods is their high computational cost and storage requirement. However, the computational efficiency of discontinuous finite element methods can be substantially improved by resorting to multilevel solution techniques. This chapter presents the application of a p-multigrid high-order accurate discontinuous finite element method to the numerical solution of compressible laminar viscous flows (compressible Navier–Stokes equations) and to compressible turbulent flows modeled with the Reynolds-Averaged Navier–Stokes equations coupled with the k-ω turbulence model.

Introduction

In discontinuous Galerkin (DG) methods, similarly to the classical "continuous" finite element method (FEM), the solution of the weak or variational form of a partial differential problem is approximated by polynomial functions over the elements of a suitably defined grid. However, unlike continuous FEM, in DG methods the approximation is in general discontinuous at the element interfaces, and the coupling of the approximate solution between neighboring elements is (weakly) enforced by

---

A. Colombo
Department of Engineering and Applied Sciences, Università degli Studi di Bergamo,
Bergamo, Italy

A. Ghidoni · G. Noventa · S. Rebay (✉)
Department of Mechanical and Industrial Engineering, Università degli Studi di Brescia,
Brescia, Italy
e-mail: stefano.rebay@unibs.it

M. Kronbichler and P.-O. Persson (eds.), *Efficient High-Order Discretizations for Computational Fluid Dynamics*, CISM International Centre for Mechanical Sciences 602, https://doi.org/10.1007/978-3-030-60610-7_4

interface (or numerical) flux functions. An appropriate definition of the numerical flux function guarantees the consistency and stability of the DG numerical approximation.

DG schemes are currently finding use in very diverse applications (Li and Shu 2005; Persson et al. 2009; Bernard et al. 2009; Bassi et al. 2015a, b; Noventa et al. 2016; Flad et al. 2016; Frère et al. 2018) since they are among the few approximation techniques that combine high-order accuracy, geometrical flexibility and robustness in a single method. However, the drawback of robustness, accuracy, and flexibility is the high computational cost and storage requirement of these methods.

As a consequence, a substantial research effort has been spent to enhance the computational efficiency of high-order DG solvers. Many computational strategies have been investigated for the efficient assembly of DG space discretization operators (Warburton et al. 1999; Cockburn et al. 2009; Kopriva and Gassner 2010; Bassi et al. 2013; Kronbichler and Wall 2018), for the time integration of the space discretized DG equations by means of implicit time, and possibly adaptive, discretization schemes (Bassi and Rebay 2000; Persson and Peraire 2008; Diosady and Darmofal 2009; Crivellini and Bassi 2011; Pazner and Persson 2018; Massa et al. 2018; Noventa et al. 2020; Bassi et al. 2020a, b) and multigrid (MG) solution strategies (Bassi and Rebay 2002; Helenbrook et al. 2003; Fidkowski et al. 2005; Luo et al. 2006; Nastase and Mavriplis 2006; Klaji et al. 2007; van der Vegt and Rhebergen 2012a, b; Franciolini et al. 2020), both in the h and p variants.

In this work, we are interested to devise an efficient approach for the solution of steady-state problems. In this context, implicit methods can decrease drastically the iterations number to reach the convergence, but require a huge amount of memory to store the Jacobian and/or the preconditioner matrix for the solution of the linear system arising at each time step, which may become prohibitive for realistic problems and higher polynomial approximations. Multigrid methods are potentially more efficient in terms of computing time and memory requirement than implicit methods. Two multigrid strategies can be considered, i.e., h- and the p-MG. While in the classical h-MG method the discrete equations are solved on a series of recursively coarsened grids, in the p-MG algorithm the equations are solved by considering a series of progressively lower order approximations on the same grid.

The basic idea of MG strategies, i.e., using low-order approximations to correct high-order solutions, was initially exploited for the solution of elliptic equation with spectral finite elements methods. In particular, Zang et al. (1982) and Streett et al. (1985) demonstrated the improvement over simple iterative smoothers given by the coupling of a multigrid approach to Fourier spectral approximations with periodic and Dirichlet boundary conditions. A more general multigrid approach was introduced by Rønquist and Patera (1989). More recently, p-MG was also applied in the DG framework, as reported by Helenbrook et al. (2003), who analyzed the coupling of p-MG and DG in one and two dimensions, and Darmofal and Fidkowski (2004), who presented p-MG results for the 2-D compressible Euler equations. The use of different smoothers, i.e., element/element-line Jacobi (Mascarenhas et al. 2007) and implicit/explicit (Luo et al. 2006; Bassi et al. 2009), has been also investigated. p-MG has been also adopted in DG solvers for the solution of the compressible Navier–Stokes (Fidkowski et al. 2005; Shahbazi et al. 2009; Bassi et al. 2011a; Ghidoni et al.

2014) and RANS (Luo et al. 2012; Ghidoni et al. 2007, 2012b; Bassi et al. 2015a; Wallraff et al. 2013; Jiang et al. 2015; Rueda-Ramírez et al. 2019) equations.

In this work, we consider the nonlinear full approximation scheme (FAS) p-MG algorithm for the solution of the Navier–Stokes and RANS (with the k-ω turbulence model) equations. Different smoothers, e.g., (i) block implicit and (ii) line implicit Runge–Kutta, (iii) linearized Backward Euler schemes, are investigated in terms of computational efficiency (CPU time and memory usage), and compared with an implicit solver. The performance of the proposed p-MG solution strategy is here investigated by computing many complex 3-D test cases: the compressible laminar flow around a Delta Wing and a 3-D streamlined body (these test cases have been defined during the ADIGMA EU project (Kroll 2009)), and the compressible turbulent flow around a Delta Wing, a train head, and through a turbine cascade (these test cases have been defined during the IDIHOM EU project (IDIHOM 2015)).

Compressible RANS Equations

The Reynolds-Averaged Navier–Stokes equations and k-$\widetilde{\omega}$ turbulence model equations can be written as

$$\frac{\partial \rho}{\partial t} + \frac{\partial}{\partial x_j}(\rho u_j) = 0, \tag{1}$$

$$\frac{\partial}{\partial t}(\rho u_i) + \frac{\partial}{\partial x_j}(\rho u_j u_i) = -\frac{\partial p}{\partial x_i} + \frac{\partial \widehat{\tau}_{ji}}{\partial x_j}, \tag{2}$$

$$\frac{\partial}{\partial t}(\rho E) + \frac{\partial}{\partial x_j}(\rho u_j H) = \frac{\partial}{\partial x_j}\left[u_i \widehat{\tau}_{ij} - \widehat{q}_j\right] - \tau_{ij}\frac{\partial u_i}{\partial x_j} + \beta^* \rho \overline{k} e^{\widetilde{\omega}_r}, \tag{3}$$

$$\frac{\partial}{\partial t}(\rho k) + \frac{\partial}{\partial x_j}(\rho u_j k) = \frac{\partial}{\partial x_j}\left[(\mu + \sigma^* \overline{\mu}_t)\frac{\partial k}{\partial x_j}\right] + \tau_{ij}\frac{\partial u_i}{\partial x_j} - \beta^* \rho \overline{k} e^{\widetilde{\omega}_r}, \tag{4}$$

$$\begin{aligned}\frac{\partial}{\partial t}(\rho \widetilde{\omega}) + \frac{\partial}{\partial x_j}(\rho u_j \widetilde{\omega}) = {} & \frac{\partial}{\partial x_j}\left[(\mu + \sigma \overline{\mu}_t)\frac{\partial \widetilde{\omega}}{\partial x_j}\right] + \\ & + \frac{\alpha}{\overline{k}}\tau_{ij}\frac{\partial u_i}{\partial x_j} - \beta \rho e^{\widetilde{\omega}_r} + \left(\mu + \sigma \overline{\mu}_t\right)\frac{\partial \widetilde{\omega}}{\partial x_k}\frac{\partial \widetilde{\omega}}{\partial x_k},\end{aligned} \tag{5}$$

where ρ denotes the density, u_j the Cartesian components of the velocity vector, $E = e + \frac{1}{2}u_k u_k$ the total energy per unit mass, $H = h + \frac{1}{2}u_k u_k$ the total enthalpy per unit mass, k the turbulent kinetic energy and $\widetilde{\omega}$ the logarithm of the turbulence dissipation rate. The pressure p, the turbulent and total stress tensors τ_{ij} and $\widehat{\tau}_{ij}$, the heat flux vector $\widehat{q}_j$, the eddy viscosity coefficient $\overline{\mu}_t$ and the limited value of turbulent kinetic energy $\overline{k}$ are given by

$$p = (\gamma - 1)\rho(E - u_k u_k/2)\,, \tag{6}$$

$$\tau_{ij} = 2\overline{\mu}_t\left[S_{ij} - \frac{1}{3}\frac{\partial u_k}{\partial x_k}\delta_{ij}\right] - \frac{2}{3}\rho\overline{k}\delta_{ij}, \tag{7}$$

$$\widehat{\tau}_{ij} = 2\mu\left[S_{ij} - \frac{1}{3}\frac{\partial u_k}{\partial x_k}\delta_{ij}\right] + \tau_{ij}, \tag{8}$$

$$\widehat{q}_j = -\left(\frac{\mu}{\mathrm{Pr}} + \frac{\overline{\mu}_t}{\mathrm{Pr}_t}\right)\frac{\partial h}{\partial x_j}, \tag{9}$$

$$\overline{\mu}_t = \alpha^*\rho\overline{k}e^{-\widetilde{\omega}_r}, \quad \overline{k} = \max(0, k)\,. \tag{10}$$

The symbol γ denotes the ratio between the constant pressure and constant volume specific heats, $\gamma \overset{\mathrm{def}}{=} C_p/C_v$, Pr and Pr_t are the molecular and turbulent Prandtl numbers, and

$$S_{ij} = \frac{1}{2}\left(\frac{\partial u_i}{\partial x_j} + \frac{\partial u_j}{\partial x_i}\right)$$

is the mean strain-rate tensor. The closure parameters α, α^*, β, β^*, σ, σ^* are those of the high-Reynolds number k-ω model of Wilcox (Wilcox 1993).

The turbulence k-$\widetilde{\omega}$ model given by Eqs. (4) and (5) is that described in Bassi et al. (2005). Its main differences with respect to the standard k-ω turbulence model can be briefly summarized as follows. First of all, the variable $\widetilde{\omega} = \log(\omega)$ is used in place of ω in order to guarantee the positivity and in order to improve the resolution of the rapid variation of this variable close to solid walls. Moreover, limited values $\overline{k}$ and $\overline{\mu}_t$ appear in Eqs. (3)–(5) to deal with possible negative values of the turbulent kinetic energy. Finally, a constrained $\widetilde{\omega}_r$ that sets a lower bound on $\widetilde{\omega}$ by fulfilling to suitably defined "realizability conditions" appears in the source terms and in the eddy viscosity equation.

The "slightly-rough-wall" boundary condition is adopted to prescribe the value of $\widetilde{\omega}$ at the wall. In particular the approach proposed in Menter (1994) is modified to introduce a dependence on the polynomial degree ℓ of the solution. The $\widetilde{\omega}$ value at the wall is computed by using the Taylor series expansion of the near wall solution $\widetilde{\omega}(y)$ as a function of the distance y in the direction normal to the wall (which gives the correct behavior $\widetilde{\omega}(y) \to \infty$ as $y \to 0$). A finite value $\widetilde{\omega}_w^\ell$ at the wall is obtained by evaluating the Taylor series truncated to ℓ terms at $y = h$, where h denotes the distance from the wall of the center of the element adjacent to the wall, thus obtaining the finite value

$$\widetilde{\omega}_w^\ell = \log\left(\frac{6\nu_w}{\beta\left(\alpha^\ell h\right)^2}\right) \qquad \text{where} \qquad \alpha^\ell = \exp\left(-\sum_{n=1}^{\ell}\frac{1}{n}\right). \tag{11}$$

Discontinuous Galerkin Approximation of the RANS and k-$\widetilde{\omega}$ Turbulence Model Equations

The system of RANS and k-$\widetilde{\omega}$ turbulence model equations can be written in compact form as

$$\frac{\partial \mathbf{q}}{\partial t} + \nabla \cdot \mathbf{F}^{\mathrm{c}}(\mathbf{q}) + \nabla \cdot \mathbf{F}^{\mathrm{v}}(\mathbf{q}, \nabla \mathbf{q}) + \mathbf{s}(\mathbf{q}, \nabla \mathbf{q}) = \mathbf{0} \ , \tag{12}$$

where $\mathbf{q} \in \mathbb{R}^m$ denotes the unknown vector of conservative variables, $\mathbf{F}^{\mathrm{c}} \in \mathbb{R}^m \otimes \mathbb{R}^d$ and $\mathbf{F}^{\mathrm{v}} \in \mathbb{R}^m \otimes \mathbb{R}^d$ are the inviscid and viscous flux functions, $\mathbf{s} \in \mathbb{R}^m$ the sum of turbulence source and volume forces vectors, m is the number of variables and d the number of space dimensions ($m = 7$ for the 3-D compressible RANS equations with a two equation turbulence model closure). Instead of solving the RANS equations for the unknown vector

$$\mathbf{q} = [\rho, \rho u_1, \rho u_2, \rho u_3, \rho E, \rho k, \rho\widetilde{\omega}]^T \ ,$$

the vector of "primitive variables"

$$\mathbf{w} = [p, u_1, u_2, u_3, T, k, \widetilde{\omega}]^T$$

is adopted. The system of governing equations (12) for the new set of variables $\mathbf{w}$ can be written as

$$\mathbf{P}(\mathbf{w}) \frac{\partial \mathbf{w}}{\partial t} + \nabla \cdot \mathbf{F}^{\mathrm{c}}(\mathbf{w}) + \nabla \cdot \mathbf{F}^{\mathrm{v}}(\mathbf{w}, \nabla \mathbf{w}) + \mathbf{s}(\mathbf{w}, \nabla \mathbf{w}) = \mathbf{0} \ , \tag{13}$$

where

$$\mathbf{P}(\mathbf{w}) = \frac{\partial \mathbf{q}(\mathbf{w})}{\partial \mathbf{w}} \in \mathbb{R}^m \otimes \mathbb{R}^m$$

denotes the Jacobian matrix of the transformation $\mathbf{q} = \mathbf{q}(\mathbf{w})$, and where for simplicity we have used the same symbols $\mathbf{F}^{\mathrm{c}}(\cdot)$, $\mathbf{F}^{\mathrm{v}}(\cdot, \cdot)$ and $\mathbf{s}(\cdot, \cdot)$ in (12) and (13), even if obviously the flux and source terms appearing in (13) are functionally different from those in (12). This will not cause any confusion since in the following our discussion will be entirely focused on system (13).

Space Discretization

In order to construct the DG discretization of the governing equations, the weak formulation

$$\int_{\Omega} \mathbf{v} \cdot \left(\mathbf{P}\left(\mathbf{w}\right) \frac{\partial \mathbf{w}}{\partial t} \right) \, \mathrm{d}\Omega - \int_{\Omega} \nabla \mathbf{v} : \left(\mathbf{F}^{\mathrm{c}}\left(\mathbf{w}\right) + \mathbf{F}^{\mathrm{v}}\left(\mathbf{w}, \nabla \mathbf{w}\right) \right) \, \mathrm{d}\Omega$$

$$+ \int_{\partial\Omega} \mathbf{v} \otimes \boldsymbol{n} : \left(\mathbf{F}^{\mathrm{c}}\left(\mathbf{w}\right) + \mathbf{F}^{\mathrm{v}}\left(\mathbf{w}, \nabla \mathbf{w}\right) \right) \, \mathrm{d}\sigma + \int_{\Omega} \mathbf{v} \cdot \mathbf{s}\left(\mathbf{w}, \nabla \mathbf{w}\right) \, \mathrm{d}\Omega = \mathbf{0} \qquad (14)$$

of system (13) is considered, where $\mathbf{v} = \{v_1, \ldots, v_m\}$ denotes an arbitrary smooth vector test functions and $\boldsymbol{n}$ the outward pointing unit normal vector to the boundary $\partial\Omega$.

A discrete version of (14) is obtained by approximating the domain Ω by a grid $\mathcal{T}_h = \{K\}$ consisting of a set of non-overlapping elements, and by restricting $\mathbf{w}$ and $\mathbf{v}$ to finite dimensional functions $\mathbf{w}_h$ and $\mathbf{v}_h$ which are piecewise polynomial inside the elements $K \in \mathcal{T}_h$ and generally discontinuous at the element interfaces. In other words, we are looking for an approximate solution $\mathbf{w}_h$ belonging to the discontinuous finite element space $\mathbf{V}_{h,d}^{m,\ell}$ spanned by d-dimensional polynomial functions of degree at most ℓ inside each element K, namely,

$$\mathbf{V}_{h,d}^{m,\ell} \stackrel{\text{def}}{=} [\mathbb{P}_d^{\ell}(\mathcal{T}_h)]^m \qquad (15)$$

$$\mathbb{P}_d^{\ell}(\mathcal{T}_h) \stackrel{\text{def}}{=} \left\{ v_h \in L^2(\Omega_h) \, : \, v_h|_K \in \mathbb{P}_d^{\ell}(K), \, \forall K \in \mathcal{T}_h \right\}. \qquad (16)$$

A convenient set of basis functions for $\mathbf{V}_h^{\ell}$, where from now on subscript d and superscript m will be omitted in $\mathbf{V}_{h,d}^{m,\ell}$ to simplify the notation, is the set of orthogonal and hierarchical polynomials defined in physical space described, for example, in Bassi et al. (2012). If $\varphi_i(\mathbf{x})$, $i = 1, \ldots, N_b$ and $\mathbf{x} \in K$, denote the basis functions of $\mathbf{V}_h^{\ell}$ for an arbitrary element K, the αth component $w_{h,\alpha}(\mathbf{x}, t)$ and $v_{h,\alpha}(\mathbf{x})$ of the vector functions $\mathbf{w}_h(\mathbf{x}, t)$ and $\mathbf{v}_h(\mathbf{x})$, respectively, can be written on K as

$$w_{h,\alpha}(\mathbf{x}, t) = \sum_{i=1}^{N_b} \varphi_i(\mathbf{x}) w_{\alpha i}(t) \qquad v_{h,\alpha}(\mathbf{x}) = \sum_{i=1}^{N_b} \varphi_i(\mathbf{x}) v_{\alpha i} \qquad \mathbf{x} \in K \ ,$$

where $w_{\alpha i}(t)$ and $v_{\alpha i}$ are the expansion coefficients, also called "degrees of freedom" (DoF), of $w_{h,\alpha}(\mathbf{x}, t)$ and $v_{h,\alpha}(\mathbf{x})$, respectively.

Since $\mathbf{w}_h(\mathbf{x}, t)$ and $\mathbf{v}_h(\mathbf{x})$ are linear in $\varphi_i(\mathbf{x})$ and the weak formulation is linear in $\mathbf{v}_h(\mathbf{x})$, Eq. (14) will hold for an arbitrary $\mathbf{v}_h \in \mathbf{V}_h^{\ell}$ if it holds for all the basis functions $\varphi_i(\mathbf{x})$ of all the elements $K \in \mathcal{T}_h$. This leads to the system of ordinary differential equations

$$
\begin{aligned}
&\sum_{K\in\mathcal{T}_h}\int_K \varphi_i P_{\alpha\beta}(\mathbf{w}_h)\,\varphi_j \frac{\mathrm{d}w_{\beta j}}{\mathrm{d}t}\,\mathrm{d}\Omega+\\
&-\sum_{K\in\mathcal{T}_h}\int_K \frac{\partial\varphi_i}{\partial x_k}\left(F^{\mathrm{c}}_{\alpha k}(\mathbf{w}_h)+F^{\mathrm{v}}_{\alpha k}\left(\mathbf{w}_h,\nabla_h\mathbf{w}_h+\mathbf{r}^g_h([\![\mathbf{w}_h]\!])\right)\right)\mathrm{d}\Omega+\\
&+\sum_{f\in\mathcal{F}}\int_f [\![\varphi_i]\!]_k\left(\widehat{F}^{\mathrm{c}}_{\alpha k}\left(\mathbf{w}_h^{\pm}\right)+\widehat{F}^{\mathrm{v}}_{\alpha k}\left(\mathbf{w}_h^{\pm},\left(\nabla_h\mathbf{w}_h+\eta_f\mathbf{r}^f_h([\![\mathbf{w}_h]\!])\right)^{\pm}\right)\right)\mathrm{d}\sigma+\\
&+\sum_{K\in\mathcal{T}_h}\int_K \varphi_i s_\alpha\left(\mathbf{w}_h,\nabla_h\mathbf{w}_h+\mathbf{r}^g_h([\![\mathbf{w}_h]\!])\right)\mathrm{d}\Omega=0\ ,
\end{aligned}
\tag{17}
$$

where repeated indices indicate an implicit sum over the repeated index range, namely, $1,\ldots,N_b$ for indices i and j, $1,\ldots,d$ for index k, and $1,\ldots,m$ for indices α and β.

The symbols f and $\mathcal{F}$ appearing in the third summation of Eq. (17) denote an edge of $\mathcal{T}_h$ (boundary or internal) and the collection of all the edges of $\mathcal{T}_h$, respectively. The superscript $\pm$ indicates the trace of the solution of the two adjacent elements sharing face f. The symbol $[\![\cdot]\!]$ denotes the jump trace operator that, for a scalar quantity z, is the vector quantity $[\![z]\!]=z^-\boldsymbol{n}^-+z^+\boldsymbol{n}^+$. The k-th component of the vector quantity $[\![z]\!]$ is denoted by $[\![z]\!]_k$.

The functions $\mathbf{r}^g_h$ and $\mathbf{r}^f_h:[\mathbb{P}^\ell_d(f)]^m\rightarrow[\mathbb{P}^\ell_d(\mathcal{T}_h)]^m$, denote the global and local lifting operators introduced in Bassi et al. (1997), respectively. The local lifting operator is defined for a generic (internal or boundary) face f and for a generic quantity $z_h\in\mathbb{R}^m\otimes\mathbb{R}^d$ defined on f as

$$
\int_{K^-}\boldsymbol{\tau}_h:\mathbf{r}^f(z_h)\,\mathrm{d}\Omega+\int_{K^+}\boldsymbol{\tau}_h:\mathbf{r}^f(z_h)\,\mathrm{d}\Omega=\int_f\{\boldsymbol{\tau}_h\}:z_h\,\mathrm{d}\sigma\ ,
$$

where $K^{\pm}$ are the two adjacent elements sharing face f, $\boldsymbol{\tau}_h\in\mathbf{V}_h^{m\times d}$ is an arbitrary test function, and $\{\cdot\}$ denotes the average trace operator

$$
\{\cdot\}\overset{\text{def}}{=}\frac{(\cdot)^-+(\cdot)^+}{2}\ .
$$

The global lifting $\mathbf{r}^g_h$ is the sum of all the local lifting operators

$$
\mathbf{r}^g_h(z_h)=\sum_{f\in\mathcal{F}}\mathbf{r}^f_h(z_h)\ .
$$

The inviscid and viscous "numerical" flux functions $\widehat{\mathbf{F}}^{\mathrm{c}}$ and $\widehat{\mathbf{F}}^{\mathrm{v}}$ appearing in the third line of (17) are introduced to uniquely define the flux on internal faces of $\mathcal{T}_h$ where the solution is discontinuous, thus ensuring conservation of the resulting numerical approximation, or to weakly prescribe the boundary conditions on the boundary faces of $\mathcal{T}_h$.

Any numerical flux function commonly considered in upwind finite volume methods can be used as inviscid numerical flux $\widehat{\mathbf{F}^{\mathrm{c}}}$. In practice we adopt either the Godunov flux, using the exact Riemann solver of Gottlieb and Groth (1988), or the modified van Leer flux introduced by Hänel et al. (1987).

The viscous numerical flux function $\widehat{\mathbf{F}^{\mathrm{v}}}$ is instead defined as

$$\widehat{\mathbf{F}^{\mathrm{v}}}\left(\mathbf{w}_h^{\pm}, \left(\nabla_h \mathbf{w}_h + \eta_f \mathbf{r}_h^f\left(\llbracket \mathbf{w}_h \rrbracket\right)\right)^{\pm}\right) \stackrel{\mathrm{def}}{=} \left\{\mathbf{F}^{\mathrm{v}}\left(\mathbf{w}_h, \nabla_h \mathbf{w}_h + \eta_f \mathbf{r}_h^f\left(\llbracket \mathbf{w}_h \rrbracket\right)\right)\right\}, \tag{18}$$

where η_f is a stabilization parameter introduced in Arnold et al. (2002) that must be greater or equal to the number of elements adjacent to element K, i.e., $\eta_f \geq 2$ in 1-D, $\eta_f \geq 3$ for 2-D grids of triangles, $\eta_f \geq 4$ for 3-D grids of tetrahedra, etc.

The adopted shock-capturing technique, based on the work presented in Bassi et al. (2010), consists in introducing for each element $K \in \mathcal{T}_h$ an artificial diffusion term that is active only where unphysical oscillations are present.

Computation of the Steady-State Solution

Assembling together all the elemental contributions, the discrete problem corresponding to Eq. (17) can be written as the ordinary differential system (ODS)

$$\mathbf{M_P}\frac{\mathrm{d}\mathbf{W}}{\mathrm{d}t} + \mathbf{R}(\mathbf{W}) = \mathbf{0}\ , \tag{19}$$

where $\mathbf{R}$ is the residuals vector and $\mathbf{M_P}$ is the global block diagonal matrix arising from the discretization of the first integral of Eq. (17). Matrix $\mathbf{M_P}$ couples the degrees of freedom of different variables within the element through the variable transformation matrix $\mathbf{P}$.

The focus of this work is however on steady-state computations, i.e., in finding the vector $\mathbf{W}$ that is the solution of the nonlinear algebraic problem

$$\mathbf{R}(\mathbf{W}) = \mathbf{0}\ . \tag{20}$$

In practice, the steady-state solution $\mathbf{W}$ can be found by integrating (19) in time starting from a suitably guessed initial solution until the steady state is reached. To avoid the CFL stability restrictions of explicit time integration methods, the ODS (19) is advanced in time by means of the linearised backward Euler (LBE) scheme

$$\left[\frac{\mathbf{M}_{\mathbf{P}}^n}{\Delta t} + \frac{\partial \mathbf{R}(\mathbf{W}^n)}{\partial \mathbf{W}}\right]\Delta \mathbf{W}^n = -\mathbf{R}\left(\mathbf{W}^n\right)\ . \tag{21}$$

The linear system arising at each time step in the LBE time discretization (21) is solved using the "matrix-explicit" GMRES (Generalized Minimal RESidual) algorithm. System preconditioning is required to make the convergence behavior of the GMRES solver acceptable for problems of practical interest. The block Jacobi method with one block per process and ILU(0) factorization or the additive Schwartz method (ASM) are usually employed as preconditioner. Linear algebra and parallelization are handled through the PETSc library, the Portable Extensible Toolkit for Scientific Computations of Balay et al. (2001), and the MPI standard for message-passing communication.

The law to compute the time step, which can significantly affect both the efficiency and the robustness of the method, is defined as

$$\Delta t_K = \mathrm{CFL}\frac{h_{K,\mathrm{CFL}}}{c_v + d_v},$$

where

$$c_v = |\mathbf{v}| + a \qquad d_v = 2\frac{\mu_e + \lambda_e}{h_{K,\mathrm{CFL}}} \qquad h_{K,\mathrm{CFL}} = d\frac{\Omega_K}{S_K}$$

denote the convective and diffusive velocities and the reference length of the generic element K, respectively. The coefficients μ_e and λ_e are the effective dynamic viscosity and conductivity, while Ω_K and S_K denote the volume and the surface of K. In the above relations, all quantities depending on $\mathbf{w}_h$ are computed from mean values for element K.

Devising an effective and robust strategy to increase the CFL number as the residual decreases is far from trivial, especially for turbulent computations, and we here describe an empirically determined "CFL law" that in practice works well for a large class of problems. It is based on the L^∞ and L^2 norms of the residual and depends on three user-defined parameters. The first parameter, denoted as $\mathrm{CFL}_{\mathrm{min}}$, simply sets the minimum possible value of CFL number. The second parameter $\mathrm{CFL}_{\mathrm{ord}}$, which introduces a dependence of the computed CFL value on the polynomial order of the DG approximation, is usually set equal to the maximum CFL number dictated by stability for explicit multistage SSP (or TVD) RK schemes. This means setting the parameter $\mathrm{CFL}_{\mathrm{ord}}$ as

$$\mathrm{CFL}_{\mathrm{ord}} = \frac{1}{2k+1},$$

where k is the polynomial order of the DG approximation. The third parameter is an exponent α governing the growth rate of the CFL number (typically $\alpha \leq 1$). The law gives CFL as

$$\mathrm{CFL} = \begin{cases} \mathrm{CFL}_{\mathrm{min}}/\xi^\alpha & \text{if } \xi \leq 1 \\ \mathrm{CFL}_{\mathrm{ord}} + \Phi(\xi)(\mathrm{CFL}_{\mathrm{min}} - \mathrm{CFL}_{\mathrm{ord}}) & \text{if } \xi > 1 \end{cases} \tag{22}$$

with parameter ξ given by

$$\xi \stackrel{\text{def}}{=} \left\{ \begin{array}{ll} \min(1, \xi_2) \text{ if } \xi_\infty \leq 1, & \xi_2 \stackrel{\text{def}}{=} \max\limits_{\alpha=1,\ldots,m} \left(||R_\alpha||_2 / ||R_{\alpha 0}||_2 \right), \\ \xi_\infty \text{ if } \xi_\infty > 1, & \xi_\infty \stackrel{\text{def}}{=} \max\limits_{\alpha=1,\ldots,m} \left(||R_\alpha||_\infty / ||R_{\alpha 0}||_\infty \right), \end{array} \right\}.$$

where $||\cdot||_2$ and $||\cdot||_\infty$ denote the L_2 and L_∞ norms, respectively, R_α denotes the residual vector of the αth equation of the RANS system, and $R_{\alpha 0}$ denotes the corresponding residual at the first iteration. The function $\Phi(\xi)$ appearing in (22) is defined as

$$\Phi(\xi) \stackrel{\text{def}}{=} \exp\left(\frac{\alpha(1-\xi)\text{CFL}_{\text{min}}}{\text{CFL}_{\text{min}} - \text{CFL}_{\text{ord}}} \right).$$

p-Multigrid Solution Strategy

The convergence rate of standard iterative solvers deteriorate after few iterations, due to the frequency content of the error in the solution. The convergence rate is fast only when the error is distributed in the high-frequency modes, but, after the first few iterations, the high frequencies of the error are smoothed out.

To remove all the error modes and enhance the convergence rate, multigrid (MG) strategies (Briggs et al. 2000; Trottenberg et al. 2001) can be adopted, both in the h and/or p variant. In h-MG methods, hierarchical set of nested grids is used to damp the solution error over the whole spectrum. In the p-MG context coarse levels are represented by a sequence of progressively lower order approximations on a single grid (see Fig. 4.1).

The relaxation scheme (smoother), at an arbitrary level, must damp all the error modes which cannot be represented on "coarser" levels, where "coarse" can be a coarser grid in the h-multigrid case or a lower order approximation in the p-multigrid case.

Different paths (V-cycle and W-cycle) can be used to visit the various levels (see Fig. 4.2). At each level, a number ν_1/ν_2 of pre-(bullets) and post-smoothing (circle) iterations are performed prior to restricting/prolongating the solution to the next coarser/finer level To enhance the basic MG algorithm, the Full Multigrid (FMG)

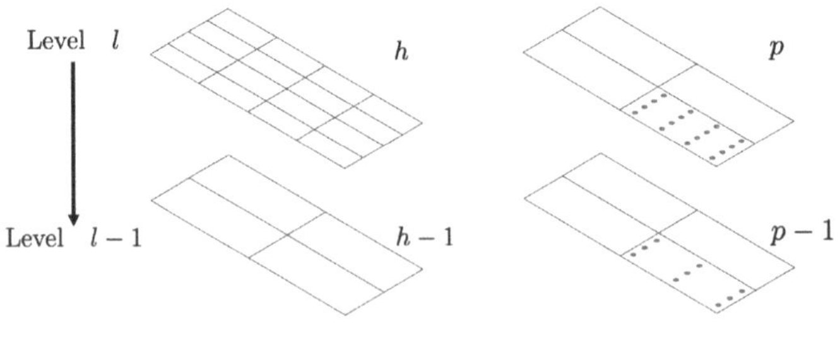

Fig. 4.1 Multigrid scheme between level l and $l-1$ for the h (left) and p (right) variants

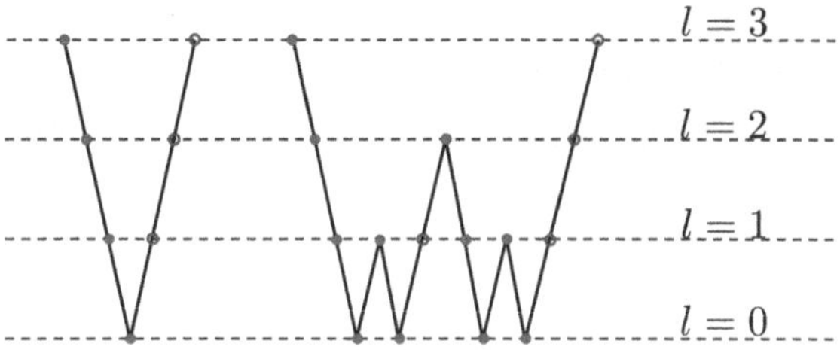

Fig. 4.2 V-cycle and W-cycle for $L = 4$ multigrid levels (•: pre-smoothing; ∘: post-smoothing)

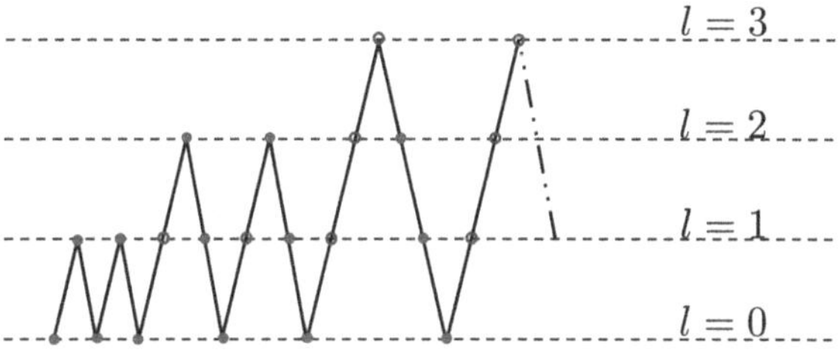

Fig. 4.3 V-cycle full multigrid for $L = 4$ multigrid levels (•: pre-smoothing; ∘: post-smoothing)

algorithm (see Fig. 4.3) can be exploited, where the coarser level solutions are used to initialize the computation of the finer grids. It is not necessary to compute the fully converged solution on each level before switching to the next finer level, since the corresponding discretization error can be relatively large. To save computing time, a residual-based criterion can be adopted, which allows prolongating the solution $\mathbf{W}^l$ to the next finer level $l + 1$ when $(\max(||\mathbf{R}^l||_2) < 10^{-2})$, where $||\mathbf{R}^l||_2$ is the L^2 norm of the residual vector.

The full approximation MG scheme (NMG) adopted to solve a generic nonlinear problem $\mathbf{A}^l(\mathbf{W}^l) = \mathbf{b}^l$ is described by the recursive Algorithm 1, where the superscript l denotes a "level" that can represent the grid size h or the polynomial order p for the h- and p-MG cases, respectively. In practice, $\mathbf{b}^l - \mathbf{A}^l(\mathbf{W}^l) \equiv \mathbf{R}^l(\mathbf{W}^l) = \mathbf{0}$ is the nonlinear algebraic problem to be solved for a steady-state flow computation, as shown in Eq. (20).

In the Algorithm 1 ν_1 and ν_2 are the number of pre- and post-iterations, l the current level, l_{min} the lowest level, $\mathbf{s}$ the forcing term, $\widetilde{\mathbf{I}}_l^{l-1}$ and $\mathbf{I}_l^{l-1}$ the solution and the residual restriction operators, $\widetilde{\mathbf{I}}_{l-1}^{l}$ the error prolongation operator. The parameter τ allows to choose between the V-cycle ($\tau = 1$), and the W-cycle ($\tau = 2$). Different relaxation schemes can be employed as nonlinear smoother ($SMOOTH$). More details about smoothers are given is Sect. 4.5.

The system matrix, $\mathbf{A}^l$, can be computed on each level or only on the finest level and projected on the lower levels with the matrix restriction operator, $\widehat{\mathbf{I}}_l^{l-1}$, depending on the type of smoother adopted at each level (implicit, block implicit, and line implicit).

Algorithm 1 $NMG(l, l_{min}, \tau, \nu_1, \nu_2, \mathbf{A}^l, \mathbf{W}^l, \mathbf{b}^l)$

if $(l = l_{min})$ **then**
  Solve $\mathbf{A}^l_{min}(\mathbf{W}^l_{min}) = \mathbf{b}^l_{min}$
  $\widetilde{\mathbf{W}}^l_{min} = \mathbf{W}^l_{min}$
else if $(l > l_{min})$ **then**
  for $i = 1$ to ν_1 **do** ▷ pre-smoothing
    $\mathbf{W}^l = SMOOTH^l(\mathbf{A}^l, \mathbf{W}^l, \mathbf{b}^l)$
  end for
  $\mathbf{W}^{l-1}_0 = \widetilde{\mathbf{I}}^{l-1}_l \mathbf{W}^l$
  $\mathbf{s}^{l-1} = \mathbf{A}^{l-1}(\widetilde{\mathbf{W}}^{l-1}_0) + \mathbf{I}^{l-1}_l(\mathbf{b}^l - \mathbf{A}^l(\mathbf{W}^l))$
  for $j = 1$ to τ **do** ▷ recursion
    $\mathbf{W}^{l-1}_j = NMG(l-1, l_{min}, \tau, \nu_1, \nu_2, \mathbf{A}^{l-1}, \mathbf{W}^{l-1}_{j-1}, \mathbf{s}^{l-1})$
  end for
  $\mathbf{e}^{l-1} = \mathbf{W}^{l-1}_\tau - \mathbf{W}^{l-1}_0$ ▷ correction
  $\widetilde{\mathbf{W}}^l = \mathbf{W}^l + \widetilde{\mathbf{I}}^l_{l-1}\mathbf{e}^{l-1}$
  for $i = 1$ to ν_2 **do** ▷ post-smoothing
    $\widetilde{\mathbf{W}}^l = SMOOTH^l(\mathbf{A}^l, \widetilde{\mathbf{W}}^l, \mathbf{b}^l)$
  end for
end if
return $\widetilde{\mathbf{W}}^l$

Solution and Error Transfer Operators

The solution restriction and prolongation operators are simply L^2 projections onto the spaces $\mathbf{V}^{l-1}_h$ and $\mathbf{V}^l_h$, respectively.

The low-order level solution is obtained from the high-order solution by requiring that

$$\int_\Omega \mathbf{w}^{l-1}_h \mathbf{v}^{l-1}_h d\Omega = \int_\Omega \mathbf{w}^{l-1}_h \mathbf{v}^p_h d\Omega, \quad \forall \mathbf{w}^{l-1}_h \in \mathbf{V}^{l-1}_h. \tag{23}$$

By introducing the matrices

$$\mathbf{M}^{l-1} = \left[M_{ij}\right]^{l-1} = \int_\Omega \phi^{l-1}_i \phi^{l-1}_j d\Omega, \qquad \mathbf{M}^{l-1}_l = \left[M_{ij}\right]^{l-1}_l = \int_\Omega \phi^{l-1}_i \phi^l_j d\Omega,$$

Equation (23) can be rewritten as

$$\mathbf{M}^{l-1}\mathbf{v}^{l-1} = \mathbf{M}^{l-1}_l \mathbf{v}^{l-1},$$

which shows that $\mathbf{v}^{l-1}$ can be obtained from $\mathbf{v}^l$ as

$$\mathbf{v}^{l-1} = \widetilde{\mathbf{I}}^{l-1}_l \mathbf{v}^l, \qquad \widetilde{\mathbf{I}}^{l-1}_l \equiv (\mathbf{M}^{l-1})^{-1}\mathbf{M}^{l-1}_l, \tag{24}$$

which represents the solution restriction operator $\widetilde{\mathbf{I}}^{l-1}_l$. In a similar fashion, the L^2 prolongation of $\mathbf{e}^{l-1}$ to the higher order space $\mathbf{V}^l_h$ can be defined as

$$\mathbf{M}^l \mathbf{e}^l = \mathbf{M}^l_{l-1} \mathbf{e}^{l-1},$$

and consequently

$$\mathbf{e}^l = \widetilde{\mathbf{I}}^l_{l-1} \mathbf{e}^{l-1}, \qquad \widetilde{\mathbf{I}}^l_{l-1} \equiv (\mathbf{M}^l)^{-1} \mathbf{M}^l_{l-1} = (\mathbf{M}^l)^{-1} (\mathbf{M}^{l-1}_l)^T. \tag{25}$$

The choice of orthogonal and hierarchical modal expansion bases for the DG discretization allows to write the operators $\widetilde{\mathbf{I}}^{l-1}_l$ and $\widetilde{\mathbf{I}}^l_{l-1}$ as

$$\widetilde{\mathbf{I}}^{l-1}_l = \delta_{l,l-1}, \qquad \widetilde{\mathbf{I}}^l_{l-1} = \delta_{l-1,l}, \tag{26}$$

where δ_{ij} is the Kronecker symbol. In practice the dofs of the restricted solution are equal to the low-order subset of their high-order representations, while the dofs of the prolongated error are the same as the low-order error with null high-order components.

Residual and Matrix Restriction Operator

The residual operator at level l can be written as $\mathbf{R}^l_j = B(\phi^l_j, \mathbf{v}^l)$, where $B(\cdot,\cdot)$ is linear in its first argument. The restricted residual $\mathbf{R}^{l-1}$ is defined as $\mathbf{R}^{l-1}_j \equiv B(\phi^{l-1}_j, \mathbf{v}^l)$. The approach proposed by Fidkowski, see, e.g., Fidkowski et al. (2005), has been used to obtain an explicit expression of the residual restriction operator $\mathbf{I}^{l-1}_l$. If the space $\mathbf{V}^{l-1}_h$ is a subspace of $\mathbf{V}^l_h$, a linear relation exists between the basis functions ϕ^{l-1} and ϕ^l, which can be exploited to define any basis function ϕ^{l-1} as a linear combination of the basis functions ϕ^l_j

$$\phi^{l-1}_i = \alpha_{ij} \phi^l_j, \tag{27}$$

where α_{ij} are constant coefficients. The matrix $\boldsymbol{\alpha} = [\alpha_{ij}]$ can be related to the solution prolongation operator $\widetilde{\mathbf{I}}^l_{l-1}$ by considering Eq. (25) and by expressing ϕ^{l-1}_i in terms of ϕ^l_i using Eq. (27), thus obtaining

$$\begin{aligned} \widetilde{\mathbf{I}}^l_{l-1} = (\mathbf{M}^l)^{-1} \mathbf{M}^l_{l-1} &= (\mathbf{M}^l)^{-1} \int_\Omega \phi^l_i \phi^{l-1}_j d\Omega, \\ &= (\mathbf{M}^l)^{-1} \alpha_{jk} \int_\Omega \phi^l_i \phi^l_k d\Omega, = (\mathbf{M}^l)^{-1} \mathbf{M}^l \alpha^T = \boldsymbol{\alpha}^T. \end{aligned} \tag{28}$$

The linearity of $B(\cdot,\cdot)$ in its first argument can be used to write

$$\mathbf{R}^{l-1}_i = B(\phi^{l-1}_i, \mathbf{v}^l) = B(\alpha_{ij}\phi^l_j, \mathbf{v}^l) = \alpha_{ij} B(\phi^l_j, \mathbf{v}^l) = \alpha_{ij} \mathbf{r}^l_j, \tag{29}$$

i.e., in matrix notation

$$\mathbf{r}^{l-1} = \alpha \mathbf{r}^{l} = \left(\widetilde{\mathbf{I}}_{l-1}^{l}\right)^{T} \mathbf{r}^{l},$$

which shows that in fact $\mathbf{I}_{l}^{l-1} = (\widetilde{\mathbf{I}}_{l-1}^{l})^{T}$.

The choice of orthogonal and hierarchical modal expansion bases for the DG discretization allows to write the operators $\mathbf{I}_{l}^{l-1}$ as

$$\mathbf{I}_{l}^{l-1} = \delta_{l,l-1}, \tag{30}$$

where δ_{ij} is the Kronecker symbol. In practice the dofs of the restricted residual are equal to the low-order subset of its high-order representation.

The Galerkin coarse grid operator is chosen as matrix restriction operator, and is defined as

$$\widehat{\mathbf{I}}_{l}^{l-1} = \mathbf{I}_{l}^{l-1} [\quad] \widetilde{\mathbf{I}}_{l-1}^{l}, \tag{31}$$

where [ ] represents the matrix to be restricted.

Smoothers

The choice of smoother heavily influences the convergence rate of the multigrid algorithm. The smoothers most commonly considered in the literature for p-multigrid DG approximations include explicit Runge–Kutta (RK) schemes (Bassi et al. 2011b), element-block Jacobi/Gauss–Seidel, element-line, and alternating-direction implicit (ADI) iterations, incomplete lower–upper (ILU) decomposition, and multistage schemes (Fidkowski et al. 2005; Helenbrook and Atkins 2005). As a general rule, explicit smoothers are the less demanding in terms of memory requirements but also show slow convergence rates, particularly for stiff problems. Fully implicit methods, on the other hand, have higher memory requirements but offer better convergence rates.

In this work, we consider three different smoothers, (1) a Line-Implicit RK scheme (LIRK), (2) a Block-Implicit RK scheme (BIRK), and (3) the linearized Backward Euler scheme (LBE). A detailed description and an assessment of the performance offered by the LIRK and BIRK RK schemes is given in Ghidoni et al. (2014).

The LIRK smoother is displayed in Algorithm 2. At each MG cycle, the matrix

$$\left[\frac{\mathbf{M_P}}{\Delta t} + \alpha_m \mathbf{T}(\mathbf{W}^0)\right],$$

where $\mathbf{T}(\mathbf{W}^0)$ is the block tridiagonal matrix associated to a line, is computed only for the first RK stage. Moreover, this smoother is used only on the finest level since, as demonstrated in Ghidoni et al. (2014), the use on coarser levels has very limited effect on the convergence rate but increases the computational cost.

Algorithm 2 The Line-Implicit Runge-Kutta (LIRK) scheme

1: $\mathbf{W}^0 = \mathbf{W}^n$
2: **for** $k = 1, m$ **do**
3: Solve $\left[\frac{\mathbf{M_P}}{\Delta t} + \alpha_m \mathbf{T}\left(\mathbf{W}^0\right)\right] \delta\mathbf{W}^k = -\alpha_k \mathbf{R}\left(\mathbf{W}^{k-1}\right)$
4: $\mathbf{W}^k = \mathbf{W}^{k-1} + \delta\mathbf{W}^k$
5: **end for**
6: return $\mathbf{u}^{n+1} = \mathbf{u}^m$

The BIRK scheme is instead described by Algorithm 3, where $\mathbf{D}(\mathbf{W}^0)$ is the block diagonal part of the full Jacobian matrix. On the finest level, the matrix

$$\left[\frac{\mathbf{M_P}}{\Delta t} + \alpha_m \mathbf{D}\left(\mathbf{W}^0\right)\right]$$

is computed only for the first stage, while on coarser levels it is projected from the finest level. At each MG iteration, the time step is computed only on the finest level for both the LIRK and BIRK smoothers, because in the intermediate levels the matrix is only projected.

Algorithm 3 The Block-Implicit Runge-Kutta (BIRK) scheme

1: $\mathbf{W}^0 = \mathbf{W}^n$
2: **for** $k = 1, m$ **do**
3: Solve $\left[\frac{\mathbf{M_P}}{\Delta t} + \alpha_m \mathbf{D}\left(\mathbf{W}^0\right)\right] \delta\mathbf{W}^k = -\alpha_k \mathbf{R}\left(\mathbf{W}^{k-1}\right)$
4: $\mathbf{W}^k = \mathbf{W}^{k-1} + \delta\mathbf{W}^k$
5: **end for**
6: return $\mathbf{W}^{n+1} = \mathbf{W}^m$

The LBE scheme is displayed in Algorithm 4, where $\mathbf{J}$ represents the full Jacobian matrix. $\mathbf{J}$ is computed only for the first smoothing iteration at each level. Even if this choice reduces the effectiveness of the implicit smoother, it has been adopted since it significantly improves the robustness of the MG algorithm.

The restarted GMRES algorithm preconditioned with the block Jacobi method (one block per process) as available in the PETSc library (Balay et al. 2001) is used to solve the linear system. The CFL law for the MG strategies is defined by the Eq. 22.

Algorithm 4 The linearized Backward Euler (LBE) scheme

1: Solve $\left[\frac{\mathbf{M_P}}{\Delta t} + \mathbf{J}\left(\mathbf{W}^n\right)\right] \delta\mathbf{W}^n = \mathbf{R}\left(\mathbf{W}^n\right)$
2: return $\mathbf{W}^{n+1} = \mathbf{W}^n + \delta\mathbf{W}^n$

Line Creation Algorithm

The convergence rate of multigrid solvers can deteriorate dramatically when highly stretched grids are adopted, i.e., for the solution of the Navier–Stokes and RANS equations. To overcome this limitation in h-multigrid the mesh is coarsened only in the direction normal to the grid stretc.hing, or, alternatively, a line solver in the direction normal to the grid stretc.hing is adopted.

In the p-multigrid context, the latter approach is easier to implement, even if the creation of the lines for complex geometries can be not trivial. The line solver should reduce the stiffness induced by grid anisotropy, and, therefore, lines are created only in regions where stretc.hed elements are present, and they must propagate along the direction of strong coupling. The lines can be created (i) as (ii) the solution of a scalar advection-diffusion problem (Fidkowski et al. 2005; Wallraff et al. 2013), or (iii) following a geometrical approach (Langer 2013).

In this work the lines are created following an algorithm based on a geometrical approach, which is described in the Algorithm 5. nb is the number of boundary faces, $\boldsymbol{n}_f$ the normal to face f, e_1 the element adjacent to the boundary face f, ne_{max} is the maximum number of elements per line, e_i the element adjacent to previous element e_{i-1} inserted in the line, AR_{e_i} the aspect ratio of an element e_i defined as

$$\mathrm{AR}_{e_i} = \frac{\min(S_{e_i,j})}{\max(S_{e_i,j})},$$

where $S_{e_i,j}$ is the surface of the j-th face of the element e_i.

Algorithm 5 Line creation algorithm

for $f = 1, nb$ **do**
  Set $l_f(ne_{max})$ ▷ Allocate a line for each face f
  Compute $\boldsymbol{n}_{l_f}, e_1$ ▷ e_1: element adjacent to f
  $l_f(1) = e_1$ ▷ Add element to line
  for $i = 2, ne_{max}$ **do**
    if $e_i \cap \boldsymbol{n}_f$ **then** ▷ Intersection between element and normal
      $l_f(i) = e_i$ ▷ Add element to line
      Compute AR_{e_i} and $\mathrm{AR}_{e_{i-1}}$
      if $\mathrm{AR}_{e_i}/\mathrm{AR}_{e_{i-1}} < 0.1 \quad or \quad i = ne_{max}$ **then**
        Exit
      end if
    end if
  end for
end for

Figure 4.4 shows the lines for a 3-D streamlined body mesh (lines are longer than normal to enhance visibility). For each line a block tridiagonal system is assembled and solved efficiently with the Thomas algorithm (Thomas 1949).

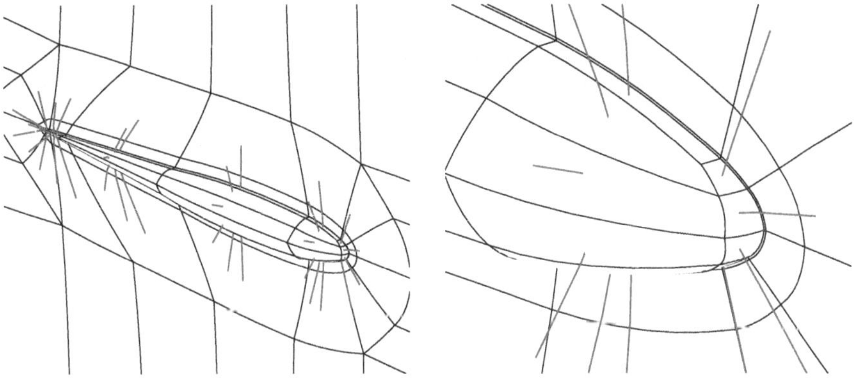

Fig. 4.4 Lines for a 3-D streamlined body mesh

The use of a line smoother in parallel computations needs an ad hoc domain partitioning procedure to avoid the cut of the lines and the loss of the line smoother effectiveness on stretc.hed meshes. An example of procedure for partitioning can be found in Ghidoni et al. (2014)

Fourier Analysis of the p-multigrid Scheme

In order to assess the performance of the p-MG algorithm with the LIRK and BIRK smoothers, this section presents the Fourier analysis of the PMG algorithm for a linear 2-D model problem, following an approach similar to that described in Bassi et al. (2011b).

In particular, we here consider the linear advection-diffusion equation on a square domain with periodic boundary conditions in both the x and y directions, that can be written as

$$w_t + aw_x + bw_y - \nu(w_{xx} + w_{yy}) = f(x, y) \quad \text{on} \quad [-1, 1] \times [-1, 1] \ , \tag{32}$$

where subscripts denotes partial differentiation, $w(x, y, t)$ is the unknown, $\mathbf{v} = (a, b)$ is a constant velocity vector, and f is a source term. Upwind numerical flux function and the BR2 scheme are used for the advective and diffusive terms, respectively.

The DG space discretization of order p of the above problem can be written as the following ordinary differential system:

$$\mathbf{M}^p \frac{\mathrm{d}\mathbf{w}^p}{\mathrm{d}t} + \mathbf{A}^p \mathbf{w}^p = \mathbf{b}^p \ ,$$

where $\mathbf{M}^p$ denotes the block diagonal mass matrix, $\mathbf{A}^p$ the DG space discretization matrix of the linear advection-diffusion equation, and $\mathbf{b}^p$ the discrete version of the source term (and of the boundary data when present). We are here interested in the steady-state solution

$$\mathbf{A}^p \mathbf{w}^p = \mathbf{b}^p \ ,$$

which is to be computed by the PMG algorithm. At each level of the PMG algorithm we adopt as smoother a multistage RK scheme, as described in more detail in the previous section. Let $\widetilde{\mathbf{W}}^{p,n}$ denote the approximate solution at iteration n, $\mathbf{w}^p$ the exact solution, and $\mathbf{e}^{p,n} = \widetilde{\mathbf{W}}^{p,n} - \mathbf{w}^p$ the error vector for polynomial order p. For a linear problem, the error at the new iteration $\mathbf{e}^{p,n+1}$ depends linearly on the error $\mathbf{e}^{p,n}$ at the previous iteration, and can be written as

$$\mathbf{e}^{n+1} = \mathbf{S}\mathbf{e}^n \ , \tag{33}$$

where $\mathbf{S}$ is the "iteration" or "smoother" matrix that, for the multistage RK schemes used as smoothers here considered, is given by

$$\mathbf{S} = \mathbf{I} + \alpha_1 \mathbf{P} + \alpha_2 \mathbf{P}(\mathbf{I} + \alpha_1 \mathbf{P}) + \alpha_3 \mathbf{P}(\mathbf{I} + \alpha_1 \mathbf{P} + \alpha_2 \mathbf{P}(\mathbf{I} + \alpha_1 \mathbf{P})) + \dots \ , \tag{34}$$

where α_k are the coefficients of the considered RK scheme. The matrix $\mathbf{P}$ appearing in (34) is given by

$$\mathbf{P} = (\mathbf{M} + \mathbf{D})^{-1} \mathbf{A} \ , \tag{35}$$

where $\mathbf{D}$ is the block diagonal part of $\mathbf{A}$ for the BIRK smoother, or the sum of the block diagonal part and of the part of $\mathbf{A}$ corresponding to the coupling along horizontal or vertical lines for the LIRK smoother.

The spectral radius of the iteration matrix, denoted by $\rho(\mathbf{S})$, determines the growth or decay of the error at each iteration. Damping of the error (i.e., a stable time integration scheme) requires that all the eigenvalues of $\mathbf{S}$ are located within the unit circle centered at the origin of the complex plane, i.e., $\rho(\mathbf{S}) \leq 1$.

By regarding the error as the sum of error modes, equal to the eigenvectors of matrix $\mathbf{S}$, Eq. (33) shows that at each iteration, each error mode (eigenvector) is amplified by a factor equal to the absolute value of the corresponding eigenvalue. At each level of the PMG algorithm, only the high-frequency modes of that level need to be effectively smoothed out, since the remaining low-frequency modes are seen as high-frequency modes on coarser levels and will therefore be effectively smoothed out by the same smoother applied at the coarser level.

For the two-level multigrid algorithm the error evolution equation can be written as

$$\mathbf{e}^{n+1} = \mathbf{TL}^p \mathbf{e}^n \ , \tag{36}$$

where $\mathbf{TL}^p$ is the two-level multigrid operator

$$\mathbf{TL}^p = (\mathbf{S}^p)^{\nu_2} \mathbf{K}^p (\mathbf{S}^p)^{\nu_1} \ , \tag{37}$$

where ν_1 and ν_2 are the number of pre- and post-iterations ($\nu_1 = 1$ and $\nu_2 = 1$ for this analysis), $\mathbf{S}^p$ is the iteration or smoother matrix described previously, and

$$\mathbf{K}^p = \mathbf{I} - \widetilde{\mathbf{I}}_{l-1}^{l} (\mathbf{A}^{p-1})^{-1} \mathbf{I}_l^{l-1} \mathbf{A}^p$$

is the two-grid correction operator. In this case the spectral radius $\rho(\mathbf{TL})$ of the two-level multigrid operator determines the growth or decay of the error at each multigrid iteration.

Similarly to the previously considered one-level case, also in the two-level case the error can be regarded as the sum of error modes equal to the eigenvectors of the matrix **TL**, and also in this case each error mode is amplified at each iteration by a factor equal to the absolute value of the corresponding eigenvalue. However, in the two-level case the entire spectrum of eigenmodes must be effectively smoothed out by the operator **TL**. In 2-D, the eigenvalues of matrix **TL** depends on the type of smoother, polynomial order p, element Reynolds number $Re = a\Delta x/\nu$, element aspect ratio $AR = \Delta x/\Delta y$, CFL number, and the flow angle $\tan(\alpha) = b/a$. The asymptotic convergence rate is bounded by the spectral radius of $\rho(\mathbf{TL}^p)$. However, since for some flow conditions isolated regions exist in the wave-number space (θ_x, θ_y) where the absolute value of some eigenvalue $\rightarrow 1$, and therefore $\rho(\mathbf{TL}^p) \rightarrow 1$, the performance of the p-MG algorithm is not quantified in terms of the spectral radius $\rho(\mathbf{TL})$ but in terms of an "average amplification factor" (AF) equal to the absolute value of the average eigenvalue over all the error modes, see, e.g., Fidkowski et al. (2005). This corresponds to tune the algorithm so as to achieve an effective smoothing of the "average error mode".

The BIRK smoother is first analyzed by comparing the performance of different numbers of RK stages and of different values of the RK coefficients. The RK schemes considered are the three-, four-, and five-stage schemes with coefficients given in Table 4.1, and the five-stage scheme, called from now on BIRK5$_{old}$, with the coefficients adopted in Bassi et al. (2009), i.e., the "standard" coefficients $\{0.2, 0.25, 0.333, 0.5, 1.0\}$. Notice that the coefficients reported in Table 4.1 have been obtained empirically by "numerical experimentation", and better convergence rates could probably be achieved with a truly optimized set of coefficients.

The investigation has been done using the following parameters: $CFL = 1$, $AR = 100$, $Re = 10000$, $\alpha = 1°$, $\alpha = 10°$, orders of the fine and coarse levels in the two-level algorithm $p_f = 2$ and $p_c = 1$, respectively. Table 4.2 summarizes the results

Table 4.1 BIRK/LIRK coefficients (α_i) for three (BIRK3/LIRK3), four (BIRK4/LIRK4) and five (BIRK5/LIRK5) stages schemes

Scheme	α_1	α_2	α_3	α_4	α_5
BIRK3/LIRK3	0.666	0.833	1		
BIRK4/LIRK4	0.625	0.750	0.875	1	
BIRK5/LIRK5	0.600	0.700	0.800	0.900	1

Table 4.2 The maximum and average amplification factor, AF_{max} and AF_{avg}, for the BIRK3 – 5 and BIRK5$_{old}$ schemes with the following parameters: the mesh aspect ratio $AR = 100$, the Reynolds number $Re = 10000$, the inflow angle $\alpha = (1°, 10°)$, $CFL = 1$, and the fine/coarse levels order of the two levels algorithm $p_f = 2/p_c = 1$

Stage	$AF_{max,1°}$	$AF_{avg,1°}$	$AF_{max,10°}$	$AF_{avg,10°}$
3	0.47088	0.08775	0.95761	0.06854
4	0.44116	0.05682	0.94633	0.04576
5	0.37738	0.04176	0.93555	0.03393
5_{old}	0.55969	0.07598	0.96021	0.07441

Table 4.3 BIRK5: the average amplification factor for a flow angle $\alpha = 1°$ and different mesh aspect ratios AR, Reynolds numbers Re, and fine/coarse levels order of the two levels algorithm

p_f/p_c	$AR = 1$		$AR = 100$	
	$Re = 10$	$Re = 100$	$Re = 10000$	$Re = 10000$
2/1	0.02595	0.02910	0.04716	0.04176
3/2	0.01525	0.01235	0.02504	0.03329
4/3	0.01154	0.00805	0.01510	0.02744

and shows that the five-stage scheme with the new set of coefficients is characterized by a better average and maximum amplification factor of $\mathbf{TL}^p$. For the five-stage schemes (BIRK5 and BIRK5$_{old}$) the influence of the stage coefficients on the average AF can be also observed. This leads to the conclusion that the new set al.lows a 45% and 55% reduction of the average amplification factor for a flow angle $\alpha = 1°$ and $\alpha = 10°$, respectively. The same analysis has been also carried out on a real test case, as described in Sect. 4.7.1.

The performance of the BIRK5/LIRK5 smoothers is next analyzed. Horizontal (LIRK5x) and vertical (LIRK5y) lines are considered. The $\mathbf{TL}^p$ average amplification factor is computed for (i) the nearly inviscid conditions $AR = 1$ and $Re = 10000$, (ii) the low Reynolds number conditions $AR = 1$ and $Re = 10$, (iii) the moderate Reynolds number conditions $AR = 1$ and $Re = 100$, and (iv) the boundary layer conditions $AR = 100$ and $Re = 10000$. The flow angle for all cases is $\alpha = 1°$. The influence of the fine/coarse orders is also investigated, and the results are reported in Tables 4.3, 4.4 and 4.5. A reduction of the average AF can be observed for all smoothers as the polynomial order of the fine and coarse levels is increased. For all the considered cases the BIRK5 smoother is the least effective scheme when $AR = 1$, even if for increasing Reynolds number both BIRK5 and LIRK5y schemes tend to be comparable. However, due to the greater coupling along the convection direction, the best performance is obtained with the LRIK5x, but for the reasons explained in Sect. 4.5.1 this type of line smoother is not considered in the proposed p-MG algorithm. As a concluding remark, for the "boundary layer" case characterized by $AR = 100$, the LIRK5y is the best scheme displaying the lowest average amplification factor.

Table 4.4 LIRK5x: the average amplification factor for a flow angle $\alpha = 1°$ and different mesh aspect ratios AR, Reynolds numbers Re, and fine/coarse levels order of the two levels algorithm

p_f/p_c	$AR = 1$		$AR = 100$	
	$Re = 10$	$Re = 100$	$Re = 10000$	$Re = 10000$
2/1	0.00839	0.01077	0.02541	0.04115
3/2	0.00458	0.00458	0.01076	0.03305
4/3	0.00452	0.00304	0.00656	0.02734

Table 4.5 LIRK5y: the average amplification factor for a flow angle $\alpha = 1°$ and different mesh aspect ratios AR, Reynolds numbers Re, and fine/coarse levels order of the two levels algorithm

p_f/p_c	$AR = 1$		$AR = 100$	
	$Re = 10$	$Re = 100$	$Re = 10000$	$Re = 10000$
2/1	0.01871	0.02871	0.04701	0.00089
3/2	0.01167	0.01066	0.02485	0.00012
4/3	0.00740	0.00604	0.01493	0.00004

Results—Laminar Test Cases

This section presents the computation of the compressible laminar flow around the Delta Wing and a streamlined three-dimensional body, which have been considered in the EU project ADIGMA (Kroll 2009).

The effect of the number of stages used in the BIRK smoother on the performance of the FMG V-cycle algorithm is first investigated by computing the flow field around the Delta Wing. Three, four, and five stages and the RK coefficients given in Table 4.1 have been considered.

In the second test case the effectiveness of the line smoother on stretc.hed grids is investigated. Two different setups have been considered: (i) the LIRK smoother only on the finest level and (ii) on all levels (with the exception of the lowest). In addition the possibility to use the BIRK smoother on the coarsest level will be also considered.

To optimize the convergence rate, the coarsest level for all computations is $p_{min} = 1$. The L^2 norm of the density residual has been used as convergence indicator. The solution of the equations is considered to be converged when the maximum value of the residuals L^2 norm is below 10^{-7}. The computing time t_{CPU} is reported in the figures as a normalized value with respect to the TauBenchmark (IPACS 2005) value, $t_{TauBench}$, obtained on a full node of the cluster used for the CFD simulation .[1] The normalized computing time is measured in IDIHOM units (IU) (Sermeus 2011).

[1]-n 250000 -s 10 define the reference TauBench workload for the hardware benchmark. The cluster used for all computations achieved the benchmark value $t_{TauBench} = 10.32$ s.

Laminar Delta Wing

The laminar flow around a Delta Wing with sharp leading-edge and a blunt-trailing edge is here considered. The flow field is computed for a Mach number $M = 0.3$, a Reynolds number $Re = 4000$ and an angle of attack $\alpha = 12.5°$ with isothermal no-slip wall boundary condition imposed on the wing. Details of the coarse (3264 hexahedrals) and fine (26112 hexahedrals) meshes (linear edges) are displayed in Fig. 4.5. To assess only the performance of the BIRK smoother, meshes are not characterized by highly stretc.hed elements.

Computed Mach number contours on the fine ($\mathbb{P}^5$ solution) and coarse ($\mathbb{P}^6$ solution) mesh are compared in Fig. 4.6, to show the capability of the p-MG solver to reach high-order solution approximations for complex viscous flows. The smoothers adopted for this test case are the LBE for the coarsest level and the BIRK for other levels. The number of pre- and post-iterations guarantees the p independent property of the p-MG algorithm, at least on the coarse mesh, i.e., the same number of multigrid iteration N_{MG} for each polynomial order, and optimizes computational efficiency. On the coarse level $\nu_c = 5$ is adopted, on the finest level $\nu_{f,1-2} = 10$, and on the intermediate levels $\nu_{1-2} = 20$.

The influence of stages number on computational efficiency to obtain a converged solution is first analyzed. Tables 4.6 and 4.7 show the MG cycles and the normalized CPU time for both meshes and different solution approximations. As the number of stages is increased, the convergence rate for higher order polynomials improves. In particular, on the coarse mesh from $\mathbb{P}^5$ the BIRK5 is the most efficient, while on lower orders similar performance can be observed. A reduction of the computing time around 7.4% and 19.6% (the BIRK3 scheme is considered as reference) is observed for the $\mathbb{P}^5$ and $\mathbb{P}^6$ solution, respectively. On the fine mesh the BIRK5 is the most efficient and shows a computational time reduction around 20% for all solution approximations. These results confirm the data of the Fourier analysis for the two-level multigrid algorithm presented in Sect. 4.6.

Figure 4.7 shows the asymptotic behavior of the residual L^2 norm, for the FMG with the BIRK5+LBE strategy. $\mathbb{P}^6$ spatial discretization on the coarse mesh and $\mathbb{P}^5$

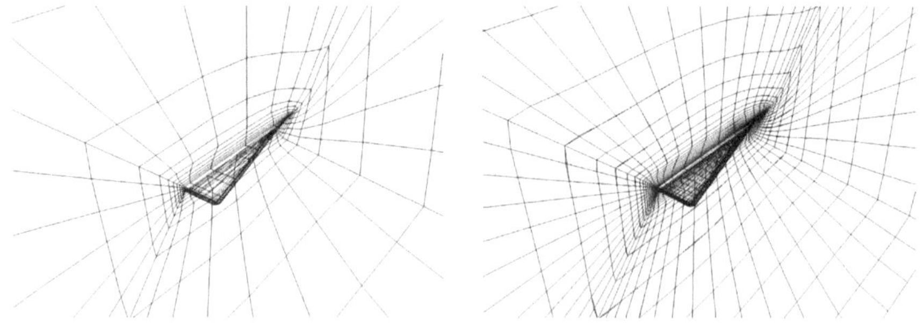

Fig. 4.5 Laminar Delta Wing: the computational mesh consisting of 3263 (left) and 26112 (right) linear hexahedral elements

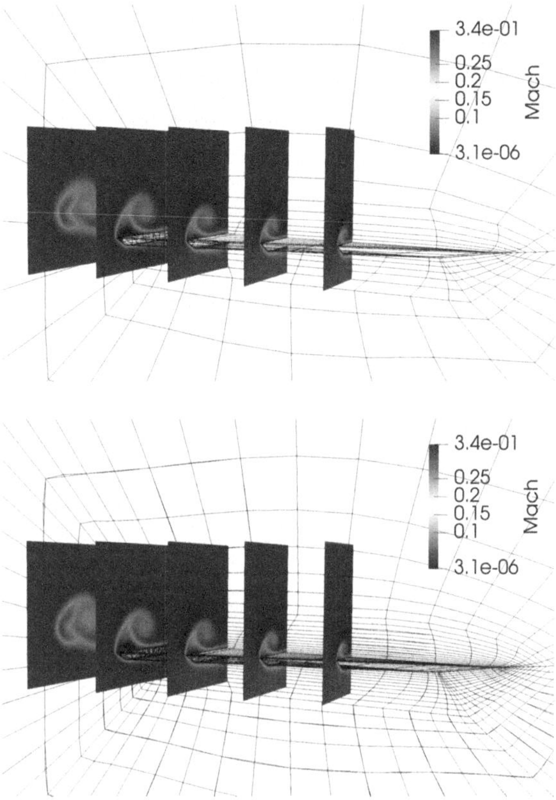

Fig. 4.6 Laminar Delta Wing: comparison of the computed Mach number contours for $\mathbb{P}^6$ solution on the coarse mesh (top) and for $\mathbb{P}^5$ solution on the fine mesh (bottom)

spatial discretization on the fine mesh with full convergence on each level has been used. Tables 4.8 and 4.9 summarize the convergence history, where the number of iterations needed to fully converge on each level, N_{MG}, the slope module of the linear regression of each convergence curve, $|s|$, have been reported for the coarse and the fine mesh, respectively. Notice that the p-multigrid order-independent property is satisfied on the coarse mesh, while on the fine mesh the convergence slope decreases for higher spatial solution approximation. To recover the p-independent property, the number of smoothing iterations should be increased, which would penalize the computing time.

Streamlined 3-D Body

In the second test case, the laminar flow around a streamlined three-dimensional body, called BTC0 in the ADIGMA EU project, based on a 10% thick airfoil with boundaries constructed by a surface of revolution is computed for a far-field Mach

Table 4.6 Laminar Delta Wing: the performance of the Block-implicit RK scheme for different stages number on the coarse mesh. $\mathbb{P}^k$ is the solution approximation, ν_c the number of iterations at the coarsest level, ν_{1-2} the number of pre and post-iterations for intermediate levels, $\nu_{f,1-2}$ the number of pre- and post-iterations at the finest level, N_{MG} the number of MG iterations needed to reach a converged solution, t_{CPU} the normalized CPU time, Δt_{CPU} the computational time reduction due to the use of a different number of stages (the CPU time of the BIRK3 is the reference)

BIRK stages	$\mathbb{P}^k$	ν_c	ν_{1-2}	$\nu_{f,1-2}$	N_{MG}	t_{CPU} [IU]	Δt_{CPU}
3	3	5	20-20	10-10	7	6.01	–
4	3	5	20-20	10-10	6	5.71	−4.9%
5	3	5	20-20	10-10	6	6.37	+5.1%
3	4	5	20-20	10-10	7	13.56	–
4	4	5	20-20	10-10	6	12.67	−6.5%
5	4	5	20-20	10-10	6	13.12	−3.2%
3	5	5	20-20	10-10	8	40.14	–
4	5	5	20-20	10-10	8	41.47	+3.3%
5	5	5	20-20	10-10	7	37.18	−7.4%
3	6	5	20-20	10-10	7	157.27	–
4	6	5	20-20	10-10	6	155.23	−1.3%
5	6	5	20-20	10-10	6	126.43	−19.6%

Table 4.7 Laminar Delta Wing: the performance of the Block-implicit RK scheme for different stages number on the fine mesh. $\mathbb{P}^k$ is the solution approximation, ν_c the number of iterations at the coarsest level, ν_{1-2} the number of pre- and post-iterations for intermediate levels, $\nu_{f,1-2}$ the number of pre- and post-iterations at the finest level, N_{MG} the number of MG iterations needed to reach a converged solution, t_{CPU} the normalized CPU time, Δt_{CPU} the computational time reduction due to the use of a different number of stages (the CPU time of the BIRK3 is the reference)

BIRK stages	$\mathbb{P}^k$	ν_c	ν_{1-2}	$\nu_{f,1-2}$	N_{MG}	t_{CPU} [IU]	Δt_{CPU}
3	2	5	–	10-10	15	45.90	–
4	2	5	–	10-10	13	39.24	−14.5%
5	2	5	–	10-10	12	37.69	−17.8%
3	3	5	20-20	10-10	20	157.17	–
4	3	5	20-20	10-10	17	146.51	−6.8%
5	3	5	20-20	10-10	15	118.51	−24.6%
3	4	5	20-20	10-10	24	433.82	–
4	4	5	20-20	10-10	20	410.66	−5.3%
5	4	5	20-20	10-10	17	334.69	−22.7%
3	5	5	20-20	10-10	28	1183.72	–
4	5	5	20-20	10-10	24	1110.46	−6.1%
5	5	5	20-20	10-10	20	947.87	−19.9%

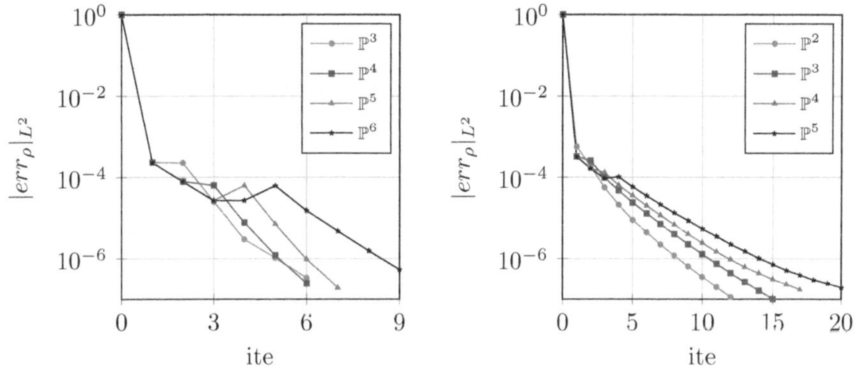

Fig. 4.7 Laminar Delta Wing: the residual convergence rate (full convergence on each level) as a function of MG cycles for $\mathbb{P}^6$ solution on the coarse mesh (left) and for $\mathbb{P}^5$ solution on the fine mesh (right)

Table 4.8 Laminar Delta Wing: MG cycles needed to converge on each level and the slope module of the convergence curves for the BIRK5+LBE smoothing strategy on the coarse mesh. $\mathbb{P}^k$ is the solution approximation, ν_c the number of iterations at the coarsest level, ν_{1-2} the number of pre- and post-iterations for intermediate levels, $\nu_{f,1-2}$ the number of pre- and post-iterations at the finest level, N_{MG} the number of MG iterations needed to reach a converged solution, $|s|$ the slope module of the linear regression of each convergence curve

$\mathbb{P}^k$	ν_c	ν_{1-2}	$\nu_{f,1-2}$	N_{MG}	$\|s\|$
3	5	20-20	10-10	7	0.704
4	5	20-20	10-10	6	0.805
5	5	20-20	10-10	7	0.841
6	5	20-20	10-10	6	0.511

Table 4.9 Laminar Delta Wing: MG cycles needed to converge on each level and the slope module of the convergence curves for the BIRK5+LBE smoothing strategy on the fine mesh. $\mathbb{P}^k$ is the solution approximation, ν_c the number of iterations at the coarsest level, ν_{1-2} the number of pre- and post-iterations for intermediate levels, $\nu_{f,1-2}$ the number of pre- and post-iterations at the finest level, N_{MG} the number of MG iterations needed to reach a converged solution, $|s|$ the slope module of the linear regression of each convergence curve

$\mathbb{P}^k$	ν_c	ν_{1-2}	$\nu_{f,1-2}$	N_{MG}	$\|s\|$
2	5	–	10-10	15	0.326
3	5	20-20	10-10	15	0.255
4	5	20-20	10-10	17	0.206
5	5	20-20	10-10	20	0.172

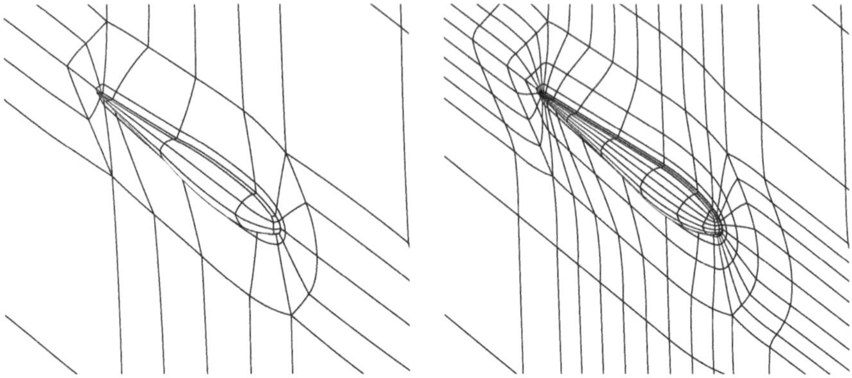

Fig. 4.8 BTC0: the computational mesh consisting of 832 (left) and 6656 (right) quadratic hexahedral elements

number M $= 0.5$, an angle of attack $\alpha = 1^\circ$, and a Reynolds number Re $= 500$ with adiabatic no-slip wall boundary condition.

The effectiveness of the line smoother on meshes characterized by elements with a severe aspect ratio (AR) in the boundary layer zone is investigated. A coarse mesh with 832 hexahedrals and $AR = 8850$, and a fine mesh with 6656 hexahedrals and $AR = 20000$, both with quadratic edges, have been used. Figure 4.8 shows some details of the coarse and fine meshes. Also in this case, the number of pre- and post-iterations guarantee the p independent property of the p-MG algorithm on the coarse mesh. The number of smoothing iterations for the coarse and the fine mesh is as follows: $\nu_c = 5$, $\nu_{f,1-2} = 5$ and $\nu_{1-2} = 10$ for the coarse mesh, and $\nu_c = 5$, $\nu_{f,1-2} = 10$ and $\nu_{1-2} = 20$ for the fine mesh.

The performance of the line smoother is investigated for different sets of smoothers: (i) the implicit smoother (LBE) on the coarsest level, the LIRK5 on the finest level, and the BIRK5 on the other levels (LIRK5+BIRK5+LBE) and (ii) the LBE on the coarsest level and the LIRK5 on the other levels (LIRK5+LBE). Previous strategies have been also compared with the BIRK5+LBE. In Tables 4.10 and 4.11 the MG cycles and the normalized CPU time t_{CPU} needed to reach a converged solution, the computational time variation Δt_{CPU} due to the use of the three strategies described above (the LIRK5+LBE is the reference) are shown for both meshes and different solution approximations. LIRK5+BIRK5+LBE guarantees the best performance on both meshes. Even if LIRK5+BIRK5+LBE and the LIRK5+LBE have the same number of iterations to reach a converged solution, MG iterations of the LIRK5+LBE are more expensive, due to the assembly and solution of the block tridiagonal system for each level. The huge AR deteriorates the BIRK5+LBE computational efficiency on the coarse mesh for higher solution polynomial approximation, while on the fine mesh it prevents the algorithm to converge. For this reason, Table 4.11 does not report the BIRK5+LBE scheme.

Table 4.10 BTC0: the effectiveness of different strategies to exploit the line smoother in the p-MG algorithm on the coarse mesh. $\mathbb{P}^k$ is the solution approximation, ν_c the number of iterations at the coarsest level, ν_{1-2} the number of pre- and post-iterations for intermediate levels, $\nu_{f,1-2}$ the number of pre- and post-iterations at the finest level, N_{MG} the number of MG iterations needed to reach a converged solution, t_{CPU} the normalized CPU time, Δt_{CPU} the computational time variation due to the use of the different solution strategies (the CPU time of the LIRK5+LBE strategy is the reference)

Scheme	$\mathbb{P}^k$	ν_c	ν_{1-2}	$\nu_{f,1-2}$	N_{MG}	t_{CPU} [IU]	Δt_{CPU} (%)
BIRK5+LBE	3	5	10-10	5-5	22	20.34	−12.3%
LIRK5+LBE	3	5	10-10	5-5	4	23.22	–
LIRK5+BIRK5+LBE	3	5	10-10	5-5	4	7.76	−66.5%
BIRK5+LBE	4	5	10-10	5-5	24	68.50	−13.6%
LIRK5+LBE	4	5	10-10	5-5	4	79.32	–
LIRK5+BIRK5+LBE	4	5	10-10	5-5	4	25.72	−67.5%
BIRK5+LBE	5	5	10-10	5-5	34	285.51	−12.3%
LIRK5+LBE	5	5	10-10	5-5	6	325.58	–
LIRK5+BIRK5+LBE	5	5	10-10	5-5	5	98.25	−69.8%
BIRK5+LBE	6	5	10-10	5-5	66	4458.62	+357.1%
LIRK5+LBE	6	5	10-10	5-5	6	975.22	–
LIRK5+BIRK5+LBE	6	5	10-10	5-5	6	567.83	−41.7%

Table 4.11 BTC0: the effectiveness of different strategies to exploit the line smoother in the p-MG algorithm on the fine mesh. $\mathbb{P}^k$ is the solution approximation, ν_c the number of iterations at the coarsest level, ν_{1-2} the number of pre- and post-iterations for intermediate levels, $\nu_{f,1-2}$ the number of pre- and post-iterations at the finest level, N_{MG} the number of MG iterations needed to reach a converged solution, t_{CPU} the normalized CPU time, Δt_{CPU} the computational time variation due to the use of the different solution strategies (the CPU time of the LIRK5+LBE strategy is the reference)

Scheme	$\mathbb{P}^k$	ν_c	ν_{1-2}	$\nu_{f,1-2}$	N_{MG}	t_{CPU} [IU]	Δt_{CPU} (%)
LIRK5+LBE	2	5	–	10-10	6	19.34	–
LIRK5+BIRK5+LBE	2	5	–	10-10	6	19.19	−0.7%
LIRK5+LBE	3	5	20-20	10-10	4	176.28	–
LIRK5+BIRK5+LBE	3	5	20-20	10-10	4	132.27	−24.9%
LIRK5+LBE	4	5	20-20	10-10	4	759.11	–
LIRK5+BIRK5+LBE	4	5	20-20	10-10	4	581.20	−23.4%
LIRK5+LBE	5	5	20-20	10-10	6	3546.51	–
LIRK5+BIRK5+LBE	5	5	20-20	10-10	5	2110.37	−40.4%

The use of the BIRK5 as smoother on the coarsest level has been also considered, comparing its performance with the LBE smoother. As summarized in Table 4.12, the p-MG performance without the implicit smoother on the coarsest level are penalized. In fact the LBE reduces the computing time about 80%.

Table 4.12 BTC0: the effectiveness of different smoothers (LBE and BIRK5) for the solution of the coarsest level problem (fine mesh). $\mathbb{P}^k$ is the solution approximation, ν_c the number of iterations at the coarsest level, ν_{1-2} the number of pre- and post-iterations for intermediate levels, $\nu_{f,1-2}$ the number of pre- and post-iterations at the finest level, N_{MG} the number of MG iterations needed to reach a converged solution, t_{CPU} the normalized CPU time, Δt_{CPU} the computational time variation due to the use of a different smoother on coarsest level (the CPU time of the LIRK5+BIRK5 strategy is the reference)

Scheme	$\mathbb{P}^k$	ν_c	ν_{1-2}	$\nu_{f,1-2}$	N_{MG}	t_{CPU} [IU]	Δt_{CPU} (%)
LIRK5+BIRK5	3	5	20-20	10-10	13	32.67	–
LIRK5+BIRK5+LBE	3	5	20-20	10-10	4	7.76	−76.2%
LIRK5+BIRK5	4	5	20-20	10-10	17	190.02	–
LIRK5+BIRK5+LBE	4	5	20-20	10-10	4	25.72	−86.5%
LIRK5+BIRK5	5	5	20-20	10-10	20	776.94	–
LIRK5+BIRK5+LBE	5	5	20-20	10-10	5	98.25	−87.3%

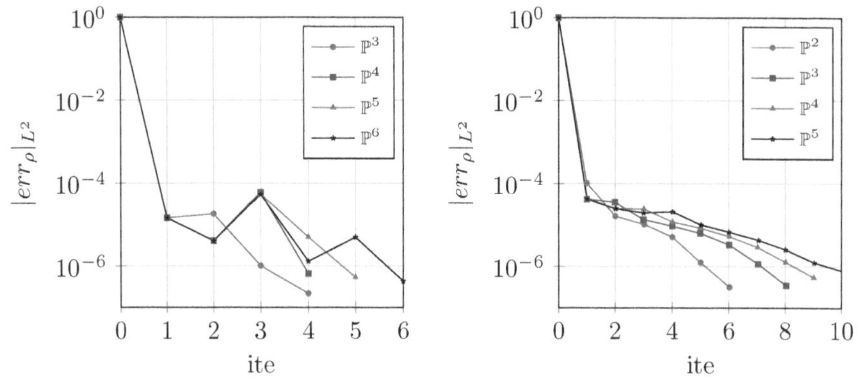

Fig. 4.9 BTC0: the residual convergence rate (full convergence on each level) as a function of MG cycles for a $\mathbb{P}^6$ spatial discretization on the coarse mesh (left) and for a $\mathbb{P}^5$ spatial discretization on the fine mesh (right)

As depicted in Fig. 4.9, the FMG solution with the LIRK5+BIRK5+LBE strategy (full convergence on each level) has been computed for a $\mathbb{P}^6$ and $\mathbb{P}^5$ spatial discretization on the coarse and fine mesh, respectively, to study the asymptotic behavior of the residuals L^2 norm. Tables 4.13 and 4.14 summarize the convergence history, where the number of iterations needed to fully converge on each level, N_{MG}, the slope module of the linear regression of each convergence curve, $|s|$, have been reported for the coarse and the fine mesh, respectively. Also in this case p-multigrid order-independent property is satisfied on the coarse mesh, while on the fine mesh the convergence slope decreases for higher spatial solution approximation. To recover the p-independent property, the number of smoothing iterations should be increased, which would penalize the computing time.

Table 4.13 BTC0: MG cycles needed to converge on each level and the slope module of the convergence curves for the LIRK5+BIRK5+LBE smoothing strategy on the coarse mesh. $\mathbb{P}^k$ is the solution approximation, ν_c the number of iterations at the coarsest level, ν_{1-2} the number of pre- and post-iterations for intermediate levels, $\nu_{f,1-2}$ the number of pre- and post-iterations at the finest level, N_{MG} the number of MG iterations needed to reach a converged solution, $|s|$ the slope module of the linear regression of each convergence curve

$\mathbb{P}^k$	ν_c	ν_{1-2}	$\nu_{f,1-2}$	N_{MG}	$\|s\|$
3	5	10-10	5-5	4	0.961
4	5	10-10	5-5	4	1.961
5	5	10-10	5-5	5	0.965
6	5	10-10	5-5	6	1.050

Table 4.14 BTC0: MG cycles needed to converge on each level and the slope module of the convergence curves for the LIRK5+BIRK5+LBE smoothing strategy on the fine mesh. $\mathbb{P}^k$ is the solution approximation, ν_c the number of iterations at the coarsest level, ν_{1-2} the number of pre- and post-iterations for intermediate levels, $\nu_{f,1-2}$ the number of pre- and post-iterations for the finest level, N_{MG} the number of MG iterations needed to reach a converged solution, $|s|$ the slope module of the linear regression of each convergence curve

$\mathbb{P}^k$	ν_c	ν_{1-2}	$\nu_{f,1-2}$	N_{MG}	$\|s\|$
2	5	–	10-10	6	0.461
3	5	20-20	10-10	8	0.305
4	5	20-20	10-10	9	0.260
5	5	20-20	10-10	10	0.231

Results—Turbulent Test Cases

This section presents the computation of the compressible turbulent flow around the Delta Wing and a train head, and through the T106A turbine cascade, which have been considered in the EU project IDIHOM (IDIHOM 2015).

In the first test case the effectiveness of the multigrid strategy, p-MG(LIRK), based on the line smoother (LIRK5+BIRK5+LBE) adopted for the laminar test cases is assessed, comparing its computational efficiency with an implicit solver (LBE). In the second test case the computational efficiency of the p-MG strategy with different sets of smoothers is investigated. In particular, p-MG(LIRK) is compared with a multigrid algorithm where LBE is adopted as smoother on every level, p-MG(LBE). Finally, the T106A test case is used to assess the performance of the p-MG(LBE) with respect to an implicit solver.

As in the previous laminar test cases the coarsest level for all computations is $p_{min} = 1$, the L^2 norm of the density residual has been used as convergence indicator and the solution of the equations is considered to be converged when the maximum value of the residuals L^2 norm is below 10^{-7}.

Turbulent Delta Wing

The NASA 65° sweep delta wing is proposed and investigated experimentally within the second international Vortex Flow Experiment (VFE-2) (VFE2, Project() 2020). The far-field conditions for this test case are $M = 0.4$, $\alpha = 13.3°$ and $Re = 3\times10^6$. The solution is computed up to $\mathbb{P}4$ on a mesh consisting of 13816 hexahedral elements with quadratic edges. The high-order mesh is generated by means of an in-house agglomeration software starting from a linear finite volume mesh (Crippa 2008) and is depicted in Fig. 4.10.

The p-MG(LIRK) strategy is based on the following smoothers: LIRK5 on the finest level, BIRK5 on the intermediate levels, and LBE on the coarsest level. The smoothing iterations on the coarsest level are $\nu_c = 5$, on the intermediate levels $\nu_{1-2} = 3 - 2$, and on the finest level $\nu_{f,1-2} = 2 - 2$. Figure 4.11 shows a comparison of the computed turbulence intensity contours for a $\mathbb{P}^3$ (left) and $\mathbb{P}^4$ (right) solution approximation. As expected, the latter case provides a sharper definition of the contours.

A comparison between p-MG(LIRK) and implict (LBE) algorithm is performed in terms of the memory usage (Mem), normalized CPU time (Time), and iterations (ite) needed to reach a converged solution. Figure 4.12 shows the computing time needed to converge for both strategy. p-MG(LIRK) strategy guarantees better performance for higher polynomial approximation. Table 4.15 summarizes the results of the comparison. For lower polynomial approximations, i.e., $\mathbb{P}^2$, the p-MG(LIRK) needs larger CPU time to converge, probably because the number of levels is not enough to fully exploit multigrid advantages, while for $\mathbb{P}^3$ solution approximation CPU time needed by p-MG(LIRK) to converge is decreased by 8%. In terms of memory usage, for both solution approximations the requirement of the p-MG(LIRK) is reduced by 80%.

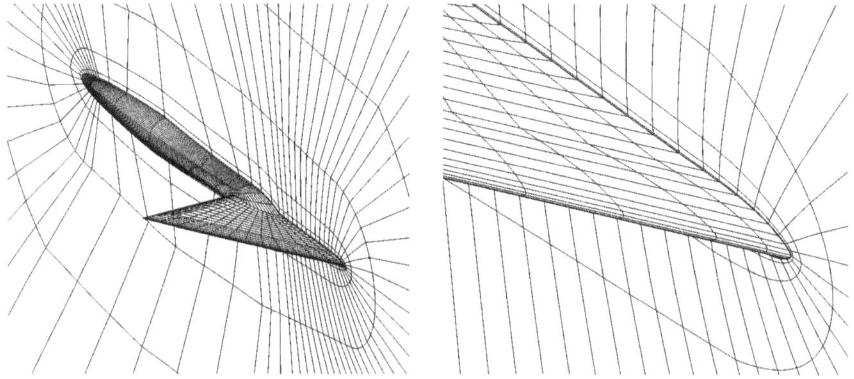

Fig. 4.10 Turbulent Delta Wing: grid consisting of 13816 hexahedral elements with quadratic edges

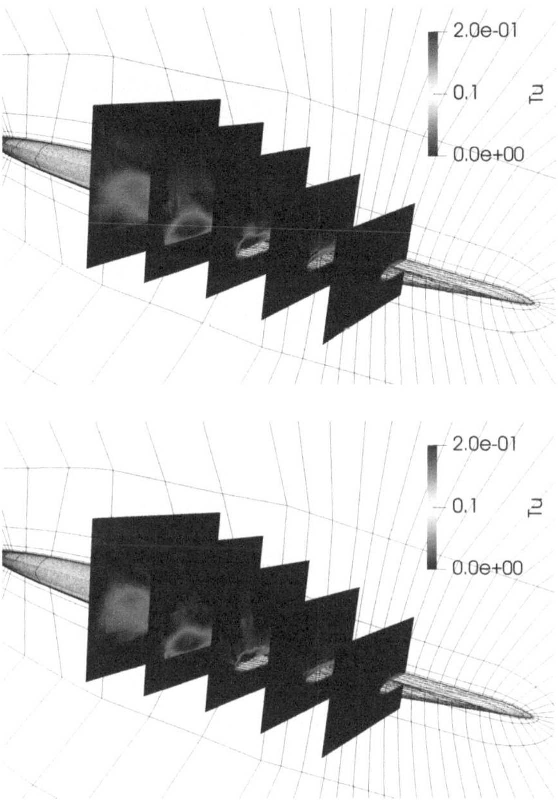

Fig. 4.11 Delta wing: comparison of the computed turbulence intensity contours for a $\mathbb{P}^3$ solution (top) and $\mathbb{P}^4$ solution (bottom)

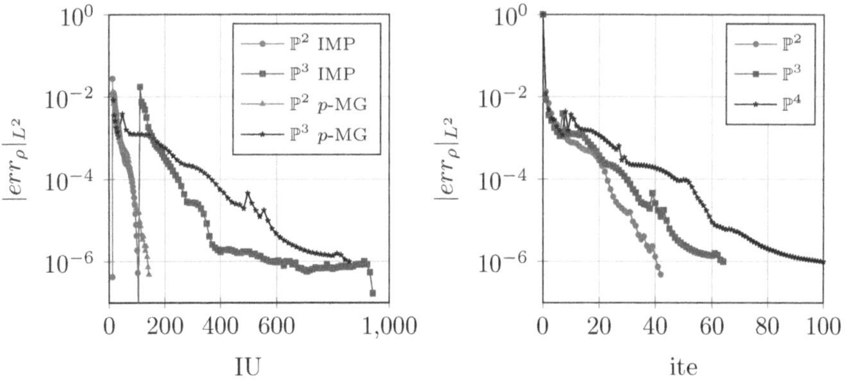

Fig. 4.12 Turbulent Delta wing: L^2 norm convergence history versus normalized CPU time for different polynomial solution approximations and strategy (right). L^2 norm convergence history versus MG cycles for $\mathbb{P}^{1\to 4}$ approximations (left)

Table 4.15 Turbulent Delta wing: comparison of p-MG(LIRK) and implict (LBE) strategies performance. $\mathbb{P}^k$ is the solution approximation, Mem the memory usage, Δ_{RAM} the memory requirement reduction due to the use of different solution strategies, t_{CPU} the normalized CPU time, Δt_{CPU} the computational time reduction due to the use of a different solution strategies (Memory usage and CPU time of the LBE scheme are the reference)

Scheme	$\mathbb{P}^k$	Mem [GB]	Δ_{RAM}	t_{CPU} [IU]	Δt_{CPU}
LBE	2	44.0	–	93	–
p-MG(LIRK)	2	6.0	−86.3%	142	+52.6%
LBE	3	112.0	–	929	–
p-MG(LIRK)	3	18.0	−83.9%	857	−7.7%

Table 4.16 Turbulent Delta Wing: MG cycles needed to converge on each level and slope module of the convergence curves for the p-MG(LIRK) strategy. $\mathbb{P}^k$ is the solution approximation, ν_c the number of iterations at the coarsest level, ν_{1-2} the number of pre- and post-iterations for intermediates levels, $\nu_{f,1-2}$ the number of pre- and post-iterations at the finest level, N_{MG} the number of MG iterations needed to reach a converged solution, $|s|$ the slope module of the linear regression of each convergence curve

$\mathbb{P}^k$	ν_c	ν_{1-2}	$\nu_{f,1-2}$	N_{MG}	$\|s\|$
2	5	–	2-2	42	0.0957
3	5	3-2	2-2	64	0.0643
4	5	3-2	2-2	100	0.0423

Figure 4.12 shows the asymptotic behavior of the residual L^2 norm. The solution has been obtained with the FMG p-MG(LIRK) strategy is computed for a $\mathbb{P}^4$ spatial discretization with full convergence on each level. Table 4.16 summarizes the convergence history, where the number of iterations at the coarsest level, ν_c, the number of pre- and post-iterations for levels, ν_{1-2}, the number of pre- and post-iterations at the finest level, $\nu_{f,1-2}$, the number of iterations needed to fully converge on each level, N_{MG}, the slope module of the linear regression of each convergence curve, $|s|$, have been shown. As shown also for the laminar test cases, the p-MG(LIRK) order-independent property is not completely satisfied and, in fact, the iterations needed to reach the converged state increase with the polynomial order.

Train Head

The flow around a simplified train (single car) with the wind velocity $U_w = 70\,\text{m/s}$ and the Reynolds number based on the reference length $\text{Re} = 1.2 \times 10^6$ is considered. Computations are performed for a yaw angle $\beta = 10°$ on a mesh with 7776 hexahedral elements with quartic edges. Figures 4.13 and 4.14 show the surface mesh and the computed turbulence intensity contours on the train symmetry plane for a $\mathbb{P}^4$ solution.

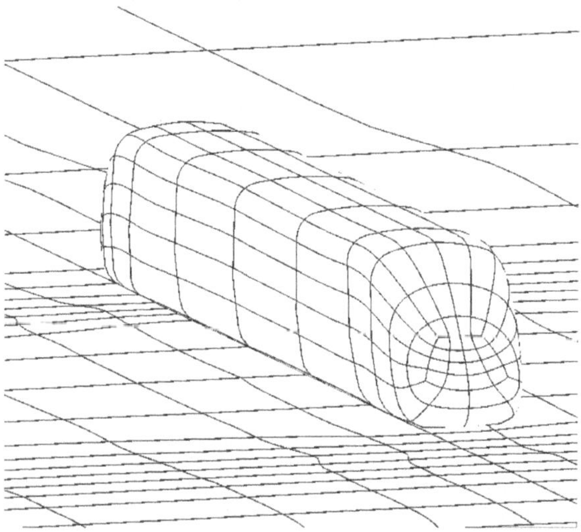

Fig. 4.13 Train: mesh consisting of 7776 hexahedral elements with quartic edges

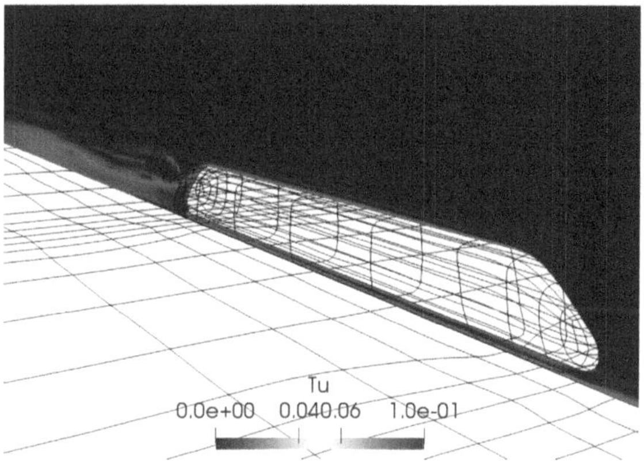

Fig. 4.14 Train: the computed turbulence intensity contours for a $\mathbb{P}^4$ solution on train symmetry plane

The effect of different smoothers on the p-MG performance is investigated in this test case in terms of the memory usage (Mem) and the normalized CPU time (Time) needed to reach a converged solution. In particular, the multigrid strategy developed for laminar test cases, p-MG(LIRK), is compared with a strategy, p-MG(LBE), where an implicit smoother (LBE) is adopted on every level (GMRES method, 120 iteration, no-restart, preconditioned with ILU(0) on each block). The smoothing iterations adopted for the p-MG algorithms are: $\nu_c = 5$ on the coarsest level, $\nu_{f,1-2} = 2 - 2$ on the finest level and $\nu_{1-2} = 3 - 3$ on intermediate levels.

Figure 4.15 shows the convergence history for the p-MG with the two sets of smoothers in terms of iterations and the normalized CPU time needed to reach a converged solution. Table 4.17 summarizes performance for both strategies. The p-MG(LBE) outperforms the p-MG(LIRK) in terms of CPU time with an average

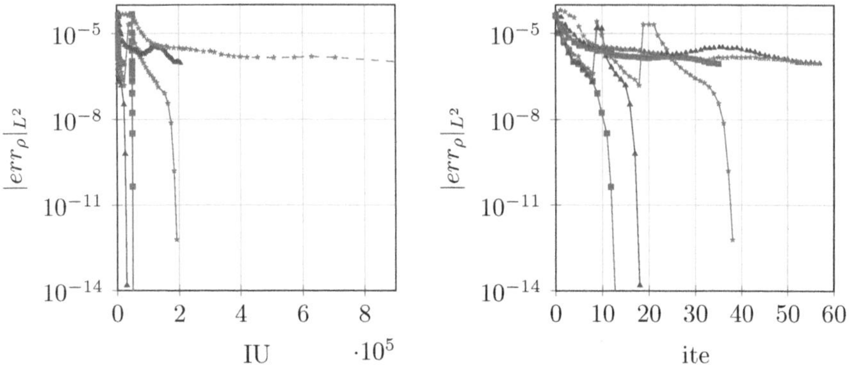

Fig. 4.15 Train: convergence history (L^2 norm of the density residual) versus normalized CPU time for different polynomial solution approximations and smoothers (left). Convergence history (L^2 norm of the density residual) versus MG cycles for different polynomial solution approximations and smoothers (right). $\mathbb{P}^2$ $p-$MG (LBE), $\mathbb{P}^3$ $p-$MG (LBE), $\mathbb{P}^4$ $p-$MG (LBE), $\mathbb{P}^2$ $p-$MG (LIRK), $\mathbb{P}^3$ $p-$MG (LIRK), $\mathbb{P}^4$ $p-$MG (LIRK)

Table 4.17 Train: comparison of p-MG(LBE) and p-MG(LIRK) strategies. $\mathbb{P}^k$ is the solution approximation, RAM the memory usage, Δ_{RAM} the memory requirement reduction due to the use of different solution strategies, t_{CPU} the normalized CPU time, Δt_{CPU} the computational time reduction due to the use of a different solution strategies (Memory usage and CPU time of the p-MG(LIRK) are the reference)

Scheme	$\mathbb{P}^k$	RAM [GB]	Δ_{RAM}	t_{CPU} [IU]	Δt_{CPU}
p-MG(LIRK)	2	4.1	–	20449	–
p-MG(LBE)	2	12	+292%	5047	−75.3%
p-MG(LIRK)	3	12	–	194647	–
p-MG(LBE)	3	37.4	+311%	30427	−84.3%
p-MG(LIRK)	4	32	–	899902	–
p-MG(LBE)	4	122.4	+382%	191512	−78.7%

reduction around 80%. However, as expected, the memory requirement is higher and increases with the polynomial order.

T106A

The T106A turbine cascade is a low-pressure turbine cascade designed by MTU Aero Engines, which has been extensively investigated in experimental and computational studies in Hoheisel (1981), Ghidoni et al. (2012a), and Bassi et al. (2016),

The computations are performed for a downstream isentropic Mach number $M_{2,is} = 0.59$, a Reynolds number based on the downstream isentropic conditions and

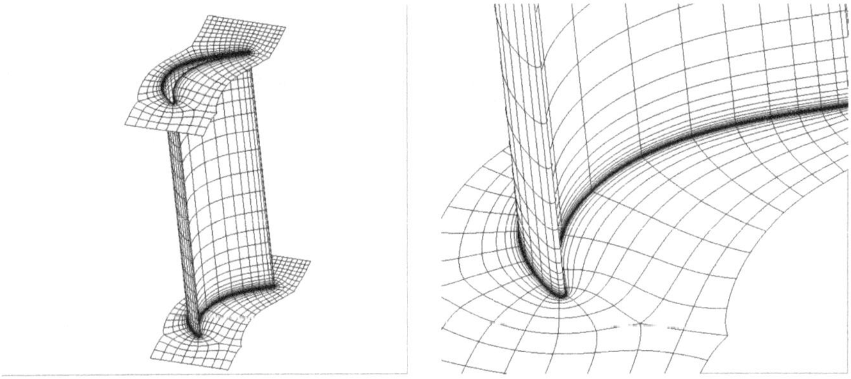

Fig. 4.16 T106A: mesh consisting of 43200 hexahedral elements with a quadratic 5representation of the boundary

on the blade chord $\mathrm{Re}_{2,is} = 0.5 \times 10^6$, an inlet turbulence intensity $Tu_1 = 4.0\%$, and an inlet angles $\alpha_1 = 37.7°$. The mesh consists of 43200 hexahedral elements with quadratic edges (see Fig. 4.16). The Mach number contours at midspan for different polynomial orders ($\mathbb{P}^{1\rightarrow3}$) are depicted in Fig. 4.17.

The objective of this test case is the comparison between a p-MG algorithm that uses only implicit smoother on all levels, p-MG(LBE), and the implicit solver, LBE, in terms of the memory usage (Mem) and the normalized CPU time (Time) needed to reach a converged solution. This is motivated by the results obtained on previous turbulent test cases. The p-MG setting, LBE smoother on each level, GMRES method (120 iteration, no-restart) preconditioned with ILU(0) on each block, can be summarized as: $\nu_c = 5$ on the coarsest level, $\nu_{f,1-2} = 2 - 2$ on the finest level and $\nu_{1-2} = 3 - 3$ on intermediate levels.

The convergence history of the p-MG(LBE) and the implicit solver (LBE) are compared as a function of the non-dimensional CPU time needed to reach a converged solution, as depicted in Fig. 4.18. Table 4.18 summarizes performance for both strategies. For every solution approximation p-MG(LBE) is better than the implicit scheme both in terms of CPU time and memory requirement. The memory requirement is reduced by 35%, while the CPU time by 79% and 82% for a $\mathbb{P}^2$ and a $\mathbb{P}^3$ solution, respectively. The different memory requirements can be explained with a different setting of the GMRES parameters, as the p-MG(LBE) adopts a small number of iteration and a higher tolerance.

Also in this case the FMG p-MG(LBE) solution is computed for a $\mathbb{P}^2$ and $\mathbb{P}^3$ spatial discretization to study the asymptotic behavior of the density residual L^2 norm, as depicted in Fig. 4.18.

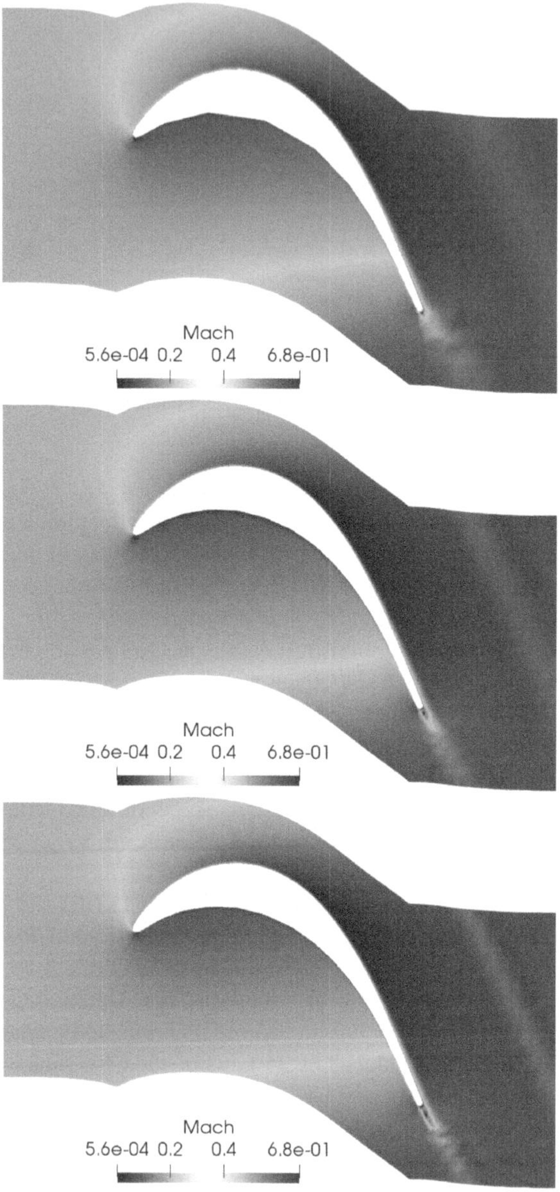

Fig. 4.17 T106A: Mach number contours for different polynomial orders, from $\mathbb{P}1$ (top) to $\mathbb{P}3$ (bottom)

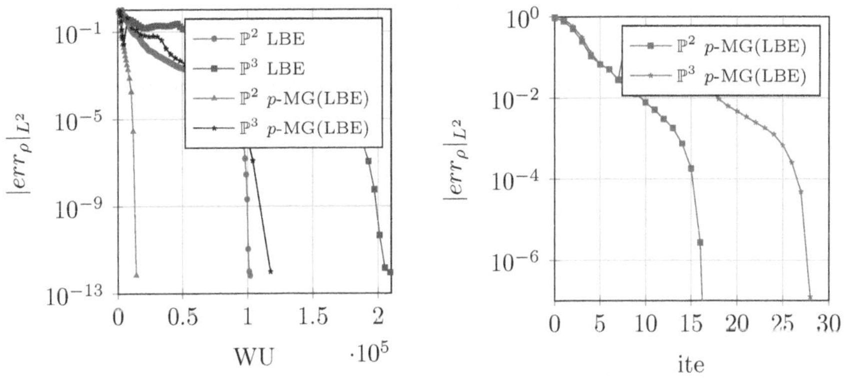

Fig. 4.18 T106A: convergence history versus normalized CPU time for different polynomial solution approximations and strategy (left) and convergence history versus MG cycles for a $\mathbb{P}^{2\to 3}$ approximation (right)

Table 4.18 T106A: comparison between p-MG(LBE) and implicit (LBE) strategies performance. $\mathbb{P}^k$ is the solution approximation, RAM the memory usage, Δ_{RAM} the memory requirement reduction due to the use of different solution strategies, t_{CPU} the normalized CPU time, Δt_{CPU} the computational time reduction due to the use of a different solution strategies (Memory usage and CPU time of the implicit scheme are the reference)

Scheme	$\mathbb{P}^k$	RAM [GB]	Δ_{RAM}	t_{CPU} [IU]	Δt_{CPU}
LBE	2	63	–	29994	–
p-MG(LBE)	2	44	−30.1%	6218	−79.2%
LBE	3	244	–	223776	–
p-MG(LBE)	3	150	−38.5%	38915	−82.6%

Conclusion

In this chapter a p-multigrid discontinuous Galerkin algorithm for the solution of the steady compressible Navier–Stokes and RANS equations has been presented and investigated. Different smoothers have been compared (block implicit and line implicit RK schemes, linearized Backward Euler) in terms of computational efficiency. The performance of the proposed approach has been evaluated by computing different 3-D laminar and turbulent shockless test cases and comparing the performance with an implicit solution strategy.

For the laminar test cases two "ingredients" have been identified for an efficient simulation: the line smoother on the finest level and the implicit smoother on the coarsest level. On the intermediate levels the block implicit smoother is adopted to reduce the memory footprint and exploit the projection of the Jacobian matrix from the finest level. The expected asymptotic convergence for different order of the finest level is not completely observed since, keeping fixed the number of pre/post-smoothing iterations, the number of iterations to reach a converged solution increases with the order of the finest level.

The strategy devised for the laminar test cases (p-MG(LIRK)) has been also used for the turbulent simulation around a Delta wing. The performance of this strategy shows a drastic reduction of the memory usage with respect to an implicit solver, while the computing time is comparable. For this reason the previous smoothers set-up has been compared with a setup based only on an implicit smoother (p-MG(LBE)) for the simulation of the turbulent flow around a train head. This choice, as expected, shows a huge increment in the memory usage, but also a reduction of the computing time around -80%.

Finally, the p-MG(LBE) has been compared with an implicit solver in the computation of the turbulent flow through a turbine cascade. Promising results have shown a reduction of the computing time around -80% and of the memory usage around -30%.

References

Arnold, D. N., Brezzi, F., Cockburn, B., & Marini, L. D. (2002). Unified analysis of discontinuous Galerkin methods for elliptic problems. *SIAM J. Numer. Anal.*, *39*(5), 1749–1779.

Balay, S., Buschelman, K., Gropp, W. D., Kaushik, D., Knepley, M. G., McInnes, L. C., Smith, B. F., & Zhang, H. (2001). PETSc Web page. http://www.mcs.anl.gov/petsc.

Bassi, F., & Rebay, S. (2000). GMRES discontinuous Galerkin solution of the compressible Navier-Stokes equations. In B. Cockburn, G. Karniadakis, & C. W. Shu (Eds.), *Discontinuous Galerkin Methods: Theory, Computation and Applications*. Number 11 in Lecture Notes in Computational Science and Engineering (pp. 197–208). New York: Springer.

Bassi, F., & Rebay, S. (2002). Numerical solution of the euler equations with a multiorder discontinuous finite element method. In Armfield, S., Morgan, P., & Srinivas, K. (Eds.), *Computational Fluid Dynamics 2002: Proceedings of the Second International Conference on Computational Fluid Dynamics* (pp. 199–204). Sydney: Springer.

Bassi, F., Rebay, S., Mariotti, G., Pedinotti, S., & Savini, M. (1997). A high-order accurate discontinuous finite element method for inviscid and viscous turbomachinery flows. In R. Decuypere & G. Dibelius (Eds.), *Proceedings of the 2nd European Conference on Turbomachinery Fluid Dynamics and Thermodynamics*, 5–7 March 1997 (pp. 99–108). Antwerpen, Belgium: Technologisch Instituut.

Bassi, F., Crivellini, A., Rebay, S., Savini, M. (2005). Discontinuous galerkin solution of the reynolds-averaged navier-stokes and k-ω turbulence model equations. *Computers & Fluids*, *34*(4), 507–540. https://doi.org/10.1016/j.compfluid.2003.08.004. Residual Distribution Schemes, Discontinuous Galerkin Schemes and Adaptation.

Bassi, F., Ghidoni, A., Rebay, S., & Tesini, P. (2009). High-order accurate p-multigrid discontinuous Galerkin solution of the Euler equations. *International Journal for Numerical Methods in Fluids*, *60*(8), 847–865. https://doi.org/10.1002/fld.1917.

Bassi, F., Botti, L., Colombo, A., Crivellini, A., Franchina, N., Ghidoni, A., & Rebay, S. (2010). Very high-order accurate discontinuous galerkin computation of transonic turbulent flows on aeronautical configurations. In N. Kroll, H. Bieler, H. Deconinck, V. Couaillier, H. Ven, and K. Sorensen (Eds.), *ADIGMA - A European Initiative on the Development of Adaptive Higher-Order Variational Methods for Aerospace Applications*, volume 113 of Notes on Numerical Fluid Mechanics and Multidisciplinary Design (pp. 25–38). Berlin, Heidelberg: Springer. https://doi.org/10.1007/978-3-642-03707-8_3.

Bassi, F., Franchina, N., Ghidoni, A., & Rebay, S. (2011a). Spectral p-multigrid discontinuous Galerkin solution of the Navier–Stokes equations. *International Journal for Numerical Methods in Fluids*, *67*(11), 1540–1558. https://doi.org/10.1002/fld.2430.

Bassi, F., Ghidoni, A., & S. Rebay. (2011b). Optimal Runge–Kutta smoothers for the p-multigrid discontinuous galerkin solution of the 1d euler equations. *Journal of Computational Physics*, *230*(11), 4153 – 4175. ISSN 0021-9991. https://doi.org/10.1016/j.jcp.2010.04.030.

Bassi, F., Botti, L., Colombo, A., Di Pietro, D. A., & Tesini, P. (2012). On the flexibility of agglomeration based physical space discontinuous Galerkin discretizations. *Journal of Computational Physics*, *231*(1), 45–65. ISSN 0021-9991.

Bassi, F., Franchina, N., Ghidoni, A., & Rebay, S. (2013). A numerical investigation of a spectral-type nodal collocation discontinuous Galerkin approximation of the Euler and Navier–Stokes equations. *International Journal for Numerical Methods in Fluids*, *71*(10), 1322–1339. https://doi.org/10.1002/fld.3713.

Bassi, F., Botti, L., Colombo, A., Crivellini, A., De Bartolo, C., Franchina, N., Ghidoni, A., Rebay, S. (2015a). Time integration in the discontinuous Galerkin code MIGALE - Steady problems. In *Notes on Numerical Fluid Mechanics and Multidisciplinary Design*, vol. 128, pp. 179–204. Berlin: Springer. https://doi.org/10.1007/978-3-319-12886-3_10.

Bassi, F., Botti, L., Colombo, A., Crivellini, A., Ghidoni, A., Nigro, A., & Rebay, S. (2015b). Time Integration in the discontinuous Galerkin Code MIGALE - Unsteady Problems. In N. Kroll, C. Hirsch, F. Bassi, C. Johnston, & K. Hillewaert (Eds.), *IDIHOM: Industrialization of High-Order Methods - A Top-Down Approach*, volume 128 of Notes on Numerical Fluid Mechanics and Multidisciplinary Design, pp. 205–230. Berlin: Springer International Publishing.

Bassi, F., Botti, L., Colombo, A., Crivellini, A., Franchina, N., & Ghidoni, A. (2016). Assessment of a high-order accurate Discontinuous galerkin method for turbomachinery flows. *International Journal of Computational Fluid Dynamics*, *30*(4), 307–328. https://doi.org/10.1080/10618562.2016.1198783.

Bassi, F., Botti, L., Colombo, A., Crivellini, A., Franciolini, M., Ghidoni, A., & Noventa, G. (2020a). A p-adaptive matrix-free discontinuous Galerkin method for the Implicit LES of incompressible transitional flows. *Flow, Turbulence and Combustion*. https://doi.org/10.1007/s10494-020-00178-2.

Bassi, F., Colombo, A., Crivellini, A., Fidkowski, K. J., Franciolini, M., Ghidoni, A., & Noventa, G. (2020b). Entropy-adjoint p-adaptive discontinuous Galerkin method for the under-resolved simulation of turbulent flows. *AIAA Journal*, *58*, 1–15. https://doi.org/10.2514/1.J058847.

Bernard, P.-E., Remacle, J.-F., Comblen, R., Legat, V., & Hillewaert, K. (2009). High-order discontinuous Galerkin schemes on general 2D manifolds applied to the shallow water equations. *Journal of Computational Physics*, *228*(17), 6514–6535. https://doi.org/10.1016/j.jcp.2009.05.046.

Briggs, W. L., Henson, V. E., & McCormick, S. F. (2000). *A Multigrid Tutorial* (2nd Edn.). Philadelphia: SIAM.

Cockburn, B., Dong, B., Guzmán, J., Restelli, M., & Sacco, R. (2009). A hybridizable discontinuous Galerkin method for steady-state convection-diffusion-reaction problems. *SIAM Journal on Scientific Computing*, *31*(5), 3827–3846. https://doi.org/10.1137/080728810.

Crippa, S. (2008). *Advances in vortical flow prediction methods for design of delta-winged aircraft*. Ph.D. thesis, KTH, Aeronautical and Vehicle Engineering. QC 20100713.

Crivellini, A., & Bassi, F. (2011). An implicit matrix-free discontinuous Galerkin solver for viscous and turbulent aerodynamic simulations. *Computers & Fluids*, *50*(1), 81–93. ISSN 0045-7930. https://doi.org/10.1016/j.compfluid.2011.06.020.

Darmofal, D., Fidkowski, K. (2004). Development of a higher-order solver for aerodynamic applications. In *42nd AIAA Aerospace Sciences Meeting and Exhibit*, AIAA 2004–112. AIAA. https://doi.org/10.2514/6.2004-436.

Diosady, L. T., & Darmofal, D. L. (2009). Preconditioning methods for discontinuous Galerkin solutions of the Navier-Stokes equations. *Journal of Computational Physics*, *228*(11), 3917–3935.

Fidkowski, K. J., Oliver, T. A., Lu, J., & Darmofal, L. (2005). *p*-multigrid solution of high-order discontinuous Galerkin discretizations of the compressible Navier–Stokes equations. *Journal of Computational Physics*, *207*(1), 92–113.

Flad, D., Beck, A., & Munz, C.-D. (2016). Simulation of underresolved turbulent flows by adaptive filtering using the high order discontinuous Galerkin spectral element method. *Journal of Computational Physics*, *313*, 1–12. https://doi.org/10.1016/j.jcp.2015.11.064.

Franciolini, M., Botti, L., Colombo, A., & Crivellini, A. (2020). *p*-multigrid matrix-free discontinuous Galerkin solution strategies for the under-resolved simulation of incompressible turbulent flows. *Computers & Fluids*, *206*, 104558. ISSN 0045-7930. https://doi.org/10.1016/j.compfluid.2020.104558.

Frère, A., Hillewaert, K., Chatelain, P., & Winckelmans, G. (2018). High reynolds number airfoil: From wall-resolved to wall-modeled LES. *Flow, Turbulence and Combustion*, *101*(2), 457–476. https://doi.org/10.1007/s10494-018-9972-9.

Ghidoni, A., Pasquale, D., Rebay, S., Colombo, A., & Bassi, F. (2007). *p*-multigrid Discontinuous Galerkin method for compressible turbulent flows. In *51st AIAA Aerospace Sciences Meeting including the New Horizons Forum and Aerospace Exposition 2013*, AIAA 2013-1002. Resto: AIAA.

Ghidoni, A., Colombo, A., Rebay, S., & Bassi, F. (2012a). Simulation of the transitional flow in a low pressure gas turbine cascade with a high-order discontinuous Galerkin method. *Journal of Fluids Engineering*, *135*(7), 1–8. https://doi.org/10.1115/1.4024107.

Ghidoni, A., Rebay, S., & Pasquale, D. (2012b). High-order accurate *p*-multigrid Discontinuous Galerkin solution for complex industrial applications. In *ECCOMAS 2012 - European Congress on Computational Methods in Applied Sciences and Engineering* (pp. 1074–1083).

Ghidoni, A., Colombo, A., Bassi, F., & Rebay, S. (2014). Efficient *p*-multigrid discontinuous Galerkin solver for complex viscous flows on stretched grids. *International Journal for Numerical Methods in Fluids*, *75*(2), 134–154. https://doi.org/10.1002/fld.3888.

Gottlieb, J. J., & Groth, C. P. T. (1988). Assessment of Riemann solvers for unsteady one-dimensional inviscid flows of perfect gases. *Journal of Computational Physics*, *78*(2), 437–458. ISSN 0021-9991. https://doi.org/10.1016/0021-9991(88)90059-9.

Hänel, D., Schwane, R., & Seider, G. (1987). On the accuracy of upwind schemes for the solution of the Navier–Stokes equations. In *8th Computational Fluid Dynamics Conference*. Reston: American Institute of Aeronautics and Astronautics. https://doi.org/10.2514/6.1987-1105, https://doi.org/10.2514/6.1987-1105.

Helenbrook, B. T., & Atkins, H. A. (2005). Application of "*p*"-multigrid to discontinuous Galerkin formulations of the poisson equation. *AIAA Journal*, *44*(3), 566–575.

Helenbrook, B. T., Mavriplis, D. J., & Atkins, H. A. (2003). Analysis of p-multigrid for continuous and discontinuous finite element discretizations. In *16th AIAA Computational Fluid Dynamics Conference*, AIAA-2003-3989, Orlando, Florida: AIAA.

Hoheisel, H. (1981). *Entwicklung neuer Entwurfskonzepte für zwei Turbinengitter, Teil III, Ergebnisse T106*. Technical report, Braunschweig: Institut für Entwurfsaerodynamik.

IDIHOM. (2015). Industrialisation of High-Order Methods—A top-down approach, Specific Targeted Research Project supported by European Commission. http://www.dlr.de/as/en/desktopdefault.aspx/tabid-7027/11654_read-27492/.

IPACS. (2005). Integrated Performance Analysis of Computer Systems—Benchmarks for distributed computer systems. http://www.ipacs-benchmark.org.

Jiang, Z., Yan, C., Yu, J., & Yuan, W. (2015). Practical aspects of *p*-multigrid discontinuous Galerkin solver for steady and unsteady RANS simulations. *International Journal for Numerical Methods in Fluids*, *78*(11), 670–690. https://doi.org/10.1002/fld.4035.

Klaji, C. M., van Raalte, M. H., van der Ven, H., & van der Vegt, J. J. W. (2007). *h*-multigrid for space-time discontinuous Galerkin discretizations of the compressible Navier–Stokes equations. *Journal of Computational Physics*, *227*(2), 1024–1045. https://doi.org/10.1016/j.jcp.2007.08.034.

Kopriva, D. A., & Gassner, G. (2010). On the quadrature and weak form choices in collocation type discontinuous Galerkin spectral element methods. *Journal of Scientific Computing*, *44*, 136–155. https://doi.org/10.1007/s10915-010-9372-3.

Kroll, N. (2009). ADIGMA - a European initiative on the development of adaptive higher-order variational methods for aerospace applications. In *47th AIAA Aerospace Sciences Meeting* (p. 498). Reston: AIAA. https://doi.org/10.2514/6.2009-176.

Kronbichler, M., & Wall, W. A. (2018). A performance comparison of continuous and discontinuous Galerkin methods with fast multigrid solvers. *SIAM Journal on Scientific Computing*, *40*(5), A3423–A3448. https://doi.org/10.1137/16M110455X.

Langer, S. (2013). Point and line implicit methods to improve the efficiency and robustness of the DLR TAU Code. In A. Dillmann, G. Heller, H.-P. Kreplin, W. Nitsche, & I. Peltzer (Eds.), *New Results in Numerical and Experimental Fluid Mechanics VIII: Contributions to the 17th STAB/DGLR Symposium Berlin, Germany 2010* (pp. 419–428). Berlin, Heidelberg: Springer. ISBN 978-3-642-35680-3. https://doi.org/10.1007/978-3-642-35680-3_50.

Li, F., & Shu, C. (2005). Locally divergence-free discontinuous Galerkin methods for MHD equations. *Journal of Scientific Computing*, *22*, 413–442. https://doi.org/10.1007/s10915-004-4146-4.

Luo, H., Baum, J. D., & Löhner, R. (2006). A p-multigrid discontinuous Galerkin method for the Euler equations on unstructured grids. *Journal of Computational Physics*, *211*(21), 767–783.

Luo, H., Segawa, H., & Visbal, M. R. (2012). An implicit discontinuous galerkin method for the unsteady compressible Navier–Stokes equations. *Computers & Fluids*, *53*, 133–144. https://doi.org/10.1016/j.compfluid.2011.10.009.

Mascarenhas, B. S., Helenbrook, B. T., & Atkins, H. A. (2007). Application of the p-multigrid algorithm to discontinuous Galerkin formulations of the compressible Euler equation. In *18th AIAA Computational Fluid Dynamics Conference*, AIAA-2007-4331, Miami, Florida, June 2007. Miami: AIAA.

Massa, F. C., Noventa, G., Lorini, M., Bassi, F., & Ghidoni, A. (2018). High-order linearly implicit two-step peer schemes for the discontinuous Galerkin solution of the incompressible Navier-Stokes equations. *Computers and Fluids*, *162*, 55–71. https://doi.org/10.1016/j.compfluid.2017.12.003.

Menter, F. R. (1994). Two-equation eddy-viscosity turbulence models for engineering applications. *AIAA Journal*, *32*(8), 1598–1605. https://doi.org/10.2514/3.12149.

Nastase, C. R., & Mavriplis, D. J. (2006). High-order discontinuous Galerkin methods using an hp-multigrid approach. *Journal of Computational Physics*, *213*(1), 330–357.

Noventa, G., Massa, F., Bassi, F., Colombo, A., Franchina, N., & Ghidoni, A. (2016). A high-order discontinuous Galerkin solver for unsteady incompressible turbulent flows. *Computers and Fluids*, *139*, 248–260. https://doi.org/10.1016/j.compfluid.2016.03.007.

Noventa, G., Massa, F., Rebay, S., Bassi, F., & Ghidoni, A. (2020). Robustness and efficiency of an implicit time-adaptive discontinuous Galerkin solver for unsteady flows. *Computers & Fluids*, *204*, 104529. https://doi.org/10.1016/j.compfluid.2020.104529.

Pazner, W., & Persson, P.-O. (2018). Approximate tensor-product preconditioners for very high order discontinuous Galerkin methods. *Journal of Computational Physics*, *354*, 344–369. ISSN 0021-9991. https://doi.org/10.1016/j.jcp.2017.10.030.

Persson, P.-O., & Peraire, J. (2008). Newton-GMRES preconditioning for discontinuous Galerkin discretizations of the Navier-Stokes equations. *SIAM Journal on Scientific Computing*, *30*(6), 2709–2733.

Persson, P.-O., Bonet, J., & Peraire, J. (2009). Discontinuous Galerkin solution of the Navier-Stokes equations on deformable domains. *Computer Methods in Applied Mechanics and Engineering*, *198*(17–20), 1585–1595. https://doi.org/10.1016/j.cma.2009.01.012.

Rønquist, E. M., & Patera, A. T. (1989). Spectral element multigrid I. Formulation and numerical results. *Journal of Scientific Computing*, *2*(4), 389–406.

Rueda-Ramírez, A. M., Manzanero, J., Ferrer, E., Rubio, G., & Valero, E. (2019). A p-multigrid strategy with anisotropic p-adaptation based on truncation errors for high-order discontinuous

Galerkin methods. *Journal of Computational Physics*, *378*, 209–233. https://doi.org/10.1016/j.jcp.2018.11.009.

Sermeus, K. (2011). IDIHOM CFD method assessment procedure. Technical report, Cassidian. http://www.idihom.de/home/.

Shahbazi, K., Mavriplis, D. J., & Burgess, N. K. (2009). Multigrid algorithms for high-order discontinuous Galerkin discretizations of the compressible Navier–Stokes equations. *Journal of Computational Physics*, *228*(21), 7917–7940. ISSN 0021-9991. https://doi.org/10.1016/j.jcp.2009.07.013.

Streett, C. L., Zang, T. A., & Hussaini, M. Y. (1985). Spectral multigrid methods with applications to transonic potential flow. *Journal of Computational Physics*, *57*(1), 43–76. https://doi.org/10.1016/0021-9991(85)90052-X.

Thomas, L.H. (1949). *Elliptic problems in linear differential equations over a network*. Technical Report, Watson Scientific Computing Lab Report, Columbia University, New York.

Trottenberg, U., Oosterlee, C. W., & Schüller, A. (2001). *Multigrid*. London, UK: Academic Press.

van der Vegt, J. J. W., & Rhebergen, S. (2012a). *hp*-multigrid as smoother algorithm for higher order discontinuous Galerkin discretizations of advection dominated flows: Part I. Multilevel analysis. *Journal of Computational Physics*, *231*(22), 7537–7563. https://doi.org/10.1016/j.jcp.2012.05.038.

van der Vegt, J. J. W., & Rhebergen, S. (2012b). *hp*-multigrid as smoother algorithm for higher order discontinuous Galerkin discretizations of advection dominated flows. Part II: Optimization of the Runge–Kutta smoother. *Journal of Computational Physics*, *231*(22), 7564–7583. ISSN 0021-9991. https://doi.org/10.1016/j.jcp.2012.05.037.

VFE2, Project (2020). The second international Vortex Flow Experiment (VFE-2). http://vfe2.dlr.de.

Wallraff, M., Leicht, T., & Lange-Hegermann, M. (2013). Numerical flux functions for Reynolds-Averaged Navier–Stokes and k-ω turbulence model computations with a line-preconditioned p-multigrid discontinuous Galerkin solver. *International Journal for Numerical Methods in Fluids*, *71*(8), 1055–1072. https://doi.org/10.1002/fld.3702.

Warburton, T. C., Lomtev, I., Du, Y., Sherwin, S., & Karniadakis, G. E. (1999). Galerkin and discontinuous Galerkin spectral/hp methods. *Computer Methods in Applied Mechanics and Engineering*, *175*, 343–359.

Wilcox, D. C. (1993). *Turbulence Modelling for CFD*. DCW industries Inc., La Cañada, CA, USA, 91011.

Zang, T. A., Wong, Y. S., & Hussaini, M. Y. (1982). Spectral multigrid methods for elliptic equations. *Journal of Computational Physics*, *48*(3), 485–501. https://doi.org/10.1016/0021-9991(82)90063-8.

Chapter 5
High-Order Accurate Time Integration and Efficient Implicit Solvers

Per-Olof Persson

Abstract This chapter introduces some of the most popular implicit time-integration methods, and a highly efficient preconditioner based on block-ILU factorizations and Minimum Discarded Fill element ordering. It also describes how to combine the benefits of both explicit and implicit time-integrators using Implicit–Explicit (IMEX) Runge–Kutta methods, which can be highly effective for problem with geometrically induced stiffness or large variations in mesh element sizes.

Introduction

After discretization of a system of conservation laws in space using, e.g., a high-order discontinuous Galerkin (DG) scheme, the result is a *semi-discrete* system of ordinary differential equations (ODEs) which needs to be integrated in time using some time-stepping scheme. In many cases, explicit time-integrators can be quite competitive for solving these systems. They are typically very easy to implement, since they only require evaluation of the residual vector. The main drawback with explicit methods is that they might need excessively small timesteps Δt to be stable. This happens when the corresponding system of ODEs is stiff, meaning it has widely varying timescales and it is not practical to resolve the fastest scales. There are several reasons why real-world problems are stiff, in particular when discretized using high-order DG methods on fully unstructured meshes, including:

- The timestep restriction for a first-order operator is given by $\Delta t \leq Ch/(p+1)^k$, where it can be shown that $k \leq 2$ or the better approximation $k \approx 1.78$, see e.g. Krivodonova and Qin (2013). This can be quite restrictive already for moderate polynomial degrees p, in particular when compared with a corresponding finite-difference method.

P.-O. Persson (✉)
Department of Mathematics, University of California, Berkeley,
Berkeley, CA 94720-3840, USA
e-mail: persson@berkeley.edu

M. Kronbichler and P.-O. Persson (eds.), *Efficient High-Order Discretizations for Computational Fluid Dynamics*, CISM International Centre for Mechanical Sciences 602, https://doi.org/10.1007/978-3-030-60610-7_5

- Fully unstructured tetrahedral meshes typically have several low-quality elements, which results in a (locally) small mesh size h which restricts the global timestep Δt.
- Many CFD problems have physically induced sources of stiffness, such as the acoustic waves of low-Mach number flows or viscous effects.
- In addition, for CFD problems with, e.g., turbulent boundary layers, thin stretched elements are used on the wall to resolve the flow at a reasonable computational cost. However, this induces high levels of stiffness. The situation is similar when shock waves are captured using h-adaptivity, with small stretched elements along the shock.

For these reasons, we consider implicit methods in this chapter. It is certainly not clear exactly which class of methods is best for a given problem, and the situation is further complicated by new generations of computer architectures such as GPUs, see, e.g., Klöckner et al. (2009), which can be very efficient, for example, in residual evaluation but typically not able to store large matrices. Nevertheless, it is clear that some sort of implicit methods will be needed for many practical problems in the future.

We will first introduce some of the most common implicit time-integration schemes. Then we will describe one of the most effective matrix-based preconditioning techniques for solving the systems that arise, the block-ILU method with Minimum Discarded Fill element ordering. We will also briefly address parallelization issues. Finally, for problems with large mesh element size variations, we will describe how to address this geometry-induced stiffness using Implicit–Explicit (IMEX) schemes.

High-Order Time-Stepping

In this section, we will review some of the main time-stepping schemes and their properties.

Semi-Discrete Formulation

We will start from a semi-discrete formulation of a system of conservation laws using the Discontinuous Galerkin method, written as a system of coupled ordinary differential equations (ODEs) of the form:

$$\boldsymbol{M}\boldsymbol{u}'(t) = \boldsymbol{r}(\boldsymbol{u}(t)), \tag{1}$$

where $\boldsymbol{u}(t)$ is a vector containing the degrees of freedom associated with the spatial solution at time t. This is typically represented using a nodal basis. The vector $\boldsymbol{u}'(t)$ denotes the component-wise time derivative of $\boldsymbol{u}(t)$, $\boldsymbol{M}$ is the mass matrix, and $\boldsymbol{r}$ is the residual vector. We consider a number of different time-integrators for (1), including explicit Runge–Kutta methods, backward differentiation formulas (BDF), diagonally implicit Runge–Kutta methods (DIRK), or fully Implicit Runge–Kutta (IRK) methods.

Explicit Methods

In many cases, explicit time-integrators can be quite competitive for solving systems of the form (1). They are typically very easy to implement, since they only require evaluation of the residual $\boldsymbol{r}(\boldsymbol{u})$. Given initial conditions $\boldsymbol{u}_0 = \boldsymbol{u}(t_0)$, the most basic method is (forward) Euler's method which steps from time t_n to t_{n+1} using the formula:

$$\boldsymbol{u}_{n+1} = \boldsymbol{u}_n + \Delta t \boldsymbol{M}^{-1} \boldsymbol{r}(\boldsymbol{u}_n). \tag{2}$$

High-order methods can be developed, and one of the most famous and widely used methods is the fourth-order Runge–Kutta method (RK4):

$$\boldsymbol{k}_1 = \boldsymbol{M}^{-1} \boldsymbol{r}(\boldsymbol{u}_i) \tag{3}$$

$$\boldsymbol{k}_2 = \boldsymbol{M}^{-1} \boldsymbol{r}(\boldsymbol{u}_i + \Delta t \boldsymbol{k}_1/2) \tag{4}$$

$$\boldsymbol{k}_3 = \boldsymbol{M}^{-1} \boldsymbol{r}(\boldsymbol{u}_i + \Delta t \boldsymbol{k}_2/2) \tag{5}$$

$$\boldsymbol{k}_4 = \boldsymbol{M}^{-1} \boldsymbol{r}(\boldsymbol{u}_i + \Delta t \boldsymbol{k}_3) \tag{6}$$

$$\boldsymbol{u}_{n+1} = \boldsymbol{u}_n + \Delta t(\boldsymbol{k}_1 + 2\boldsymbol{k}_2 + 2\boldsymbol{k}_3 + \boldsymbol{k}_4)/6. \tag{7}$$

However, as discussed above, for many problems the timestep restriction due to stiffness can be prohibitive. Therefore, we will consider implicit methods.

Backward-Differentiation Formulas (BDF)

The most basic implicit method is the backward Euler's method:

$$\boldsymbol{M}\boldsymbol{u}_{n+1} = \boldsymbol{M}\boldsymbol{u}_n + \Delta t\, \boldsymbol{r}(\boldsymbol{u}_{n+1}). \tag{8}$$

We note that since the unknown $\boldsymbol{u}_{n+1}$ appears in the right-hand side as an argument to the residual, systems of equations need to be solved to take a step. This is typically done with Newton's method, possibly with inexact Jacobians, which require solutions of large sparse systems of linear equations. Note that we do not explicitly invert the mass matrix, since it can be incorporated directly into the linear systems.

Backward Euler's method can be derived by fitting a straight line between $\boldsymbol{u}_n$ and $\boldsymbol{u}_{n+1}$, and imposing that the slope matches the equation at time t_{n+1}. The Backward-

Differentiation Formulas (BDF methods), see Shampine and Gear (1979), generalizes this to higher order accuracy by fitting polynomials of higher degree. One popular method is the BDF2 method:

$$\boldsymbol{M}\boldsymbol{u}_{n+1} = \frac{4\boldsymbol{M}}{3}\boldsymbol{u}_n - \frac{\boldsymbol{M}}{3}\boldsymbol{u}_{n-1} + \frac{2\Delta t}{3}\boldsymbol{r}(\boldsymbol{u}_{n+1}), \tag{9}$$

which is second-order accurate and L-stable. Unfortunately, higher degree BDF methods are not L-stable, and a theorem by Dahlquist states that no A-stable linear multistep method can have order of convergence greater than 2.

(Diagonally) Implicit Runge–Kutta Methods

The implicit one-step, multi-stage, Runge–Kutta methods can obtain both L-stability and high-order accuracy. A general s-stage, pth-order Runge–Kutta method for advancing the solution from t_n to t_{n+1} has the form:

$$\boldsymbol{M}\boldsymbol{k}_i = \boldsymbol{r}\left(t_n + \Delta t c_i, \boldsymbol{u}_n + \Delta t \sum_{j=1}^{s} a_{ij}\boldsymbol{k}_j\right), \tag{10}$$

$$\boldsymbol{u}_{n+1} = \boldsymbol{u}_n + \Delta t \sum_{i=1}^{s} b_i \boldsymbol{k}_i. \tag{11}$$

The coefficients a_{ij}, b_i, and c_i can be expressed in the form of the *Butcher tableau*,

$$\begin{array}{c|ccc} c_1 & a_{11} & \cdots & a_{1s} \\ \vdots & \vdots & \ddots & \vdots \\ c_s & a_{s1} & \cdots & a_{ss} \\ \hline & b_1 & \cdots & b_s \end{array} = \begin{array}{c|c} c & A \\ \hline & b^T \end{array}.$$

For an *explicit* Runge–Kutta method, the coefficient matrix A is strictly lower triangular. Since each stage only depends on previous stages, the method can be implemented using only residual evaluations. Otherwise, the method is an *implicit* Runge–Kutta method. If A is lower triangular, these methods are called *diagonally implicit* Runge–Kutta (DIRK) methods, see Alexander (1977). Although they are implicit, each stage only depends on itself and previous stages and can be implemented by a sequence of implicit solutions similar to a standard backward Euler method. For general coefficient matrices A, the method is called (fully) implicit Runge–Kutta methods (IRK). Then all the stages depend on each other and they require a fully coupled solution procedure.

One commonly used DIRK scheme is the following L-stable, third-order accurate, three-stage method:

$$\begin{array}{c|c} c & A \\ \hline & b^T \end{array} = \begin{array}{c|ccc} \alpha & \alpha & 0 & 0 \\ \tau_2 & \tau_2 - \alpha & \alpha & 0 \\ 1 & b_1 & b_2 & \alpha \\ \hline & b_1 & b_2 & \alpha \end{array} \qquad \begin{aligned} \alpha &= 0.435866521508459 \\ \tau_2 &= (1+\alpha)/2 \\ b_1 &= -(6\alpha^2 - 16\alpha + 1)/4 \\ b_2 &= (6\alpha^2 - 20\alpha + 5)/4 \end{aligned}$$

Implicit Solvers

For both the BDF and the DIRK methods described above, we solve the nonlinear systems that arise using Newton's method. This requires the solution of linear systems involving matrices of the form

$$\boldsymbol{A} \equiv \alpha_0 \boldsymbol{M} - \Delta t \frac{\partial \boldsymbol{r}}{\partial \boldsymbol{u}} = \alpha_0 \boldsymbol{M} - \Delta t \boldsymbol{K}, \tag{12}$$

where the constant α_0 depends on the method. For simplicity of presentation, we assume here that $\alpha_0 = 1$ in (12), which is the case for the first-order accurate backward Euler method. Other values of α_0, as required for higher order methods, simply correspond to a scaling of the timestep Δt.

Jacobian Matrices

The system matrix $\boldsymbol{A} = \boldsymbol{M} - \Delta t \boldsymbol{K}$ is sparse with a block-wise structure corresponding to the element connectivities. An example of a small triangular mesh with polynomial degrees $p = 2$ within each element is shown in Fig. 5.1. Note that the number of nonzero blocks in each row is equal to the number of neighbor elements plus the self-term. This mesh has triangular elements, which have three neighbors for a total of four nonzero blocks in each row. Quadrilateral elements would need five nonzero blocks, and in 3-D, tetrahedral and hexahedral elements would give five and seven nonzero blocks, respectively.

To be able to use machine optimized dense linear algebra routines, such as the BLAS/LAPACK libraries described in Anderson et al. (1999), the matrix $\boldsymbol{A}$ should be represented and stored in an efficient dense block format, see Persson and Peraire (2008) for details. A general-purpose sparse matrix format such as the Compressed Sparse Column format would result in a performance loss due to cache effects. We note that some schemes, such as the highly sparse Compact DG (CDG) scheme in Peraire and Persson (2008), have significant sparsity in the off-diagonal blocks, which we take advantage of in our implementation. However, in the presentation here, we assume for simplicity that all nonzero blocks are full dense matrices. The block with element indices $1 \le i, j \le n$ will be denoted by $\boldsymbol{A}_{ij}$, where n is the total number of elements.

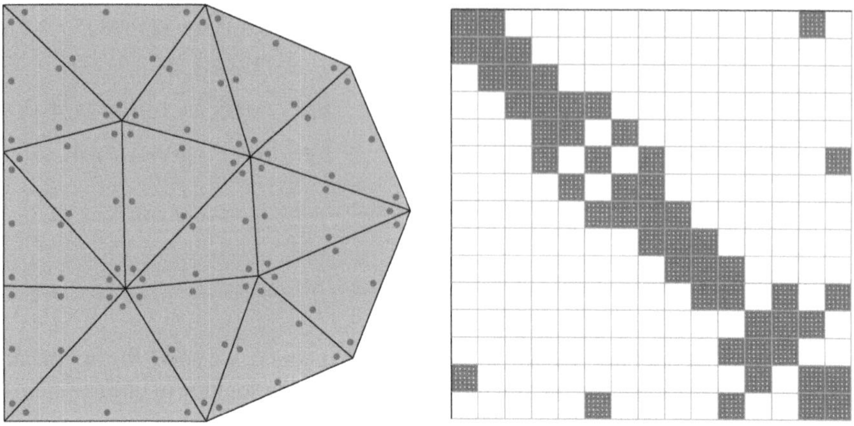

Fig. 5.1 An example mesh with elements of polynomial order $p = 2$, and the corresponding block matrix for a scalar problem

Incomplete LU Preconditioning

It is clear that the performance of the iterative solvers will depend strongly on the timestep Δt. As $\Delta t \to 0$, the matrix $\boldsymbol{A}$ reduces to the mass matrix, which is block-diagonal and inverted exactly by all preconditioners that we consider. However, as $\Delta t \to \infty$, the problem becomes harder and often not well-behaved. Physically, a small Δt means that the information propagation is local, while a large Δt means information is exchanged over large distances during the timestep. This effect, which is important when designing iterative methods, is even more important when we consider parallel algorithms since algorithms based on local information exchanges usually scale better than ones with global communication patterns.

When solving the system $\boldsymbol{A}\boldsymbol{u} = \boldsymbol{b}$ using Krylov subspace iterative methods such as GMRES, it is essential to use a good preconditioner. This amounts to finding an approximation $\tilde{\boldsymbol{A}}$ to $\boldsymbol{A}$ which allows for a relatively inexpensive computation of $\tilde{\boldsymbol{A}}^{-1}\boldsymbol{p}$ for arbitrary vectors $\boldsymbol{p}$. One of the simplest choices that performs reasonably well for many problems is the block-diagonal, or the block-Jacobi, preconditioner

$$\tilde{\boldsymbol{A}}^{\mathrm{J}}_{ij} = \begin{cases} \boldsymbol{A}_{ij} & \text{if } i = j\,, \\ \boldsymbol{0} & \text{if } i \neq j\,. \end{cases} \tag{13}$$

Clearly, $\tilde{\boldsymbol{A}}^{J}$ is cheap to invert compared to $\boldsymbol{A}$, since all the diagonal blocks are independent. However, unlike the point-Jacobi iteration, there is a significant pre-processing cost in the factorizations of the diagonal blocks $\tilde{\boldsymbol{A}}^{J}_{ij}$, which is comparable to the cost of more complex factorizations (Persson and Peraire 2008).

A minor modification of the block-diagonal preconditioner is the block Gauss–Seidel preconditioner, which keeps the diagonal blocks plus all the upper triangular blocks:

$$\tilde{\boldsymbol{A}}_{ij}^{\mathrm{GS}} = \begin{cases} \boldsymbol{A}_{ij} & \text{if } i \leq j\,, \\ \boldsymbol{0} & \text{if } i > j\,. \end{cases} \tag{14}$$

The preprocessing cost is the same as before, and solving $\tilde{\boldsymbol{A}}^{\mathrm{GS}}\tilde{\boldsymbol{u}} = \boldsymbol{p}$ is only a constant factor more expensive than for $\tilde{\boldsymbol{A}}^{J}$. The Gauss–Seidel preconditioner can perform well for some simple problems, such as scalar convection problems, but in general it only gives a small factor of improvement over the block-diagonal preconditioner. Furthermore, the sequential nature of the triangular back-solve with $\tilde{\boldsymbol{A}}^{\mathrm{GS}}$ makes the Gauss–Seidel preconditioner hard to parallelize.

A more ambitious preconditioner with similar storage and computational cost is the block incomplete LU factorization $\tilde{\boldsymbol{A}}^{\mathrm{ILU}} = \tilde{\boldsymbol{L}}\tilde{\boldsymbol{U}}$ with zero fill-in. This block-ILU(0) algorithm corresponds to block-wise Gaussian elimination, where no new blocks are allowed into the matrix. This factorization can be computed with the following simple algorithm:

function $[\tilde{\boldsymbol{L}}, \tilde{\boldsymbol{U}}] \leftarrow \mathit{IncompleteLU}(\boldsymbol{A}, \mathit{mesh})$
$\tilde{\boldsymbol{U}} = \boldsymbol{A}, \tilde{\boldsymbol{L}} = \boldsymbol{I}$
for $j = 1, \ldots, n-1$
  for neighbors $i > j$ of j in *mesh*
    $\tilde{\boldsymbol{L}}_{ij} = \tilde{\boldsymbol{U}}_{ij}\tilde{\boldsymbol{U}}_{jj}^{-1}$
    $\tilde{\boldsymbol{U}}_{ii} = \tilde{\boldsymbol{U}}_{ii} - \tilde{\boldsymbol{L}}_{ik}\tilde{\boldsymbol{U}}_{ki}$

We also note here that the upper triangular blocks of $\tilde{\boldsymbol{U}}$ are identical to those in $\boldsymbol{A}$, which reduces the storage requirements for the factorization. The back-solve using $\tilde{\boldsymbol{L}}$ and $\tilde{\boldsymbol{U}}$ has the same sequential nature as for Gauss–Seidel, but it turns out that the performance of the block-ILU(0) preconditioner can be fundamentally better.

Minimum Discarded Fill Element Ordering

It is clear that the Gauss–Seidel and the incomplete LU factorizations will depend strongly on the ordering of the blocks, or the elements, in the mesh. This is because the mesh ordering determines which element connections are kept and which are discarded when calculating $\tilde{A}$. In Persson and Peraire (2008), Persson (2009), we proposed a simple heuristic algorithm for finding appropriate element orderings. Our algorithm considers the fill that would be ignored if element j' was chosen as the pivot element at step j:

$$\Delta\tilde{\boldsymbol{U}}_{ik}^{(j,j')} = -\tilde{\boldsymbol{U}}_{ij'}\tilde{\boldsymbol{U}}_{j'j'}^{-1}\tilde{\boldsymbol{U}}_{j'k}, \qquad \text{for neighbors } i \geq j, k \geq j \text{ of element } j'. \tag{15}$$

The matrix $\Delta\tilde{\boldsymbol{U}}^{(j,j')}$ corresponds to fill that would be discarded by the ILU algorithm. In order to minimize these errors, we consider a set of candidate pivots $j' \geq j$ and pick the one that produces the smallest fill. As a measurement of the magnitude of the fill, or the corresponding weight, we take the Frobenius matrix norm of the fill matrix:

$$w^{(j,j')} = \|\Delta\tilde{\boldsymbol{U}}^{(j,j')}\|_{\mathrm{F}}\,. \tag{16}$$

As a further simplification, we note that

$$\|\Delta\tilde{\boldsymbol{U}}_{ik}^{(j,j')}\|_{\mathrm{F}} = \| - \tilde{\boldsymbol{U}}_{ij'}\tilde{\boldsymbol{U}}_{j'j'}^{-1}\tilde{\boldsymbol{U}}_{j'k}\|_{\mathrm{F}} \leq \|\tilde{\boldsymbol{U}}_{ij'}\|_{\mathrm{F}}\|\tilde{\boldsymbol{U}}_{j'j'}^{-1}\tilde{\boldsymbol{U}}_{j'k}\|_{\mathrm{F}}\,, \tag{17}$$

which means we can estimate the weight by simply multiplying the norms of the individual matrix blocks. By pre-multiplication of the block-diagonal, we can also avoid the matrix factor $\tilde{\boldsymbol{U}}_{j'j'}^{-1}$ above. The pseudocode of the final algorithm is given below.

function $p = \mathit{OrderingMDF}(\boldsymbol{A}, \mathit{mesh})$
for all neighbors i, j in *mesh*
    $C_{ij} = \|\boldsymbol{A}_{ii}^{-1}\boldsymbol{A}_{ij}\|_{\mathrm{F}}$ — *Reduce to scalars*
for $k = 1, \ldots, n$
    $w_k = \mathit{ComputeWeight}(k, w, C, \mathit{mesh})$ — *Compute all weights*
elements $= \{\}$ — *List of candidates*
for $i = 1, \ldots, n$ — *Main loop*
    if is_empty(elements)
        elements $= \mathrm{argmin}_j w_j$ — *Choose smallest fill*
    $p_i = \mathrm{argmin}_j w_j$, $j \in$ elements — *Choose best candidate*
    $w_{p_i} = \infty$ — *Do not use p_i again*
    elements $=$ elements $\setminus p_i$ — *Remove p_i from list*
    for neighbors k of p_i in *mesh*, $w_k \neq \infty$ — *Update weights*
        $w_k = \mathit{ComputeWeight}(k, w, C, \mathit{mesh})$
        elements $=$ elements $\cup\, k$

function $w_k = \mathit{ComputeWeight}(k, w, C, \mathit{mesh})$
$\Delta C = 0$
for neighbors i, j of element k in *mesh*,
    if $i \neq j$, $w_i \neq \infty$, $w_j \neq \infty$
        $\Delta C_{ij} = C_{ik}C_{kj}$ — *Discarded fill matrix*
$w_k = \|\Delta C\|_{\mathrm{F}}$ — *Fill weight*

We note that for a pure upwinded scalar convective problem, the MDF ordering is optimal since at each step it picks an element that either does not affect any other elements (downwind) or does not depend on any other elements (upwind), resulting in a perfect factorization. But the algorithm works well for other problems too, including multivariate and viscous problems, since it tries to minimize the error between the exact and the approximate LU factorizations. It also takes into account the effect of the discretization (e.g., highly anisotropic elements) on the ordering. These aspects are harder to account for with methods that are based on physical observations, such as line-based orderings (Nastase and Mavriplis 2006; Fidkowski et al. 2005; Kanschat 2008; Diosady and Darmofal 2009).

Performance

We demonstrate the performance of the various preconditions on a typical CFD problem of flow around a 2-D airfoil, at a range of Reynolds numbers, Mach numbers, timesteps Δt, and preconditioners (see Table 5.1). Both the block Gauss–Seidel and the block ILU0 preconditioners use the MDF ordering algorithm. Apart from the inviscid problem with $M = 0.2$ and $\Delta t = 10^{-3}$, all problems have CFL numbers higher than 100,000 which makes an explicit solution procedure impractical. We can make the following observations:

- The block-Jacobi preconditioner performs very poorly in general, only showing decent performance for small timesteps and large Mach numbers.
- The block GS preconditioner behaves similarly to the block Jacobi preconditioner, only with about a factor of two faster convergence.
- The ILU0 preconditioner shows very good performance for almost all of the test cases, except for large time steps and small Mach numbers. In these cases, multigrid / coarse grid corrections are required for faster convergence, for details see Persson and Peraire (2008).

Parallelization

Finally, we give a few comments on the parallelization of the presented algorithms. For more details, see Persson (2009). It is non-trivial to parallelize the schemes, due to the highly serial structure of the Gauss–Seidel and the ILU preconditioners. We parallelize using standard domain decomposition, where the mesh is divided into a set of partitions that are distributed to different computational processes (see Fig. 5.2). Our preferred approach for parallelization of the ILU factorization is to apply it according to the element orderings determined by the MDF algorithm, but ignoring any contributions between elements in different partitions. In standard domain decomposition terminology, this essentially amounts to a non-overlapping Schwartz

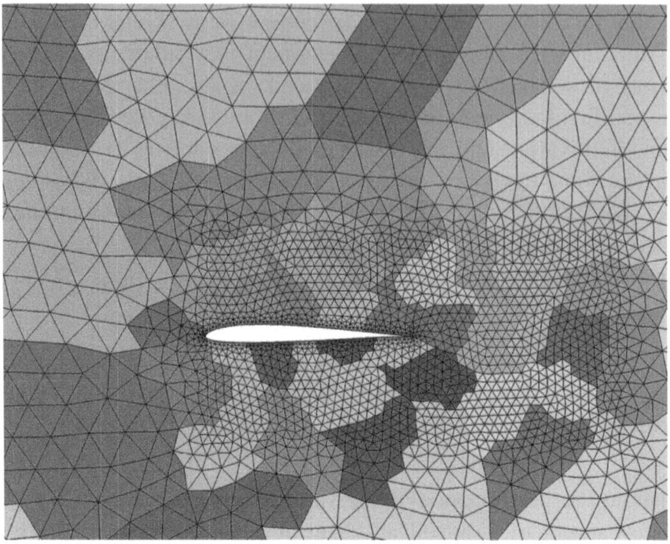

Fig. 5.2 Partitioning of the mesh elements for parallelization using domain decomposition

algorithm with incomplete solutions (Toselli and Widlund 2004). It is clear that this approach will scale well in parallel, since each process will compute a partition-wise ILU factorization independent of all other processes.

To minimize the error introduced by separating the ILU factorizations, we use the ideas from the MDF algorithm to obtain information about the weights of the connectivities between the elements. By computing a weighted partitioning using the weight

$$C_{ij} = \|\boldsymbol{A}_{ii}^{-1}\boldsymbol{A}_{ij}\|_F \tag{18}$$

between element i and j, we obtain partitions that are less dependent on each other and reduce the error from the decomposition. The drawback is that the total communication volume might be increased, but if desired, a trade-off between these two effects can be obtained by adding a constant C_0 to the weights. In practice, since the METIS software (Karypis and Kumar 1997) used for the partitioning requires integer weights, we scale and round the C_{ij} values to integers between 1 and 100. It is clear that this method reduces to the block-Jacobi method as the number of partitions approaches the number of elements. However, in any practical setting, each partition will have at least 100's of elements, in which the difference between partition-wise block-ILU and block-Jacobi is quite significant.

Implicit–Explicit (IMEX) Time-Integration

For problems involving turbulent flows modeled by Large Eddy Simulation (LES), the most critical reason for employing implicit methods is often the small and stretched mesh elements that are required to resolve the thin boundary layers. For these simulations, the stiffness is caused by the equations corresponding to the degrees of freedom in the stretched region, and the mesh elements away from the bodies might be better handled by explicit time integration. This effect will likely be even more important on the future generation of multi-core and GPU computer architectures, which appear to favor the local explicit methods over the more memory-intensive implicit ones (Klöckner et al. 2009).

In an attempt to take advantage of the fact that the problems are nonstiff in most of the computational domain, we will here present a way to use so-called IMEX schemes (Ascher et al. 1997; Kennedy and Carpenter 2003) to obtain a combination of the best properties of the implicit and the explicit solvers. These methods are based on a splitting of the residual vector into a stiff and a nonstiff part, and an additive Runge–Kutta method creates a combined method that can be made high-order accurate in time. Many of the original applications of the IMEX schemes used splittings of the actual equations (for example, into nonstiff advective terms and stiff diffusive terms), but here we use a splitting based on the size of the elements in the mesh. The resulting scheme can be highly efficient, and the Jacobians that have to be computed and used for solving nonlinear equations might only be a fraction of the size of the fully implicit ones. In addition, we re-use both the computed Jacobian matrices as well as the incomplete factorizations, which brings down the cost of the implicit solvers further. More details on the technique can be found in Persson (2011).

Implicit–Explicit Runge–Kutta Methods

The IMEX schemes are based on a splitting of the residual vector in a system of ODEs of the form

$$\frac{du}{dt} = f(u) + g(u), \tag{19}$$

where $f(u)$ is considered nonstiff terms and $g(u)$ stiff terms. The schemes are of Runge–Kutta type, with one scheme c, A, b for the implicit treatment of $g(u)$ and another scheme $\hat{c}, \hat{A}, \hat{b}$ for the explicit treatment of $f(u)$. These are standard DIRK and Explicit Runge–Kutta schemes by themselves. However, the schemes are also designed in such a way that they can be combined for integration of ODEs of the split form (19).

To integrate from step n to step $n+1$ using the timestep Δt, the first stage is always explicit, and the remaining s stages are done pairwise implicit/explicit. The solution at timestep $n+1$ is then a linear combination of the stage derivatives of both schemes, and the method can be written as

$\hat{k}_1 = f(u_n)$
for $i = 1$ **to** s
Solve for k_i in $k_i = g(u_{n,i})$, where

$$u_{n,i} = u_n + \Delta t \sum_{j=1}^{i} a_{i,j} k_j + \Delta t \sum_{j=1}^{i} \hat{a}_{i+1,j} \hat{k}_j$$

Evaluate $\hat{k}_{i+1} = f(u_{n,i})$
end for

$$u_{n+1} = u_n + \Delta t \sum_{i=1}^{s} b_j k_j + \Delta t \sum_{i=1}^{s+1} \hat{b}_j \hat{k}_j$$

A number of IMEX schemes of this form have been proposed, with various orders of accuracy and stability properties (Ascher et al. 1997). Here we consider three typical schemes:

IMEX1:
2nd order accurate: 2-stage, 2nd order DIRK + 3-stage, 2nd order ERK

$$\begin{array}{c|c} c & A \\ \hline & b^T \end{array} = \begin{array}{c|cc} \alpha & \alpha & 0 \\ 1 & 1-\alpha & \alpha \\ \hline & 1-\alpha & \alpha \end{array} \qquad \begin{array}{c|c} \hat{c} & \hat{A} \\ \hline & \hat{b}^T \end{array} = \begin{array}{c|ccc} 0 & 0 & 0 & 0 \\ \alpha & \alpha & 0 & 0 \\ 1 & \delta & 1-\delta & 0 \\ \hline & 0 & 1-\alpha & \alpha \end{array}$$

with $\alpha = 1 - \frac{\sqrt{2}}{2}$ and $\delta = -2\sqrt{2}/3$. This DIRK scheme is stiffly accurate, and while the ERK is only second-order accurate it has the same stability region as a third-order ERK which is appropriate for problems with eigenvalues close to the imaginary axis.

IMEX2:
3rd order accurate: 2-stage, 3rd order DIRK + 3-stage, 3rd order ERK

$$\begin{array}{c|c} c & A \\ \hline & b^T \end{array} = \begin{array}{c|cc} \alpha & \alpha & 0 \\ 1-\alpha & 1-2\alpha & \alpha \\ \hline & \frac{1}{2} & \frac{1}{2} \end{array} \qquad \begin{array}{c|c} \hat{c} & \hat{A} \\ \hline & \hat{b}^T \end{array} = \begin{array}{c|ccc} 0 & 0 & 0 & 0 \\ \alpha & \alpha & 0 & 0 \\ 1-\alpha & \alpha-1 & 2(1-\alpha) & 0 \\ \hline & 0 & \frac{1}{2} & \frac{1}{2} \end{array}$$

with $\alpha = (3+\sqrt{3})/6$. The resulting scheme is third-order accurate with the same number of stages as the previous scheme, at the cost of losing the L-stability of the DIRK scheme.

IMEX3:
3rd order accurate: 3-stage, 3rd order DIRK + 4-stage, 3rd order ERK

$$\begin{array}{c|c} c & A \\ \hline & b^T \end{array} =$$

$$\begin{array}{c|ccc} 0.4358665215 & 0.4358665215 & 0 & 0 \\ 0.7179332608 & 0.2820667392 & 0.4358665215 & 0 \\ 1 & 1.208496649 & -0.644363171 & 0.4358665215 \\ \hline & 1.208496649 & -0.644363171 & 0.4358665215 \end{array}$$

$$\begin{array}{c|c} \hat{c} & \hat{A} \\ \hline & \hat{b}^T \end{array} =$$

$$\begin{array}{c|cccc} 0 & 0 & 0 & 0 & 0 \\ 0.4358665215 & 0.4358665215 & 0 & 0 & 0 \\ 0.7179332608 & 0.3212788860 & 0.3966543747 & 0 & 0 \\ 1 & -0.105858296 & 0.5529291479 & 0.5529291479 & 0 \\ \hline & 0 & 1.208496649 & -0.644363171 & 0.4358665215 \end{array}$$

This three-stage DIRK scheme is L-stable and third-order accurate, and the four-stage ERK scheme is third-order accurate but has the larger stability region of a fourth-order ERK.

Mesh-Size-Based Splitting of Residual

For our system of ODEs (1), we associate each component of the solution vector $\boldsymbol{u}$ with an equation in the residual $\boldsymbol{r}(\boldsymbol{u})$. Our splitting is based on identifying stiff components $\boldsymbol{u}_{\text{im}}$ that are located in mesh elements smaller than a given size, and the remaining nonstiff components $\boldsymbol{u}_{\text{ex}}$. This produces a splitting of the residual vector as

$$\boldsymbol{r}(\boldsymbol{u}) = \begin{bmatrix} \boldsymbol{r}_{\text{im}}(\boldsymbol{u}) \\ \boldsymbol{r}_{\text{ex}}(\boldsymbol{u}) \end{bmatrix} = \begin{bmatrix} 0 \\ \boldsymbol{r}_{\text{ex}}(\boldsymbol{u}) \end{bmatrix} + \begin{bmatrix} \boldsymbol{r}_{\text{im}}(\boldsymbol{u}) \\ 0 \end{bmatrix} = \boldsymbol{f}(\boldsymbol{u}) + g(\boldsymbol{u}). \tag{20}$$

The idea behind this splitting is that the stiffness from the implicit equations should not affect the explicit ones much, and it should be possible to use a timestep limited by the equations in the nonstiff region only. But note that depending on the equations and on the splitting this might not be the case, since the two schemes are coupled between each integration stage and it is unclear how this affects the stability properties of the full scheme. However, as we show in our numerical results below, it is often the case that the IMEX schemes can produce stable results with the larger timestep dictated by the element sizes in the explicit region only.

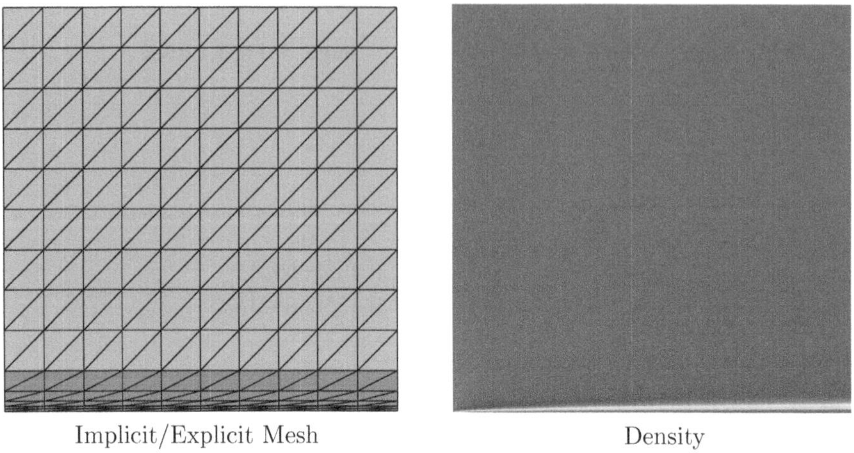

Fig. 5.3 The flow over a flat plate model problem with a thin boundary layer. A typical computational mesh (left) and the steady-state solution (right)

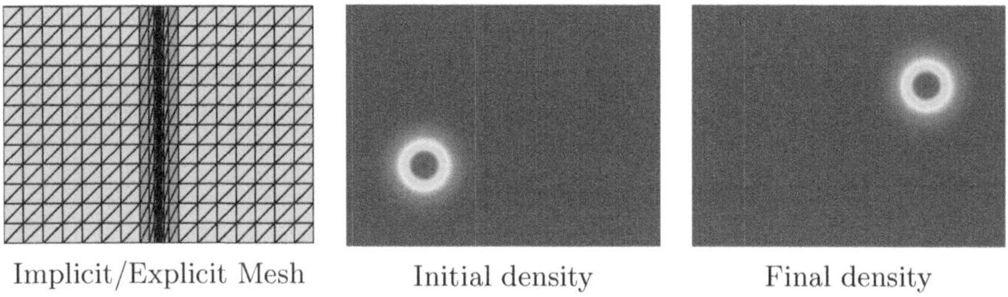

Fig. 5.4 Unsteady Euler vortex problem, computational mesh (left) with implicit elements blue and explicit elements green, and the initial/final solutions (center and right)

Quasi-Newton and Preconditioned Krylov Methods

For the implicit part of the IMEX scheme, nonlinear systems of equations $\boldsymbol{M}\boldsymbol{k}_i = g(\boldsymbol{u}_{n,i})$ must be solved. For this, we use the solution techniques presented in the previous section. In addition, we use a so-called Jacobian recycling approach where the Jacobian matrix is computed and stored explicitly but re-used between the iterations as well as between the timesteps. This turns out to work very well for these types of computations, and with the exception of the first initial transients we essentially never have to recompute the Jacobian matrix. We can also re-use the incomplete factorization and the total implicit solution time is dominated by the matrix-vector products and the backsolves.

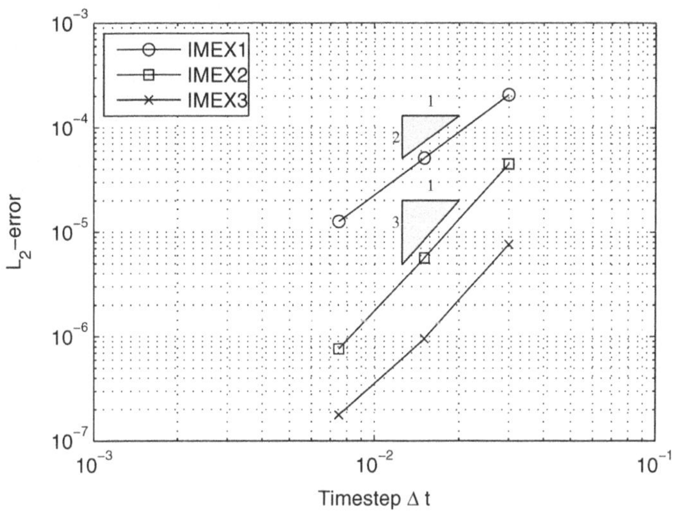

Fig. 5.5 Temporal convergence of the three IMEX schemes for the Euler vortex problem on a mesh with stretched elements. The orders of convergence 2, 3, 3, agree with the expected orders

Numerical Results

We demonstrate our method on three test problems. First, we study the stability of the schemes using a model problem of flow over a flat plate in a rectangular domain. Next, we use an Euler vortex model problem to determine the orders of accuracy of the schemes. Finally, we apply the technique on a more realistic simulation of turbulent flow over an airfoil at a high angle of attack. All simulations are done using our software package 3DG (Persson et al. 2010), which is a general-purpose toolkit for discretization of arbitrary systems of conservation laws (Peraire and Persson 2008). It produces fully analytical Jacobian matrices, and it includes efficient parallel Newton–Krylov solvers (Persson and Peraire 2008; Persson 2009). Due to the modular and general design of our software, it was straight-forward to incorporate the implicit–explicit capabilities using the existing solvers.

Absolute Stability, Flow Over a Flat Plate: Our first problem is a simple flat plate model problem that we use to to identify the feasibility of the approach, and in particular to determine if the CFL condition for the explicit portion of the domain is affected by the implicit portion. The domain is a square of unit length, with free-stream boundary conditions at left/top, no-slip wall conditions at bottom, and an outflow condition at the right boundary. We set the Mach number to 0.2 and the Reynolds number to 10,000 based on the domain width. A mesh and a steady-state solution are shown in Fig. 5.3.

A series of meshes of increasing anisotropy is generated in the following way: The initial mesh has 10-by 10 uniformly sized squares (of size 0.1-by-0.1). The bottom row is then split horizontally, and this process is repeated n_{ref} times to generate

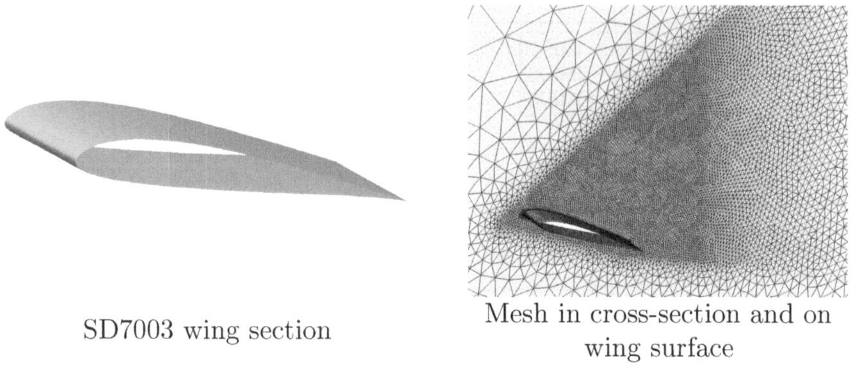

Fig. 5.6 Geometry and the extruded hybrid mesh for the ILES problem, with 312,000 tetrahedral elements

an anisotropic boundary layer mesh. Finally, we split each quadrilateral into two triangles, since our code is based on simplex elements. We note that the smallest element height is $h_{\min} = 0.1/2^{n_{\text{ref}}}$, and the highest element aspect ratio is $2^{n_{\text{ref}}}$.

For the IMEX1 and IMEX3 schemes, we first determine the largest stable timestep $\Delta t_{\max}$ if the problem was solved using the ERK method only. This is done in an automated way, using a bisection method applied to a function that determines stability numerically. In particular, we define the timestep on the coarse unrefined initial mesh by $\Delta t^0_{\max}$, which is also the timestep we hope to be able to use for our IMEX schemes on each of the stretched meshes.

We then run the test problem using the full IMEX scheme, where all split elements are considered implicit and the remaining (square) elements are considered explicit. To confirm that the stability of this scheme is determined by the mesh size in the explicit portion of the domain, we verify that the method is stable on any of the refined meshes using the timestep $\Delta t^0_{\max}$.

The results are presented in Table 5.2. We make the following observations:

- The timestep on the unrefined mesh $\Delta t^0_{\max}$ is almost equal for ERK1 but about 25% smaller for ERK3. This is unexpected since ERK3 has a larger linear stability region, but for this highly nonlinear problem it appears to be more sensitive than the other two.
- As the boundary layer is refined, the timestep $\Delta t_{\max}$ scales first linearly with $h_{\min}$ (ratio of about 2 between successive values of n_{ref}), and then quadratically (ratio of about 4). This is expected because the inviscid timestep restrictions are dominant for the under-resolved meshes, but eventually the viscous terms are limiting the timesteps.
- Both IMEX schemes are stable with the timestep $\Delta t^0_{\max}$, independently of the number of refinements n_{ref} and, therefore, of $h_{\min}$.

Order of Accuracy, Euler vortex problem: To validate the accuracy of the IMEX schemes, we solve an inviscid model problem consisting of a compressible vortex in

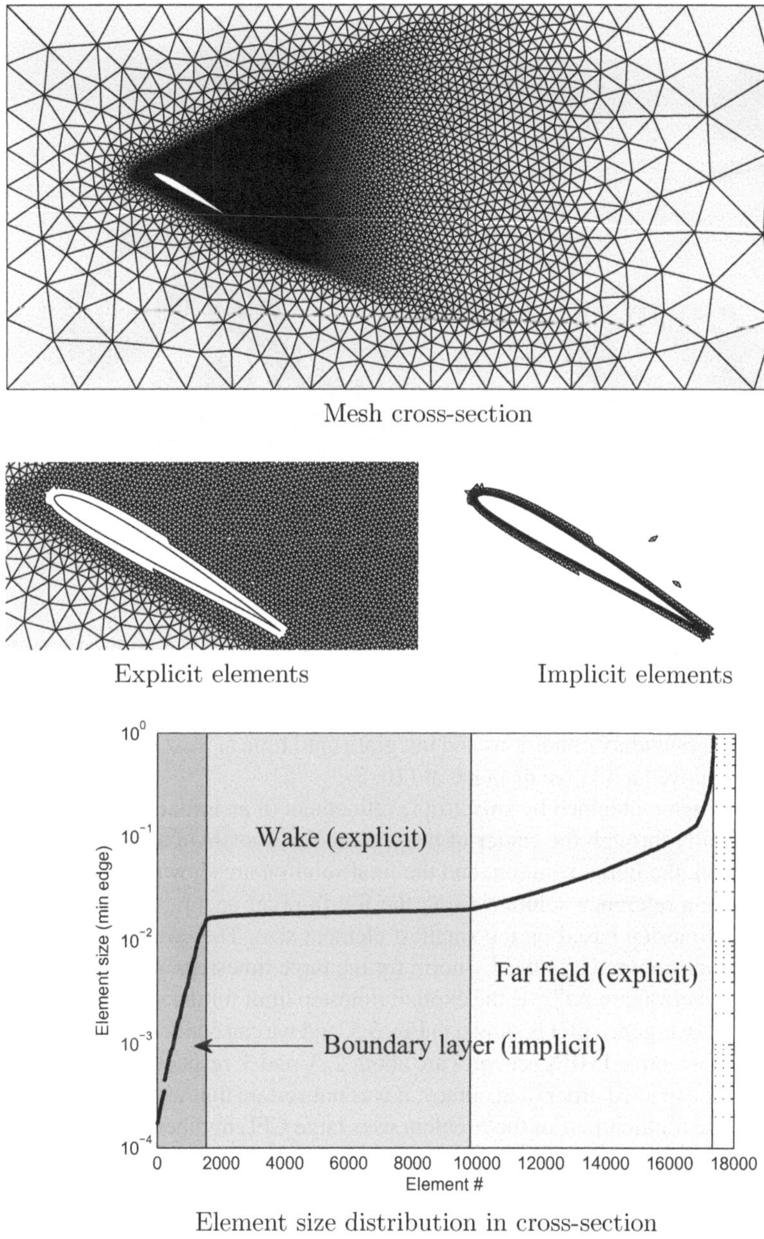

Fig. 5.7 Splitting of the mesh around an SD7003 airfoil into implicit and explicit elements. Less than 9% of the elements are in the boundary layer and integrated with the implicit scheme

Fig. 5.8 Four instantaneous solutions to the flow over an SD7003 airfoil, shown by the Mach number as color on an isosurface of the entropy

a rectangular domain (Persson et al. 2009). We use a domain of size 20-by-15, with the vortex initially centered at $(x_0, y_0) = (5, 5)$ with respect to the lower left corner. The Mach number is $M_\infty = 0.5$ and the free-stream velocity angle $\theta = \arctan(1/2)$. We use periodic boundary conditions and integrate until time $t_0 = \sqrt{10^2 + 5^2}$, when the vortex has moved a relative distance of $(10, 5)$.

Our mesh is again obtained by anisotropic refinement of an initial Cartesian grid. We split vertically through the center of the rectangular domain, a total of $n_{\text{ref}} = 5$ times. The mesh, the initial solution, and the final solution are shown in Fig. 5.4.

We compute a reference solution using the fourth-order accurate RK4 scheme, using a stable timestep based on the smallest element size. The error in the IMEX solutions are then computed in the L_2-norm for the three timesteps $\Delta t_{\max}^0$, $\Delta t_{\max}^0/2$, and $\Delta t_{\max}^0/4$, where again $\Delta t_{\max}^0$ is the explicit timestep limit for the unrefined mesh. The resulting convergence plot is shown in Fig. 5.5, and we can confirm that the order of accuracy for the three IMEX schemes are about 2, 3, and 3, respectively. Note that while this is the expected order of accuracy, it was not certain that we would observe it here, since the implicit part of the problem uses large CFL numbers and therefore might not be in the convergent regime.

Unsteady Large Eddy Simulations of Flow over Airfoil: As an example of a realistic problem where the IMEX schemes can make a significant difference in performance, we study the flow over an SD7003 foil at Reynolds number 100,000 and 30° angle of attack. Our mesh is a typical LES-type mesh, with somewhat stretched elements for resolving the boundary layer profile, and an almost uniform mesh in the wake that captures the large scale features of the unsteady flow, see Fig. 5.6. It is generated using a hybrid approach, where the DistMesh mesh generator (Persson and Strang 2004) is used to create an unstructured mesh for most of the domain, and

Table 5.1 Convergence of the compressible Navier–Stokes test problem using the three preconditioners and varying Reynolds number, Mach number, and timesteps. A cross in the GMRES iterations column indicates that the method did not converge to a relative error norm of 10^{-3} in less than 1000 iterations

Problem	Parameters		Preconditioner/Iterations		
	Δt	M	Block Jacobi	Block G-S	Block ILU0
Inviscid	10^{-3}	0.2	24	14	5
	10^{-1}	0.2	187	73	12
	∞	0.2	840	456	40
	10^{-3}	0.01	200	111	15
	10^{-1}	0.01	×	×	94
	∞	0.01	×	×	374
Laminar	10^{-3}	0.2	50	25	4
Re = 1,000	10^{-1}	0.2	×	477	11
	∞	0.2	×	×	37
	10^{-3}	0.01	98	51	8
	10^{-1}	0.01	×	×	27
	∞	0.01	×	×	135
Laminar	10^{-3}	0.2	26	14	4
Re = 20,000	10^{-1}	0.2	456	219	16
	∞	0.2	×	×	236
	10^{-3}	0.01	160	61	12
	10^{-1}	0.01	×	×	80
	∞	0.01	×	×	×
RANS	10^{-3}	0.2	76	33	8
Re = 10^6	10^{-1}	0.2	×	×	35
	∞	0.2	×	×	70
	10^{-3}	0.01	411	174	14
	10^{-1}	0.01	×	×	46
	∞	0.01	×	×	132

the boundary points are connected to the airfoil in a structured pattern that allows for stretched elements with aspects ratios up to 50 along the wing surface. The elements are curved to align with the boundaries using a nonlinear elasticity approach (Persson and Peraire 2009). Finally, the triangular mesh is extruded in the span-wise direction to generate six layers of prismatic elements, which are each split into three tetrahedral elements. The total number of elements in the mesh is 312,000, which corresponds to about 31 million degrees of freedom for the Navier–Stokes equations and polynomial orders of $p = 3$.

We split into implicit and explicit equations element-wise based on the smallest edge sizes. In Fig. 5.7 we show the two-dimensional cross section of the mesh and the corresponding element size distribution based on the smallest edge length, since this

Table 5.2 The flat plate test problem, using two ERK schemes and two IMEX schemes. This confirms that the CFL condition for the IMEX schemes is not affected by the element sizes in the implicit boundary layer region

Scheme	ERK1		IMEX1	ERK3		IMEX3
$\Delta t^0_{\max}$	$3.26 \cdot 10^{-4}$			$2.61 \cdot 10^{-4}$		
n_{ref}	$\frac{\Delta t_{\max}}{\Delta t^0_{\max}}$	Ratio		$\frac{\Delta t_{\max}}{\Delta t^0_{\max}}$	Ratio	
0	1.0000		Stable	1.0000		Stable
1	0.6612	1.51	Stable	0.4958	2.02	Stable
2	0.1747	3.79	Stable	0.1298	3.82	Stable
3	0.0457	3.82	Stable	0.0337	3.85	Stable
4	0.0118	3.89	Stable	0.0086	3.92	Stable
5	0.0032	3.62	Stable	0.0023	3.72	Stable
6	0.0009	3.82	Stable	0.0006	3.89	Stable

is what will likely dictate the CFL condition for an element (at least for well-shaped meshes). Less than 9% of the elements are considered boundary layer elements, and by excluding them from the explicit region we bring up the smallest explicit element size by about a factor of 100.

For simplicity we consider only the IMEX1 scheme, and we obtain the following stability results:

- Using the ERK1 scheme on the explicit portion only, the largest stable timestep is about $\Delta t = 1.2 \cdot 10^{-4}$.
- Using the ERK1 scheme on the entire mesh, the largest stable timestep is about $\Delta t = 1.8 \cdot 10^{-8}$. This large difference shows that the timestep is restricted by the viscous effects, which leads to a factor of almost $100^2 = 10^4$.
- Using the IMEX1 scheme on the entire mesh with the splitting shown in Fig. 5.7, the scheme is stable with the ERK1-based timestep for the explicit portion of the mesh, that is, $\Delta t = 1.2 \cdot 10^{-4}$.

This ratio of about 10,000 comes at the cost of solving nonlinear systems of equations. However, these only involve 9% of the unknowns and can be solved efficiently by re-using the Jacobians. After the initial transients have decayed, our solvers use an average of six Newton iterations per Runge–Kutta stage and the number of Krylov iterations per linear system is less than 10. In our test implementation, this leads to a total cost per stage (implicit plus explicit) that is about 3 times higher than a fully explicit stage, which corresponds to a performance improvement of about a factor of 3,000. The computed solutions are illustrated in Fig. 5.8.

We did not perform a comparison with a fully implicit method, but we estimate that it would be about a magnitude slower than the IMEX solver due to ten times larger Jacobian matrices. In addition, the fully implicit scheme would require about 10 times as much memory for storing the full Jacobians.

References

Alexander, R. (1977). Diagonally implicit Runge-Kutta methods for stiff o.d.e.'s. *SIAM Journal on Numerical Analysis*, *14*(6), 1006–102.

Anderson, E., et al. (1999). *LAPACK Users' Guide* (3rd ed.). Society for Industrial and Applied Mathematics, Philadelphia.

Ascher, U. M., Ruuth, S. J., & Spiteri, R. J. (1997). Implicit-explicit Runge-Kutta methods for time-dependent partial differential equations. *Applied Numerical Mathematics*, *25*(2–3), 151–167. ISSN 0168-9274. Special issue on time integration (Amsterdam, 1996).

Diosady, L. T., & Darmofal, D. L. (2009). Preconditioning methods for discontinuous Galerkin solutions of the Navier-Stokes equations. *Journal of Computational Physics*, *228*(11), 3917–3935.

Fidkowski, K. J., Oliver, T. A., Lu, J., & Darmofal, D. L. (2005). p-multigrid solution of high-order discontinuous Galerkin discretizations of the compressible Navier-Stokes equations. *Journal of Computational Physics*, *207*(1), 92–113.

Kanschat, G. (2008). Robust smoothers for high-order discontinuous Galerkin discretizations of advection-diffusion problems. *Journal of Computational and Applied Mathematics*, *218*(1), 53–60.

Karypis, G., & Kumar, V. (1997). METIS serial graph partitioning and fill-reducing matrix ordering. http://glaros.dtc.umn.edu/gkhome/metis/metis/overview.

Kennedy, C. A., & Carpenter, M. H. (2003). Additive Runge-Kutta schemes for convection-diffusion-reaction equations. *Applied Numerical Mathematics*, *44*(1–2), 139–181. ISSN 0168-9274.

Klöckner, A., Warburton, T., Bridge, J., & Hesthaven, J. S. (2009). Nodal discontinuous Galerkin methods on graphics processors. *Journal of Computational Physics*, *228*(21), 7863–7882. ISSN 0021-9991.

Krivodonova, L., & Qin, R. (2013). An analysis of the spectrum of the discontinuous Galerkin method. *Applied Numerical Mathematics*, *64*, 1–18.

Nastase, C. R., & Mavriplis, D. J. (2006). High-order discontinuous Galerkin methods using an hp-multigrid approach. *Journal of Computational Physics*, *213*(1), 330–357.

Peraire, J., & Persson, P.-O. (2008). The compact discontinuous Galerkin (CDG) method for elliptic problems. *SIAM Journal on Scientific Computing*, *30*(4), 1806–1824.

Persson, P.-O. (2009). Scalable parallel Newton-Krylov solvers for discontinuous Galerkin discretizations. In *47th AIAA Aerospace Sciences Meeting and Exhibit, Orlando, Florida*. AIAA-2009-606.

Persson, P.-O. (2011). High-order LES simulations using implicit-explicit Runge-Kutta sche mes. In *49th AIAA Aerospace Sciences Meeting, Orlando, FL*. AIAA-2011-684.

Persson, P.-O., & Peraire, J. (2008). Newton-GMRES preconditioning for discontinuous Galerkin discretizations of the Navier-Stokes equations. *SIAM Journal on Scientific Computing*, *30*(6), 2709–2733.

Persson, P.-O. & Peraire, J. (2009). Curved mesh generation and mesh refinement using Lagrangian solid mechanics. In *47th AIAA Aerospace Sciences Meeting and Exhibit, Orlando, Florida*. AIAA-2009-949.

Persson, P.-O. & Strang, G. (2004). A simple mesh generator in Matlab. *SIAM Review*, *46*(2), 329–345. ISSN 0036-1445.

Persson, P.-O., Bonet, J., & Peraire, J. (2009). Discontinuous Galerkin solution of the Navier-Stokes equations on deformable domains. *Computer Methods in Applied Mechanics and Engineering*, *198*(17–20), 1585–1595.

Persson, P.-O., & Peraire, J. et al. (2010). The 3DG project. http://threedg.mit.edu.

Shampine, L. F., & Gear, C. W. (1979). A user's view of solving stiff ordinary differential equations. *SIAM Review*, *21*(1), 1–17.

Toselli, A., & Widlund, O. (2004). *Domain Decomposition Methods - Algorithms and Theory* (vol. 34). Springer Series in Computational Mathematics. Berlin: Springer.

Chapter 6
An Introduction to the Hybridizable Discontinuous Galerkin Method

Sonia Fernández-Méndez

Abstract This chapter in intended to be a didactical introduction to the Hybridizable Discontinuous Galerkin (HDG) method, including the formulation and its implementation. The Laplace and Stokes equations are considered as representative problems with self-adjoint operators, accounting for the incompressibility in the second one.

Introduction

Even though the Hybridizable Discontinuous Galerkin (HDG) method is a novel method proposed just a few years ago (see Cockburn et al. 2009, 2008), it has nowadays been successfully applied to all kinds of problems, specially in the field of Computational Fluid Dynamics (CFD); see, for instance, (Cockburn et al. 2011; Nguyen et al. 2010, 2011) for its application to the Stokes and Navier–Stokes equations, or Kirby et al. (2011), Giorgiani et al. (2013b), Huerta et al. (2013) for an efficiency study in front of Continuous Finite Elements (CFE) in the context of elliptic problems and wave problems.

HDG inherits all the advantages of high-order Discontinous Galerkin (DG) methods (see, for instance, Cockburn 2004; Hesthaven and Warburton 2002; Peraire and Persson 2008; Montlaur et al. 2008) that have made them so popular in CFD in the last decade, such as local conservation of quantities of interest, intrinsic stabilization thanks to a proper definition of numerical fluxes at element boundaries, suitability for code vectorization and parallel computation, and suitability for adaptivity.

But, HDG outperforms other DG methods for problems involving self-adjoint operators, due to two main peculiarities: hybridization and superconvergence properties. The hybridization process drastically reduces the number of degrees of freedom in the discrete problem, similarly to static condensation in the context of high-order

S. Fernández-Méndez (✉)
Laboratori de Càlcul Numèric (LaCàN), Universitat Politècnica de Catalunya (UPC-BarcelonaTech), Barcelona, Spain
e-mail: sonia.fernandez@upc.edu
URL: https://www.lacan.upc.edu/user/sonia-fernandez/

M. Kronbichler and P.-O. Persson (eds.), *Efficient High-Order Discretizations for Computational Fluid Dynamics*, CISM International Centre for Mechanical Sciences 602, https://doi.org/10.1007/978-3-030-60610-7_6

CFE, see, for instance, (Giorgiani et al. 2013b). More precisely, in a Laplace equation the unknowns reduce to the approximation of the trace of the solution at the mesh skeleton, i.e., the sides (or faces in 3-D) of the mesh; and in incompressible flow problems, the final unknowns correspond to just the trace of the velocity at the mesh skeleton plus one scalar representing the mean of the pressure at every element. On other hand, HDG is based on a mixed formulation that, differently to CFE or other DG methods, is stable even when all variables (primal unknowns and derivatives) are approximated with polynomials of the same degree k. Consequently, convergence of order $k+1$ in $\mathcal{L}_2$ norm is proved not only for the primal unknown, but also for its derivatives. In addition, a simple element-by-element postprocess of the derivatives leads to a superconvergent approximation of the primal variables, with convergence of order $k+2$ in $\mathcal{L}_2$ norm. The superconvergent solution can also be used to compute an efficient error estimator and define an adaptivity procedure as proposed by Giorgiani et al. (2013a), Giorgiani et al. (2014).

This document presents an introduction to HDG methods. The presentation aims to be didactical (including implementation) more than exhaustive, thus it does not cover completely the current state of the art of HDG methods and its applications. The Laplace and Stokes equations are considered as representative problems with self-adjoint operators, accounting for the incompressibility in the second one.

Laplace Equation

Let $\Omega \subset \mathbb{R}^d$ be a bounded domain with boundary $\partial\Omega$. The following problem is considered:

$$\begin{aligned} -\boldsymbol{\nabla} \cdot (\nu \boldsymbol{\nabla} u) &= f \text{ in } \Omega \\ -\nu \boldsymbol{\nabla} u \cdot \boldsymbol{n} &= g \text{ on } \Gamma_N \\ u &= u_D \text{ on } \Gamma_D \end{aligned} \tag{1}$$

where u is the solution, ν is a material coefficient, f is a given source term, u_D are prescribed values on the Dirichlet boundary Γ_D, and g is a prescribed flux on the Neumann boundary Γ_N, with $\overline{\Gamma_D} \cup \overline{\Gamma_N} = \overline{\partial\Omega}$.

The domain Ω is now assumed to be split by a finite element mesh with $\mathtt{n_{el}}$ disjoint elements K_i, such that

$$\overline{\Omega} \subset \bigcup_{i=1}^{\mathtt{n_{el}}} \overline{K}_i, \quad K_i \cap K_j = \emptyset \text{ for } i \neq j.$$

The union of all $\mathtt{n_{fc}}$ faces Γ_i (sides for 2D) is denoted as

$$\Gamma := \bigcup_{i=1}^{\mathtt{n_{el}}} \partial K_i = \bigcup_{f=1}^{\mathtt{n_{fc}}} \Gamma_f.$$

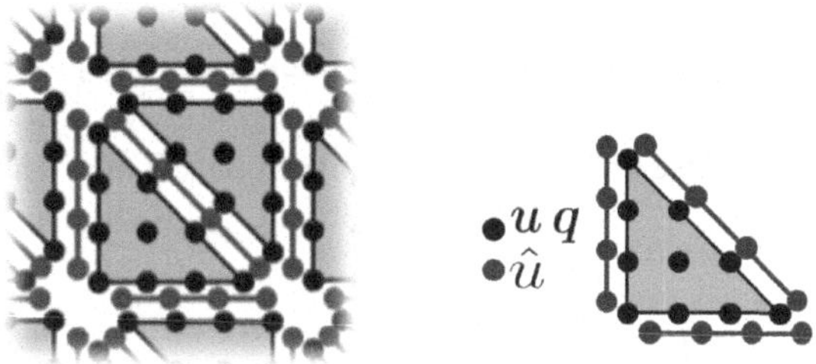

Fig. 6.1 The left picture shows a representation of an HDG discretization on a finite element mesh. Elemental variables (u and q) are element-by-element approximated from nodal values (black dots). Trace variables ($\hat{u}$) are approximated on the mesh skeleton Γ in red. The right picture shows a representation of the local problem: a pure Dirichlet problem taking $\hat{u}$ as data on the boundary of the element (in red). It is solved in each element to express the elemental variables, u and q, in black, in terms of the trace $\hat{u}$

The key idea of the HDG method is introducing a new unknow, $\hat{u}$, corresponding to the solution u on the mesh skeleton Γ, see Fig. 6.1.

The new unknown $\hat{u}$, which is usually referred to as *trace* variable, allows stating a Dirichlet problem in each element, that is, the so-called *local problems*,

$$\left.\begin{aligned} \nabla \cdot \boldsymbol{q} &= f \text{ in } K_i \\ \boldsymbol{q} + \nu \nabla u &= 0 \text{ in } K_i \\ u &= \widehat{u} \text{ on } \partial K_i \end{aligned}\right\} \text{ for } i = 1, \ldots \mathtt{n_{el}}. \tag{2}$$

The solution of the local problem, in each element, allows expressing u, and also the flux $\boldsymbol{q}$, in terms of the trace variable $\hat{u}$. Thus, now the actual unknown of the problem is the trace $\hat{u}$. It is then determined, closing the problem, imposing the *conservativity conditions*,

$$[\![\boldsymbol{q} \cdot \boldsymbol{n}]\!] = 0 \quad \text{on } \Gamma \backslash \partial\Omega, \quad \boldsymbol{q} \cdot \boldsymbol{n} = g \quad \text{on } \Gamma_N, \tag{3}$$

and the Dirichlet boundary conditions

$$\widehat{u} = \mathbb{P}_2(u_D) \quad \text{on } \Gamma_D, \tag{4}$$

where $\mathbb{P}_2(u_D)$ is the $\mathcal{L}_2$ projection of the Dirichlet data u_D onto the finite element space on the boundary. The *jump* $[\![\cdot]\!]$ operator is defined at a face Γ_f as

$$[\![\circledcirc]\!] = \circledcirc_{L(f)} + \circledcirc_{R(f)} \quad \text{on } \Gamma_f,$$

where $R(f)$ and $L(f)$ are the numbers of the left and right elements sharing the face, that is, $\Gamma_f = \overline{K_{L(f)}} \cap \overline{K_{R(f)}}$, and the subindex $\circledcirc_i$ denotes the value of function $\circledcirc$ from element K_i. In particular, $[\![\boldsymbol{q} \cdot \boldsymbol{n}]\!] = \boldsymbol{q}_{L(f)} \cdot \boldsymbol{n}_{L(f)} + \boldsymbol{q}_{R(f)} \cdot \boldsymbol{n}_{R(f)} = (\boldsymbol{q}_{L(f)} - \boldsymbol{q}_{R(f)}) \cdot \boldsymbol{n}_{L(f)}$.

It is important noting that the continuity of the solution u across Γ is imposed by the Dirichlet boundary condition in the local problems (2) and the fact that $\widehat{u}$ is single valued on Γ.

The discretization of the local problems and the global equations leads to the HDG discrete problem: find $u_h \in \mathcal{V}^h$, $\boldsymbol{q}_h \in [\mathcal{V}^h]^d$ and $\widehat{u}_h \in \Lambda^h$ such that $\widehat{u}_h = \mathbb{P}_2(u_D)$ on Γ_D and

$$\begin{aligned} \int_{K_i} v \nabla \cdot \boldsymbol{q}_h \, dV + \int_{\partial K_i} \tau \nu v (u_h - \widehat{u}_h) \, dS &= \int_{K_i} v f \, dV \\ \int_{K_i} \boldsymbol{q}_h \cdot \boldsymbol{w} \, dV - \int_{K_i} \nu u_h \nabla \cdot \boldsymbol{w} \, dV + \int_{\partial K_i} \nu \widehat{u}_h \boldsymbol{w} \cdot \boldsymbol{n} \, dS &= 0 \end{aligned} \tag{5}$$

for $i = 1, \dots \mathtt{n_{el}}$, and

$$\begin{aligned} \int_{\Gamma} \widehat{v} [\![\boldsymbol{q}_h \cdot \boldsymbol{n}]\!] \, dS + 2 \int_{\Gamma} \tau \widehat{v} (\{\nu u_h\} - \{\nu\} \widehat{u}_h) \, dS &= 0, \\ \int_{\Gamma_N} \widehat{v} \boldsymbol{q}_h \cdot \boldsymbol{n} \, dS + \int_{\Gamma_N} \tau \nu \widehat{v} (u_h - \widehat{u}_h) &= \int_{\Gamma_N} \widehat{v} g \, dS, \end{aligned} \tag{6}$$

$\forall \; v \in \mathcal{V}^h$, $\boldsymbol{w} \in [\mathcal{V}^h]^d$ and $\widehat{v} \in \Lambda^h$ such that $\widehat{v} = 0$ on Γ_D, where $\{\cdot\}$ is the mean operator on the interior faces,

$$\{\circledcirc\} = \frac{1}{2} \left(\circledcirc_{L(f)} + \circledcirc_{R(f)} \right) \quad \text{on } \Gamma_f.$$

The discrete spaces for elemental variables, u and $\boldsymbol{q}$, and for the trace variable, $\widehat{u}$, are

$$\begin{aligned} \mathcal{V}^h &:= \left\{ v \in \mathcal{L}^2(\Omega) \; : v|_{K_i} \in \mathcal{P}_k(K_i) \; \text{ for } \; i = 1, \dots, \mathtt{n_{el}} \right\} \\ \Lambda^h &:= \left\{ \hat{v} \in \mathcal{L}^2(\Gamma) \; : \hat{v}|_{\Gamma_f} \in \mathcal{P}_k(\Gamma_f) \; \text{ for } \; f = 1, \dots, \mathtt{n_{fc}} \right\}, \end{aligned} \tag{7}$$

where $\mathcal{P}_k$ denotes the space of polynomials of degree less or equal to k.

Remark 6.1 The parameter τ is a non-negative stabilization parameter usually taken of order $\mathcal{O}(1)$. For each element, it may be taken as a positive constant on all faces, or positive on one arbitrary face and zero at the rest (single face). Both options lead in practice to stable and optimally convergent solutions, with superconvergent post-processed solutions. See Sect. 6.2.1 and, for instance, (Giorgiani et al. 2013a; Cockburn et al. 2008) for details on the influence of this parameter on the solution behavior.

Remark 6.2 Different degree of approximation can be considered in each element with a straight-forward implementation. Based on this advantage of HDG, and the superconvergence properties that will be commented later, Giorgiani et al. (2013a) proposed an automatic degree-adaptive procedure.

Equations (5) correspond to the discretization of the local problem in each element. The first equation can be derived from the first equation in (2) applying integration by parts, replacing the flux by the numerical flux

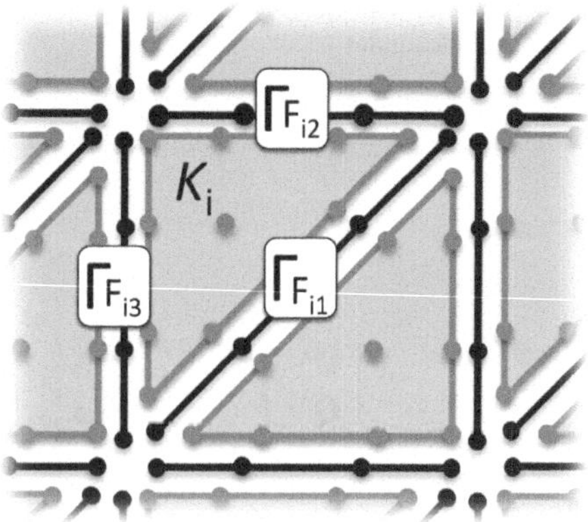

Fig. 6.2 Representation of the connectivity matrix for faces, **F**: the three faces for the i-th element, K_i, correspond to faces F_{i1}, F_{i2} and F_{i3}

$$\widehat{\boldsymbol{q}} := \boldsymbol{q} + \tau\nu(u - \widehat{u})\boldsymbol{n}, \tag{8}$$

and undoing the integration by parts. In fact, it can also be interpreted as the weighted residual of the PDE plus an stabilization term (that is zero for the analytical solution) with the parameter τ. The second equation in (5) is obtained from the weak form of the second equation in (2), applying integration by parts, and replacing the boundary condition $u = \widehat{u}$ on the element boundary.

The discretization of the local problem (5), for each element, can also be written in matrix from as

$$\begin{cases} \mathbf{A}_{uu}^{K_i}\mathbf{u}^i + \mathbf{A}_{uq}^{K_i}\mathbf{q}^i + \mathbf{A}_{u\widehat{u}}^{K_i}\Lambda^i = \mathbf{f}_u^{K_i} \\ \mathbf{A}_{qu}^{K_i}\mathbf{u}^i + \mathbf{A}_{qq}^{K_i}\mathbf{q}^i + \mathbf{A}_{q\widehat{u}}^{K_i}\Lambda^i = \mathbf{0} \end{cases} \tag{9}$$

where $\mathbf{u}^i$ and $\mathbf{q}^i$ are the vectors of nodal values of u and $\boldsymbol{q}$ in element K_i, and Λ^i is the vector of nodal values of $\widehat{u}$ on the n faces of the element ($n = 3$ for triangles, $n = 4$ for tetrahedra or quads, etc). That is,

$$\Lambda^i := \begin{bmatrix} \widehat{\mathbf{u}}^{\mathbf{F}_{i1}} \\ \vdots \\ \widehat{\mathbf{u}}^{\mathbf{F}_{in}} \end{bmatrix}, \tag{10}$$

where $\widehat{\mathbf{u}}^f$ denotes the nodal values of $\widehat{u}$ on face Γ_f, and $\mathbf{F}_{ij}$ is the number of the j-th face of element K_i; see an example in Fig. 6.2. Note also that the subindexes in the **A** matrices refer to the space for the weighting function and the test function.

System (9) can be solved for $\mathbf{u}^i$ and $\mathbf{q}^i$ in each element, obtaining the so-called *local solver* in the element K_i

$$\mathbf{u}^i = \mathbf{U}^{K_i}\Lambda^i + \mathbf{f}_U^{K_i}, \quad \mathbf{q}^i = \mathbf{Q}^{K_i}\Lambda^i + \mathbf{f}_Q^{K_i}, \tag{11}$$

with

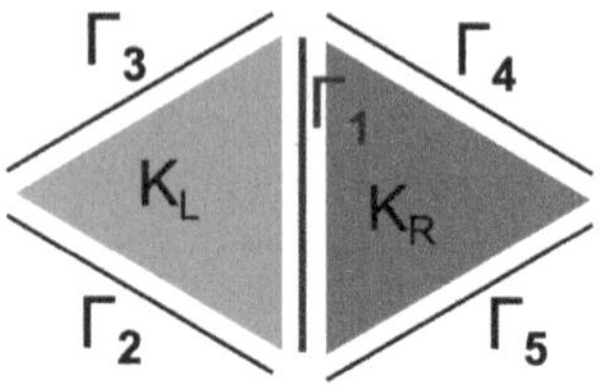

Fig. 6.3 Example of face Γ_1 shared by 2 triangular elements

$$\begin{bmatrix} \mathbf{U}^{K_i} \\ \mathbf{Q}^{K_i} \end{bmatrix} = -\mathbb{A}^{-1} \begin{bmatrix} \mathbf{A}_{u\widehat{u}}^{K_i} \\ \mathbf{A}_{q\widehat{u}}^{K_i} \end{bmatrix}, \quad \begin{bmatrix} \mathbf{f}_U^{K_i} \\ \mathbf{f}_Q^{K_i} \end{bmatrix} = \mathbb{A}^{-1} \begin{bmatrix} \mathbf{f}_u^{K_i} \\ \mathbf{0} \end{bmatrix} \tag{12}$$

and

$$\mathbb{A} = \begin{bmatrix} \mathbf{A}_{uu}^{K_i} & \mathbf{A}_{uq}^{K_i} \\ \mathbf{A}_{qu}^{K_i} & \mathbf{A}_{qq}^{K_i} \end{bmatrix}$$

That is, for each element, the elemental values of the solution, $\mathbf{u}^i$ and $\mathbf{q}^i$, can be explicitly expressed in terms of the trace on its faces, Λ^i.

On other hand, the equations (6) correspond to the discretization of the global equations, (3), to determine the trace variable $\widehat{u}$, imposing in weak form continuity of the normal flux and Neumann boundary conditions. They can also be written as

$$\sum_{i=1}^{\mathrm{n_{el}}} \int_{\partial K_i} \left(\widehat{v}\boldsymbol{q}_h \cdot \boldsymbol{n} + \tau \widehat{v}\,(u_h - \widehat{u}_h)\right) dS = \int_{\Gamma_N} \widehat{v} g \, dS, \tag{13}$$

where $\boldsymbol{n}$ denotes the normal vector exterior to K_i. For an interior face Γ_f, the equation can be written in matrix form as

$$\mathbf{A}_{\widehat{u}u}^{f,L} \mathbf{u}^{L(f)} + \mathbf{A}_{\widehat{u}q}^{f,L} \mathbf{q}^{L(f)} \; + \; \mathbf{A}_{\widehat{u}u}^{f,R} \mathbf{u}^{R(f)}(f) + \mathbf{A}_{\widehat{u}q}^{f,R} \mathbf{q}^{R(f)} \; + \; \mathbf{A}_{\widehat{u}\widehat{u}}^{f} \widehat{\mathbf{u}}^f = 0. \tag{14}$$

Then, replacing the local solver (11), for the elements $K_{L(f)}$ and $K_{R(f)}$, in (14) for every face Γ_f, leads to a system of equations involving only the trace variables $\{\widehat{\mathbf{u}}^f\}_{f=1}^{\mathrm{n_{fc}}}$.

For instance, the block equation corresponding to the face Γ_1, shared by elements K_L and K_R, as represented in Fig. 6.3, would be

$$\begin{aligned} \left[\mathbf{A}_{\widehat{u}u}^{1,L} \mathbf{U}^{K_L} + \mathbf{A}_{\widehat{u}q}^{1,L} \mathbf{Q}^{K_L}\right] \begin{bmatrix} \widehat{\mathbf{u}}^1 \\ \widehat{\mathbf{u}}^2 \\ \widehat{\mathbf{u}}^3 \end{bmatrix} + \left[\mathbf{A}_{\widehat{u}u}^{1,R} \mathbf{U}^{K_R} + \mathbf{A}_{\widehat{u}q}^{1,R} \mathbf{Q}^{K_R}\right] \begin{bmatrix} \widehat{\mathbf{u}}^1 \\ \widehat{\mathbf{u}}^4 \\ \widehat{\mathbf{u}}^5 \end{bmatrix} \\ + \mathbf{A}_{\widehat{u}\widehat{u}}^{1} \widehat{\mathbf{u}}^1 = -\mathbf{A}_{\widehat{u}u}^{1,L} \mathbf{f}_U^{K_L} - \mathbf{A}_{\widehat{u}q}^{1,L} \mathbf{f}_Q^{K_L} - \mathbf{A}_{\widehat{u}u}^{1,R} \mathbf{f}_U^{K_R} - \mathbf{A}_{\widehat{u}q}^{1,R} \mathbf{f}_Q^{K_R} \end{aligned}$$

The only lost of generalization in the example is the numbering of the faces in the two triangles. The structure of the equations would be the same for any interior face, just involving the faces in the two elements sharing the face. Note that the contributions from each element to the the block equation can be easily identified. That is, the face equations can be in fact computed as an assembly of the contributions from each element sharing the face.

Thus, similarly to the assembly in CFE, the computation of the HDG system is implemented with a loop over elements. For each element, the matrices and vectors for the local solver (11) are computed, and the contribution to the Eq. (14) is assembled for each one of the faces of the element. Once the system is assembled for all elements, and Dirichlet boundary conditions (4) are imposed, the system can be solved. Then, given the trace variables $\{\widehat{\mathbf{u}}^f\}_{f=1}^{\mathtt{n_{fc}}}$, the solution, $\mathbf{u}^i$ and $\boldsymbol{q}^i$, can be computed for each element using (11).

A MATLAB code with the same notation used in this document can be found at https://www.lacan.upc.edu/user/sonia-fernandez/.

*Convergence and Postprocess for Superconvergent Approximation u**

The HDG approximation, with degree k, provides an approximation with convergence of order $k+1$ for both u and for the approximation of the derivative $\boldsymbol{q}$. In addition, the mean of the solution in the elements is superconvergent with order $k+2$. See Cockburn et al. (2009) and Cockburn et al. (2008) for mathematical proofs. Thanks to these exceptional convergence properties, a cheap element-by-element postprocess can be computed to get a new approximation u_h^* with superconvergence of order $k+2$.

The problem to be solved in each element is

$$
\begin{aligned}
&\nabla \cdot \nabla u_h^* = \nabla \cdot \boldsymbol{q}_h && \text{in } K_i\\
&\nabla u_h^* \cdot \boldsymbol{n} = \boldsymbol{q}_h \cdot \boldsymbol{n} && \text{on } \partial K_i\\
&\textstyle\int_{K_i} u_h^* \, dV = \int_{K_i} u_h \, dV &&
\end{aligned}
$$

Superconvergence is ensured when the stabilization parameter τ is defined in each element as null in all faces except one, with an arbitrary positive constant value. This choice is referred to by some authors as *single face*. In any case, in practice, the superconvergence postprocess always provides a better approximation, with a convergence rate at least $k+1$ and in most cases close to $k+2$. The reader is invited to do some tests with the MATLAB code available at https://www.lacan.upc.edu/user/sonia-fernandez/.

Figure 6.4 shows an example on a coarse finite element mesh, with degree $k=2$. Discontinuities can be clearly seen in u_h, whereas u_h^* shows much smaller discontinuities, which can be understood as an indicator of improved accuracy. More precisely, in this example the postprocess reduces the $\mathcal{L}_2$ error from $0.9\ 10^{-2}$ to $0.7\ 10^{-3}$.

Giorgiani et al. (2013a), Giorgiani et al. (2014) proposed an authomatic adaptive algorithm based on the superconvergent solution u^*. The error is estimated in each

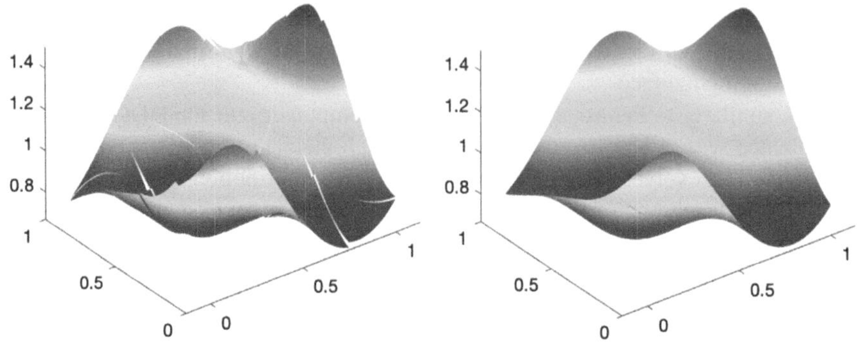

Fig. 6.4 Example of an HDG solution u_h (left) and u_h^* (right) on a coarse mesh of triangles with degree 2. The improvement in the post-processed superconvergent solution can be clearly seen

element simply as the difference of the two available approximations, u and u^*, and the degree is each element is consequently adapted.

Sparsity Pattern of the HDG Matrix and Computational Efficiency

Compared to other DG methods (such as Interior Penalty Method (IPM), Compact Discontinuous Galerkin (CDG), among others), the HDG global system has much less degrees of freedom, mainly thanks to the static condensation of elemental variables in terms of trace variables. In fact, the final number of degrees of freedom is close to CFE for high degree, as can be observed in the example in Fig. 6.5 and in the work by Huerta et al. (2013). Moreover, HDG matrices have a special block structure, because every interior face is connected to the same number of faces. For instance, in a mesh of triangles, every face is connected to itself and four more faces, corresponding to the faces in the two elements sharing the face, as can be seen in Fig. 6.3. Thus, every block of rows has five non-null blocks.

This special structure seems to be advantageous for linear solvers. A detailed comparison of HDG and CG in 2-D and 3-D can be found in the work by Kirby et al. (2011), Yakovlev et al. (2015), and also in Giorgiani et al. (2013b). Numerical experiments show that the computational time for assembly of HDG is greater than for CFE, but it may be compensated by a smaller CPU time in the linear solver, and by the higher accuracy (in part thanks to superconvergence) of HDG. Thus, HDG exhibits computational efficiency similar to CFE (in terms of CPU time for a given level of accuracy) but with the classical advantages of DG methods, such as easy adaptivity, suitability for parallel computing, stability through numerical fluxes and conservativity.

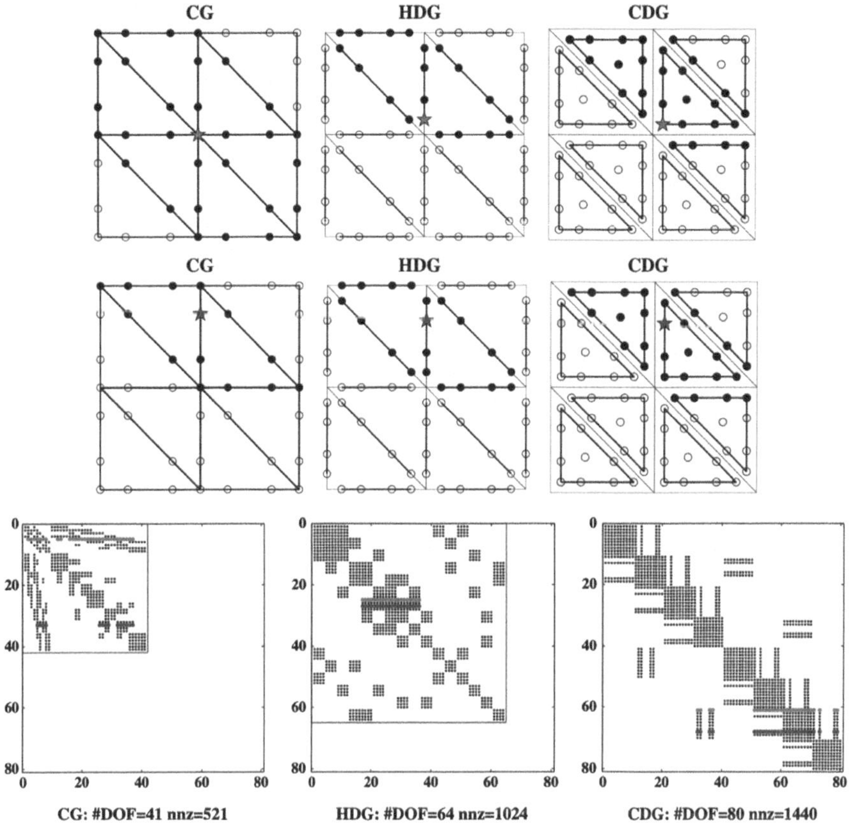

Fig. 6.5 Figures from Giorgiani et al. (2013b). The rows in red and blue correspond to the equations of the nodes marked with a red star (vertex node) and a blue star (side node), respectively. The stencil of these nodes is shown as black nodes on the mesh

Incompressible Flow

The following Stokes problem is considered:

$$
\begin{aligned}
-\nabla \cdot (\nu \nabla \boldsymbol{u}) + \nabla p &= \boldsymbol{f} \text{ in } \Omega \\
\nabla \cdot \boldsymbol{u} &= \boldsymbol{0} \text{ in } \Omega \\
(-\nu \nabla \boldsymbol{u} + p\mathbf{I}) \cdot \boldsymbol{n} &= g \text{ on } \Gamma_N \\
\boldsymbol{u} &= \boldsymbol{u}_D \text{ on } \Gamma_D
\end{aligned}
\tag{15}
$$

where $\boldsymbol{u}$ is the velocity, p is the pressure, ν is the material viscosity, $\mathbf{I}$ is the second-order identity tensor, $\boldsymbol{f}$ is an external force, $\boldsymbol{u}_D$ are prescribed values on the Dirichlet boundaries Γ_D, and g is a prescribed traction on the Neumann boundary Γ_N.

The domain Ω is again assumed to be covered by a finite element mesh, with the same notation as in the previous section. And, following the HDG rationale, problem

(15) is split in a set of local problems, for each one of the elements, and some global equations defined on element faces. The local problems are pure Dirichlet problems. That is, for each element

$$\boldsymbol{L} - \nabla \boldsymbol{u} = \boldsymbol{0}, \quad \nabla \cdot (-\nu \boldsymbol{L} + p\mathbf{I}) = \boldsymbol{f} \text{ and } \nabla \cdot \boldsymbol{u} = \boldsymbol{0} \text{ in } K_i \tag{16a}$$

$$\boldsymbol{u} = \hat{\boldsymbol{u}} \text{ on } \partial K_i \tag{16b}$$

$$\frac{1}{|\partial K_i|} \int_{\partial K_i} p \, dS = \rho_i. \tag{16c}$$

The variable $\boldsymbol{L}$ is the gradient of $\boldsymbol{u}$, allowing the splitting of the PDE in two first-order PDEs, $\hat{\boldsymbol{u}}$ is the trace of $\boldsymbol{u}$ at the mesh faces Γ, and ρ_i is the mean of the pressure at the boundary of the element, which is only a scalar for each element. The new variables, $\hat{\boldsymbol{u}}$ and $\{\rho_i\}_{i=1}^{\mathtt{n_{el}}}$, are assumed to be data in the local problems. Note that the Stokes problem with only Dirichlet data does not have a unique solution, and (16c) closes the problem, setting the mean of the pressure on the element boundary to ρ_i.

The local problems (16) can be solved element-by-element to determine $\boldsymbol{u}$, $\boldsymbol{L}$ and p, given $\hat{\boldsymbol{u}}$ and ρ_i. Thus, now the problem reduces to find $\hat{\boldsymbol{u}}$ and $\{\rho_i\}_{i=1}^{\mathtt{n_{el}}}$ with the global equations

$$[\![(-\nu \boldsymbol{L} + p\mathbf{I}) \cdot \boldsymbol{n}]\!] = 0 \quad \text{on } \Gamma \backslash \partial\Omega, \quad (-\nu \boldsymbol{L} + p\mathbf{I}) \cdot \boldsymbol{n} = g \quad \text{on } \Gamma_N, \tag{17a}$$

$$\int_{\partial K_i} \hat{\boldsymbol{u}} \cdot \boldsymbol{n} \, dS = 0 \quad \text{for } i = 1, \ldots \mathtt{n_{el}}, \tag{17b}$$

$$\hat{\boldsymbol{u}} = \mathbb{P}_2(\boldsymbol{u}_D) \quad \text{on } \Gamma_D. \tag{17c}$$

Equation (17a) is the conservativity condition, imposing equilibrium of the traction on element faces and on the Neumann boundary. Equation (17b) imposes the incompressibility condition on the boundary of the elements, ensuring well-posedness of the Dirichlet local problems (16).

If the problem is a pure Dirichlet problem, i.e., $\Gamma_N = \emptyset$, the solution of (15) is determined up to a constant for the pressure. In this case, an additional constraint for the pressure should be imposed at the global level, for instance, setting the mean of the pressure on the boundary of the first element to 0, that is $\rho_1 = 0$, or to a given constant.

The discretization spaces in (7) are now considered for elemental variables $\boldsymbol{u}$, $\boldsymbol{L}$ and p, and for the trace variable $\hat{\boldsymbol{u}}$, respectively. The discretization of the local problems (16) and the global equations (17) lead to the complete HDG formulation detailed next.

The HDG local problem for each element K_i is: given $\hat{\boldsymbol{u}}_h \in [\Lambda^h]^d$ and $\rho_i \in \mathbb{R}$, find $\boldsymbol{u}_h \in [\mathcal{P}_k(K_i)]^d$, $\boldsymbol{L}_h \in [\mathcal{P}_k(K_i)]^{d \times d}$ and $p_h \in \mathcal{P}_k(K_i)$ such that

$$
\begin{aligned}
\int_{K_i} (-\nabla \cdot (\nu \boldsymbol{L}_h) + \nabla p_h) \cdot \boldsymbol{v}\, dV + \int_{\partial K_i} \tau\nu(\boldsymbol{u}_h - \hat{\boldsymbol{u}}_h) \cdot \boldsymbol{v}\, dS = \int_{K_i} \boldsymbol{f} \cdot \boldsymbol{v}\, dV \\
\int_{K_i} \boldsymbol{L}_h : \mathbf{Q}\, dV + \int_{K_i} (\nabla \cdot \mathbf{Q}) \cdot \boldsymbol{u}_h\, dV - \int_{\partial K_i} (\mathbf{Q} \cdot \boldsymbol{n}) \cdot \hat{\boldsymbol{u}}_h\, dS = 0 \\
\int_{K_i} \boldsymbol{u}_h \cdot \nabla q\, dV - \int_{\partial K_i} (\hat{\boldsymbol{u}}_h \cdot \boldsymbol{n}) q\, dS = 0 \\
\frac{1}{|\partial K_i|} \int_{\partial K_i} p_h\, dS = \rho_i,
\end{aligned}
\tag{18}
$$

for all $\boldsymbol{v} \in [\mathcal{P}_k(K_i)]^d$, $\mathbf{Q} \in [\mathcal{P}_k(K_i)]^{d\times d}$ and $q \in \mathcal{P}_k(K_i)$. The first equation in (18) can be derived from the first equation in (16a) by applying integration by parts, replacing the velocity gradient by the numerical velocity gradient,

$$
\widehat{\boldsymbol{L}} := \boldsymbol{L} + \tau(\hat{\boldsymbol{u}} - \boldsymbol{u}) \otimes \boldsymbol{n}, \tag{19}
$$

and undoing the integration by parts. The stabilization parameter τ can be taken as $\tau = 1$, see Nguyen et al. (2009), Giorgiani et al. (2014) for details. The second and third equations are obtained from the weak form of the second and third equations in (16a) by simply applying integration by parts and replacing the boundary condition (16c) on the element boundary.

The discretization of the local problem leads now to a system of equations of the form

$$
\begin{bmatrix}
\mathbf{A}^i_{uu} & \mathbf{A}^i_{uL} & \mathbf{A}^i_{up} & \mathbf{0} \\
\mathbf{A}^i_{Lu} & \mathbf{A}^i_{LL} & \mathbf{0} & \mathbf{0} \\
\mathbf{A}^i_{pu} & \mathbf{0} & \mathbf{0} & \mathbf{A}^{i\ T}_{\rho p} \\
\mathbf{0} & \mathbf{0} & \mathbf{A}^i_{\rho p} & \mathbf{0}
\end{bmatrix}
\begin{bmatrix} \mathbf{u}^i \\ \mathbf{L}^i \\ \mathbf{p}^i \\ \lambda \end{bmatrix}
=
\begin{bmatrix} \mathbf{f}^i_u \\ \mathbf{f}^i_L \\ \mathbf{f}^i_p \\ \mathbf{0} \end{bmatrix}
+
\begin{bmatrix} \mathbf{0} \\ \mathbf{0} \\ \mathbf{0} \\ \mathbf{1} \end{bmatrix} \rho_i
-
\begin{bmatrix} \mathbf{A}^i_{u\hat{u}} \\ \mathbf{A}^i_{L\hat{u}} \\ \mathbf{A}^i_{p\hat{u}} \\ \mathbf{0} \end{bmatrix} \Lambda^i
\tag{20}
$$

where the vectors $\mathbf{u}^i$, $\mathbf{L}^i$ and $\mathbf{p}^i$ are elemental vectors of nodal values, and Λ^i is a vector with the nodal values of the trace of the velocity $\hat{\boldsymbol{u}}^f$ for the faces of the element. Note that the constraint for the mean of the pressure, $\frac{1}{|\partial K_i|} \int_{\partial K_i} p\, dS = \rho_i$, is imposed with a Lagrange multiplier λ.

This system can be solved for each element, leading to the so-called *local solver* in the element, that is, an explicit expression of the elemental variables $\mathbf{u}^i$, $\mathbf{L}^i$ and $\mathbf{p}^i$, in terms of the trace of the velocity at the faces Λ^i and the mean of the pressure ρ_i.

$$
\begin{aligned}
\mathbf{u}^i &= \mathbf{S}^i_u \Lambda^i + \mathbf{r}^i_u \rho_i + \mathbf{f}^i_U, \\
\boldsymbol{L}^i &= \mathbf{S}^i_L \Lambda^i + \mathbf{r}^i_L \rho_i + \mathbf{f}^i_L, \\
\mathbf{p}^i &= \mathbf{S}^i_p \Lambda^i + \mathbf{r}^i_p \rho_i + \mathbf{f}^i_p,
\end{aligned}
\tag{21}
$$

where the matrices $\mathbf{S}^i_*$ and the vectors $\mathbf{r}^i_*$, $\mathbf{f}^i_*$ depend on the matrices and vectors in (20).

The HDG global problem corresponding to the discretization of (17) with the numerical velocity gradient (19) is: find $\hat{\boldsymbol{u}}_h \in [\Lambda^h]^d$ and $\rho^i \in \mathbb{R}$ for $i = 1, \ldots, \mathtt{n_{el}}$

satisfying

$$\sum_{i=1}^{\mathtt{n_{el}}} \int_{\partial K_i} \widehat{\boldsymbol{v}} \cdot \left((-\nu \boldsymbol{L}_h + p_h \mathbf{I}) \cdot \boldsymbol{n} + \tau \nu \widehat{\boldsymbol{v}} \cdot \left(\boldsymbol{u}_h - \hat{\boldsymbol{u}}_h \right) \right) dS = \int_{\Gamma_N} \widehat{\boldsymbol{v}} \cdot g \, dS, \tag{22}$$

for all $\widehat{\boldsymbol{v}} \in [\Lambda^h]^d$ such that $\widehat{\boldsymbol{v}} = 0$ on Γ_D,

$$\int_{\partial K_i} \hat{\boldsymbol{u}}_h \cdot \boldsymbol{n} \, dS = 0 \quad \text{for } i = 1, \ldots \mathtt{n_{el}} \tag{23}$$

and

$$\hat{\boldsymbol{u}}_h = \mathbb{P}_2(\boldsymbol{u}_D) \quad \text{on } \Gamma_D, \tag{24}$$

where $\boldsymbol{u}_h$, $\boldsymbol{L}_h$ and p_h are solutions to the local problem for each element, i.e., the solution of (18). Equations (22) and (23) can be written in matrix form as

$$\begin{aligned} &\mathbf{A}_{\widehat{u}u}^{f,L} \mathbf{u}^{L(f)} + \mathbf{A}_{\widehat{u}L}^{f,L} \boldsymbol{L}^{L(f)} + \mathbf{A}_{\widehat{u}p}^{f,L} \mathbf{p}^{L(f)} \\ &+ \mathbf{A}_{\widehat{u}u}^{f,R} \mathbf{u}^{R(f)} + \mathbf{A}_{\widehat{u}L}^{f,R} \boldsymbol{L}^{R(f)} + \mathbf{A}_{\widehat{u}p}^{f,R} \mathbf{p}^{R(f)} + \mathbf{A}_{\widehat{u}\widehat{u}}^{f} \widehat{\mathbf{u}}^f = 0 \end{aligned} \tag{25}$$

for $f = 1, \ldots, \mathtt{n_{fc}}$, and

$$\mathbf{A}_{\rho\widehat{u}}^i \Lambda^i = \mathbf{0} \tag{26}$$

for $i = 1, \ldots, \mathtt{n_{el}}$. Replacing the local solver—that is, Eq. (21) for the elements $K_{L(f)}$ and $K_{R(f)}$—in (25) and (26), and applying the Dirichlet boundary condition (24), the global problem leads to a system of equations involving only the trace variable $\{\widehat{\mathbf{u}}^f\}_{f=1}^{\mathtt{n_{fc}}}$ and the mean of the pressure on the boundary of the elements $\{\rho_i\}_{i=1}^{\mathtt{n_{el}}}$. After solving this global system, elemental variables can be obtained by simply plugging in the solution, $\{\widehat{\mathbf{u}}^f\}_{f=1}^{\mathtt{n_{fc}}}$ and $\{\rho_i\}_{i=1}^{\mathtt{n_{el}}}$, in the local solver for each element.

Analogously to Laplace, an element-by-element postprocessing provides a superconvergent solution, $\boldsymbol{u}_h^*$,with order $k+2$ in $\mathcal{L}_2$ norm: given $\boldsymbol{u}_h \in [\mathcal{P}_k(K_i)]^d$ and $\boldsymbol{L}_h \in [\mathcal{P}_k(K_i)]^{d \times d}$, find $\boldsymbol{u}_h^* \in [\mathcal{P}_{k+1}(K_i)]^d$ such that

$$\begin{aligned} \int_{K_i} \nabla \boldsymbol{u}_h^* : \nabla \boldsymbol{v} \, dV &= \int_{K_i} \boldsymbol{L}_h : \nabla \boldsymbol{v} \, dV \quad \forall \boldsymbol{v} \in [\mathcal{P}_{k+1}(K_i)]^d, \\ \int_{K_i} \boldsymbol{u}_h^* \, dV &= \int_{K_i} \boldsymbol{u}_h \, dV. \end{aligned} \tag{27}$$

A MATLAB code implementing HDG for Stokes can be found at https://www.lacan.upc.edu/user/sonia-fernandez/.

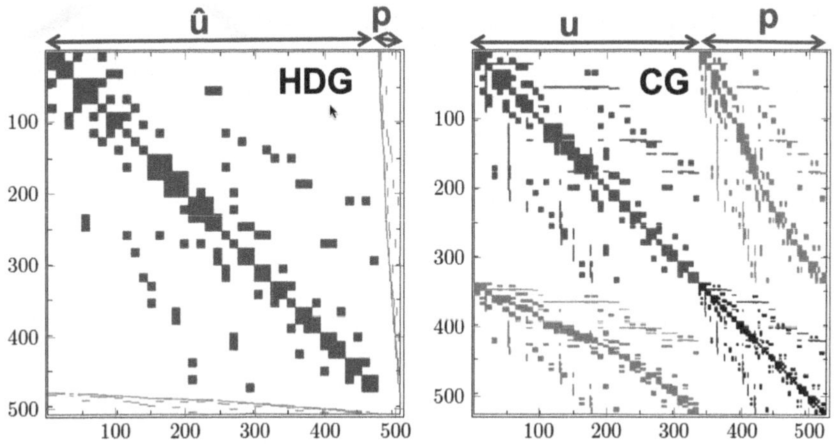

(a) Sparsity pattern of HDG, DOFs = 511, $nnz = 1830$. (b) Sparsity pattern of CG, DOFs = 530, $nnz = 3580$.

Fig. 6.6 Figures from Paipuri et al. (2018). Example of sparsity pattern for HDG and CFE (CG)

Matrix Structure and Computational Efficiency

The static condensation of elemental variables is even more advantageous in incompressible flow problems, since in this case the degrees of freedom corresponding to the pressure are reduced to just one scalar per element, ρ_i, regardless the degree of approximation k. Figure 6.6 shows an example of the sparsity pattern of the global matrix for CFE, with a Taylor-Hook approximation (with degree $k - 1$ for pressure and k for velocity) and static condensation, and for HDG with degree $k = 5$.

Looking to the blue block, corresponding to velocity unknowns, we can again observe that for the same mesh and same degree, HDG has more degrees of freedom, but it also exhibits the nice block structure that is advantageous for linear solvers. Now, in addition, the number of degrees of freedom for pressure is drastically reduced in HDG, with just one unknown ρ_i per element, for any degree of approximation k. Moreover, with these approximations the HDG solution is expected to be more accurate than the CFE one, since the velocity u_h^* converges with order $k + 2$, in front of the order k in CFE, and the pressure p_h converges with order $k + 1$, instead of k for CFE. A critical comparison, in terms of CPU time for similar accuracy, for incompressible flow can be found in Paipuri et al. (2018).

Some Comments on Navier–Stokes

Similarly to Stokes, the non-linear Navier–Stokes equations can be solved in an efficient way, just applying the same static condensation tricks for the linearized system in the non-linear solver iterations, see, for instance, (Paipuri et al. 2018) for details.

The stabilization parameter τ can play an important role in the presence of sharp fronts for high Reynolds number, see, for instance, (Paipuri et al. 2018; Giacomini et al. 2020).

References

Cockburn, B. (2004). Discontinuous Galerkin methods for computational fluid dynamics. *Encyclopedia of computational mechanics* (Vol. 3 (Fluids), chapter 4). New York: Wiley.

Cockburn, B., Dong, B., & Guzmán, J. (2008). A superconvergent LDG-hybridizable Galerkin method for second-order elliptic problems. *Mathematics of Computation*, *77*(264), 1887–1916. ISSN 0025-5718.

Cockburn, B., Gopalakrishnan, J., & Lazarov, R. (2009). Unified hybridization of discontinuous Galerkin, mixed, and continuous Galerkin methods for second order elliptic problems. *SIAM Journal on Numerical Analysis*, *47*(2), 1319–1365.

Cockburn, B., Gopalakrishnan, J., Nguyen, N. C., Peraire, J., & Sayas, F.-J. (2011). Analysis of HDG methods for Stokes flow. *Mathematics of Computation*, *80*(274), 723–760.

Giacomini, M., Sevilla, R., & Huerta, A. (2020). Tutorial on Hybridizable Discontinuous Galerkin (HDG) formulation for incompressible flow problems. In De Lorenzis, L., & Düster, A. (Eds.) *Modeling in Engineering Using Innovative Numerical Methods for Solids and Fluids* (Vol. 599, pp. 163–201). CISM International Centre for Mechanical Sciences. Springer International Publishing. https://doi.org/10.1007/978-3-030-37518-8_5.

Giorgiani, G., Fernández-Méndez, S., & Huerta, A. (2013a). Hybridizable Discontinuous Galerkin p-adaptivity for wave propagation problems. *International Journal for Numerical Methods in Fluids*, *72*(12), 1244–1262.

Giorgiani, G., Modesto, D., Fernández-Méndez, S., & Huerta, A. (2013b). High-order continuous and discontinuous Galerkin methods for wave problems. *International Journal for Numerical Methods in Fluids*, *73*(10), 883–903.

Giorgiani, G., Fernández-Méndez, S., & Huerta, A. (2014). Hybridizable Discontinuous Galerkin with degree adaptivity for the incompressible Navier-Stokes equations. *Computers & Fluids*, *98*, 196–208.

Hesthaven, J. S., & Warburton, T. (2002). Nodal high-order methods on unstructured grids: I. Time-domain solution of Maxwell's equations. *Journal of Computational Physics*, *181*(1), 186–221.

Huerta, A., Angeloski, A., Roca, X., & Peraire, J. (2013). Efficiency of high-order elements for continuous and discontinuous Galerkin methods. *International Journal for Numerical Methods in Engineering*, *96*(9), 529–560.

Kirby, R., Sherwin, S. J., & Cockburn, B. (2011). To CG or to HDG: A comparative study. *Journal of Scientific Computing*, *51*(1), 183–212. ISSN 0885-7474.

Montlaur, A., Fernández-Méndez, S., & Huerta, A. (2008). Discontinuous Galerkin methods for the Stokes equations using divergence-free approximations. *International Journal for Numerical Methods in Fluids*, *57*(9), 1071–1092.

Nguyen, N. C., Peraire, J., & Cockburn, B. (2009). An implicit high-order hybridizable discontinuous Galerkin method for linear convection-diffusion equations. *Journal of Computational Physics*, *228*(9), 3232–3254.

Nguyen, N. C., Peraire, J., & Cockburn, B. (2010). A hybridizable discontinuous Galerkin method for Stokes flow. *Computer Methods in Applied Mechanics and Engineering*, *199*(9–12), 582–597. ISSN 0045-7825.

Nguyen, N. C., Peraire, J., & Cockburn, B. (2011). An implicit high-order hybridizable discontinuous Galerkin method for the incompressible Navier-Stokes equations. *Journal of Computational Physics*, *230*(4), 1147–1170.

Paipuri, M., Fernández-Méndez, S., & Tiago, C. (2018). Comparison of high-order continuous and hybridizable discontinuous Galerkin methods in incompressible fluid flow problems. *Mathematics and Computers in Simulation*, *153*, 35–58. https://doi.org/10.1016/j.matcom.2018.05.012.

Peraire, J., & Persson, P. O. (2008). The compact discontinuous Galerkin (CDG) method for elliptic problems. *SIAM Journal on Scientific Computing*, *30*(4), 1806–1824. ISSN 1064-8275.

Yakovlev, S., Moxey, D., Kirby, R., & Sherwin, S. (2015). To CG or to HDG: A comparative study in 3D. *Journal of Scientific Computing*, *67*(1), 192–220. https://doi.org/10.1007/s10915-015-0076-6.

Chapter 7
High-Order Methods for Simulations in Engineering

Rainald Löhner

Abstract Engineers create new things and hence always have to deal with incomplete information. A critical review is made of the accuracy of the available physical and modeling parameters. It shows that in many cases key physical and modeling parameters such as viscosities, boundary conditions, geometry, and even basic physics are not known to within 1%. This raises the question as to whether any numerical method needs to be of even higher quality than a fraction of this threshold. Thereafter work estimates for traditional high-order elements are derived. The comparison of error and work estimates shows that even for relative accuracy in the 0.1% range, which is one order below the typical accuracy of engineering interest (1% range), linear elements may outperform all higher order elements. The chapter concludes with some recent LES results that indicate that eight-order schemes on a grid of size $2h$ are similar to second-order schemes on a grid of size h, and some open questions.

Introduction

The last decades have seen widespread interest in and funding for high-order (finite difference, finite volume, finite element, discontinuous Galerkin, spectral volume, spectral difference, isogeometric, etc.) methods (see, e.g., Cockburn et al. 1990; Lin and Chin 1993; Bey et al. 1996; Bassi and Rebay 1997; Atkins and Shu 1998; Cockburn et al. 2000; Schwab 2004; Karniadakis and Sherwin 2005; Nastase and Mavriplis 2006; Kroll 2006; Klaij et al. 2007; Wang 2007; Hartmann and Houston 2008; Luo et al. 2008; Cottrell et al. 2009; Kannan and Wang 2009; Liang et al. 2009; Persson et al. 2009; Nigro et al. 2010; Vos et al. 2010; Brown 2011; Cantwell et al. 2011 and the references cited thererin). Besides the basic exploration of new/unknown methods, the hope was that higher order methods would lead to faster solution of

R. Löhner (✉)
Center for Computational Fluid Dynamics, George Mason University, Fairfax, VA 22030-4444, USA
e-mail: rlohner@gmu.edu

M. Kronbichler and P.-O. Persson (eds.), *Efficient High-Order Discretizations for Computational Fluid Dynamics*, CISM International Centre for Mechanical Sciences 602, https://doi.org/10.1007/978-3-030-60610-7_7

traditional aerodynamics problems and also enable the accurate transport of vortices over very long distances, opening the way to reliable LES runs (Slotnick et al. 2014). Although these methods seem to overcome many of the difficulties encountered with traditional low-order methods, the lingering question still remains as to when these schemes pay off. The empirical evidence indicates that at present most of the production Euler and Reynolds-Averaged Navier–Stokes (RANS) cases in aerodynamics and hydrodynamics are still run using traditional second/third-order (i.e., low-order) methods. Surprisingly, most of the LES runs performed by the automotive industry are also carried out using these (Navier–Stokes) or Lattice-Boltzmann (LBM) (Duncan et al. 2010; Geier et al. 2017) based low-order methods. Naturally, one can argue that (human) inertial effects are always present, two decades is a time too short to bring codes to production status, unforseen numerical difficulties that appeared, etc. Nevertheless, two decades should also have been sufficient to conduct at least a few of the AIAA drag prediction workshop test cases (Laflin 2004) in order to see if these schemes pay off.

These observations led to the question as to whether one cannot derive from analytical extrapolations similar trends as those seen in the production environment. Work estimates for high-order reported in Löhner (2013) and Huerta et al. (2013), showed that for LES (i.e., explicit time stepping with nonlinear terms built at every Runge–Kutta stage) the advantages of high-order schemes cannot be discerned unless extremely low levels of errors are a requirement.

This result in turn led to the question of the degree of uncertainty commonly encountered in engineering practice. This is the subject of the first part of the chapter. As will be seen, boundary conditions, physical parameters, and in many cases even geometry are not known to 1% relative error, and in some cases exceed this level by an order of magnitude. Nonlinearity, butterfly effects, and rogue loads also add uncertainty. These findings weaken considerably the insistence on relative error levels that are below 10^{-4}—the realm where high-order schemes begin to be advantageous.

The chapter closes with some recent LES results and a series of fundamental open questions.

Computation in the Engineering Context

An old saying states that: "Physicists comprehend and describe what is there, engineers create what was not there before". This implies that there is always a degree of uncertainty in any new machine, process, or product that is being designed. In some industries, creating a new product leads to development costs that can easily exceed yearly profits. A new commercial airplane costs more than $\$ 5 \cdot 10^9$, a new car more than $\$ 2 \cdot 10^9$ (these and subsequent quotations in 2020 US$). This implies that if it turns out that the new product has some inherent flaw or is not able to perform as advertised, the economic viability of the entire enterprise may be at risk. The aircraft industry is littered with companies that went bankrupt after a failed product. Engineers have always sought to mitigate the risk of these potential product failures

Table 7.1 Life cycle of products

Stage	Information	Model
Specification	Global	0-D
Prelim design	Partial	1/2D
Detailed design	Local	3D
Prototype	Measured	3D
Test/evaluation	Measured	–
Production	Measured	–

by gathering as much information as possible when designing new products. The traditional repository and sources for this information included data from similar designs and products (either from in-house designs or those of the competition), experimental data from scaled models or prototypes, or computations based on engineering or mathematical physics models. The advent of supercomputers together with the ensuing development of numerical methods and large-scale codes has brought an unprecedented degree of realism to simulations done with mathematical physics models and has shifted the emphasis of information gathering during the design and analysis process of new products towards computation. It led to the emergence of computational sciences as a third pillar of the empirical sciences. Presently, any new airplane, car, train, turbine, or any consumer product of value has been extensively computed before the first prototype is ever built. Computation has become so pervasive in the design and analysis workflow that sometimes the very assumptions and limits of the models used are forgotten. Even more, for some high-value products this so-called "digital twin" is kept throughout the life-cycle and is updated regularly when maintenance is carried out or parts are being replaced. The industries that led these developments were those where failure would lead to catastrophic consequences: nuclear weapons and airplanes. In both areas, massive institutional investments funded research and development centers, with proper experimental and computing facilities to measure physical properties, carry out fundamental experimental studies, and properly debug, validate and field the computing codes that emerged from them.

The Life Cycle of Products

In broad terms, any product goes through the phases outlined in Table 7.1. The modus operandi is to maximize the information that can be extracted at each stage given the data available. As an example, it makes no sense to waste an expensive LES simulation of a complete aircraft in the preliminary design stage when even the topology of the vehicle is not yet frozen.

Information and Calculation

One can also consider the design of a new product as a process whereby information is created. At each stage, the known information is matched with an appropriate model to increase the information. This implies that in the preliminary stages, where only global information is available (e.g., payload, range, cost, etc.), global models that encapsulate the (global) data of decades of experiments and products are employed. As more information becomes available and the finer details "emerge" from the preliminary design, the models shift to partial (e.g., lifting line theory in aerodynamics, beam and plate models in structural dynamics) and then local (e.g., RANS in aerodynamics, 3-D solids in structural dynamics).

Mathematical Models of Nature and Their Uncertainty/Errors

The mathematical description of nature and, as a consequence, also engineering products or processes, is performed via models. Models are abstractions of reality, encapsulating key physical phenomena or processes, and are typically given in the form of ordinary or partial differential equations (PDEs). For the temporal and spatial scales that are encountered in most areas of mechanical and civil engineering the continuum assumption of "physically infinitesimal" (i.e., dx, dt are small compared to the overall dimensions of the system but large compared to the atomic scales) holds, leading to the typical conservation laws of mass, momentum, and energy. As an example consider air at ambient pressure and temperature: there are $2.7 \cdot 10^{25}$ molecules/m^3, and the mean free path and average time between molecular collisions is $68 \cdot 10^{-9}$ m and 10^{-10} s respectively. Nevertheless, one should always bear in mind that "all models are wrong, but some are useful", and that "There are more things in heaven and Earth, Horatio, than are dreamt of in your (our) philosophy [science]" (Hamlet, 1.5.167-8).

Given that the aim of this book is to solve the partial differential equations emanating from these models in an optimal way, it is necessary to estimate the uncertainty of all the elements that comprise a model: the assumptions that led to the partial differential equations, the physical properties/parameters required, the geometry/domain being considered, the boundary conditions and source/forcing terms that specify the problem, and the data that is extracted from the computation. In the following, a short summary of the uncertainties encountered for each of these in realistic settings is given.

Basic Model Uncertainty/Errors: Physics

Given that "all models are wrong, but some are useful", and that in some cases "the right model" may be too expensive in terms of man-hours (data preparation) and/or computing time to be useful for design, engineers have learned to extract useful information from incomplete or "inferior" models. Therefore the key question engineers have to answer before doing any kind of simulation is: Is the ODE/PDE adequate?, in other words: Does it describe the physics/phenomena sought? Examples of underlying assumptions in computational fluid dynamics (CFD) include: potential flow (nonlinear Laplacian, applicable only to attached flow), incompressible flow (uniform density, infinite speed of sound, pressure-Poisson equation), Reynolds-Averages Navier–Stokes (RANS, attached flow, stretched grids) versus Large Eddy Simulations (LES, separated flow, isotropic grids), equations of state for gases, liquids and solids, Newtonian flow (linear relation between strain rates and shear stresses), various spray models, the number of important chemical reactions for complex carbohydrates, etc.

The assumptions made are often forgotten, but in some cases are much larger than the "super-converged" numerical solutions often displayed for high-order methods. Consider the following simple examples:

(a) Incompressible Flow

The variation of density ρ in any flow is of the order of $\Delta\rho/\rho \approx Ma^2$, where Ma denotes the Mach-number. This implies that if we compute a car driving at $v = 100\ km/hr$ (approximately $Ma = 0.1$) using an incompressible flow assumption/model, the error in forces and pressures will already be of $O(0.01)$, i.e., 1%.

(b) Chemically Reacting Flows

Many carbohydrates have such complex and long molecules that a complete description of the combustion chemistry would require in excess of 600 species and 1000 reactions, placing such simulations outside the realm of useful turnaround times. The solution is to develop a so-called simplified or reduced model that captures the essential species and reactions. This reduced model may contain O(30) species and O(100) reactions. O(100) reactions imply O(300) physical input parameters that need to be provided for the Arrhenius coefficients. Many of these are not known to better than 10%. So on top of the error due to simplification another 10% needs to be factored in for some of the reactions.

(c) Atomization of Liquids

Many internal combustion engines or turbines inject liquid fuel into a chamber. The bulk liquid disintegrates into droplets, which in turn break down due to shear or shocks, eventually evaporating and combusting. Each one of these stages needs to be modeled. A large variety of models exists for each of them—an indication of basic uncertainty, with many parameters not known to even 10% accuracy.

(d) Haemolysis
When designing artificial heart leaflets or blood pumps the very basic question: what leads to the destruction of red blood cells? needs to be answered. Is it shear? Is it the integral of shear over time? Are there other factors? A number of models exist for hemolysis (destruction of red blood cells) and thrombosis (formation of blood clots)—again, an indication of basic uncertainty, with many parameters not known to even 10% accuracy.

(e) Concrete Failure Under Severe Loading
The blast resistance of structures has been the focus of decades of research. A very basic uncertainty stems from the fact that concrete properties are highly dependent on the ambient conditions when poured (i.e., temperature, humidity, aging, etc.). A number of concrete failure models have been proposed, with parameters whose variability exceeds 10%. Given the nonlinearity of the physics, it is not surprising to observe that different groups performing a safety analysis on the same structure with the same loading reach vastly different conclusions (fail, safe, marginal, etc.).

Basic Model Uncertainty/Errors: Numerical Analysis

Another source of errors can emanate from the PDEs themselves. There are three main aspects that can lead to large errors and uncertainties: (a) multiple solutions, (b) butterfly effects, and (c) rogue loads.

(a) Multiple Solutions
Given that the PDEs are in most cases nonlinear, they may admit multiple solutions. What is the number of possible solutions? If we obtain one: is this "the right one"? This may seem irrelevant at first sight, but multiple solutions were first discovered for simulations of transonic airfoils with full potential flow assumptions (Steinhoff and Jameson 1981; Salas et al. 1984). At first they were dismissed with the argument that "when solved with Euler, the unique solution will emerge". A decade later, multiple solutions were documented for Euler solvers (Jameson 1991). These were dismissed again: "when solved with RANS, the unique solution will emerge". And when multiple solutions were documented for RANS solvers (Hafez 2003), the argument shifted to LES. Hysteresis has been observed in wind tunnel experiments, so perhaps one should finally accept the possibility of multiple solutions. Is is highly uncomfortable and undesirable for an engineer, but if it describes reality it has to be taken into consideration.

(b) Butterfly Effects
For time-dependent, nonlinear equations (such as the Navier–Stokes equations that describe fluids) small changes in input parameters, geometry, equations of state, or any other physical parameters can lead to solutions that diverge exponentially in time of each other. Statistically the flows are the same. But an individual measurement

probe at a location inside the flowfield can yield vastly different values at any particular time. These so-called butterfly effects (Löhner et al. 2014a) (named after the famous "butterfly in Asia that unleashes a storm in South America a week later") are far more common than previously thought and have been observed in many wind engineering settings.

(c) Rogue Loads

As the achievable realism of simulation increases and with it the physical complexity, the interaction of different fields and the superposition of many transient modes can lead to the emergence of short-lived, high-intensity phenomena called "rogue loads". These were first documented for so-called "freak waves" in the North Sea (Draper 1964, 1971; Haver 2003) and subsequently seen in lightweight structures immersed in complex vortical wind fields (Michalski et al. 2011; Löhner et al. 2014a). As usual, in retrospect, these phenomena are explainable: after all, it stands to reason that if a lightweight structure with many length/time scales in eigenmodes/stiffness/response is immersed in a flowfield with many length/time scales in vortical structures at some point in time a superposition of effects/modes will occur, leading to rogue loads. These loads are much higher than the steady/average loads and are also much higher than the unsteady standard deviations. Furthermore, they may be much higher than the values allowed in building codes/norms. When first discovered via computational methods (Michalski et al. 2011) they were greeted with disbelief. However, subsequent wind tunnel experiments confirmed their existence.

Model Parameter Uncertainty/Errors

The key question here is: to what extent are parameters known? Examples of physical parameters needed for simulations of fluids include: viscosity (particularly turbulent viscosity), conductivity, diffusivities, chemical reactions, atomization of liquids in gases, vaporization, and droplet/wall interactions. Examples from structural dynamics include: concrete, behavior of materials under high strain rates, aging, delamination, and tissue materials (e.g., arterial walls). All of these have uncertainties that are above 1%, with some in excess of 10%.

Model Boundary Condition Uncertainty/Errors

Any solution to a partial differential equation requires proper boundary conditions. Therefore the question: to what extent are boundary conditions known? Among the many examples from CFD one may mention: turbulent inflows (e.g., wind), humidity, distribution of particulates (size, velocity, ...), in/outflow conditions for vascular and pulmonary flow, and concentration of pollutants or other dispersed species. As before, for many realistic runs the uncertainties associated with the data required for these boundary conditions exceed 1%.

Model Geometry Uncertainty/Errors

The geometry plays a fundamental role in the results obtained. So: to what extent is geometry known? It is clear that for the manufacturing industries geometries are known to a very high degree of precision. But suppose one is trying to obtain wind loads (Michalski et al. 2011; Löhner et al. 2014a) or the dispersion of some contaminant (Camelli and Löhner 2004; Löhner and Camelli 2005; Camelli and Löhner 2006) for a building: is the upstream geometry really known? And if so: how many blocks upstream need to be included for an accurate result? Or suppose one is trying to assess the effectiveness of a stent in a cerebral artery (Castro et al. 2006): to what degree is the vessel geometry known? And how many diameters upstream need to be included for an accurate result? It is not difficult to see that in many realistic settings the geometric uncertainties will exceed 1% of the characteristic dimensions (e.g., vessel diameters, upstream obstacles, etc.).

Objections to High-Order Methods

Over the last decade, a number of objections to high-order methods have arisen. Some of them are admittedly subjective. But some stem from physical intuition and are therefore worthy of consideration.

Objection 1: Monotonicity

Monotonicity and the principle of total variation diminishing solutions is an important property of transport and fluids. It is particularly important in combustion: an overshoot in temperature might trigger an early combustion, an overly damped temperature may never lead to combustion. For traditional high-order methods based on polynomial expansions (either continuous or discontinuous Galerkin finite element approximations and their variants) the enforcement of monotonicity in the degrees of freedom does not imply monotonicity in the solution (e.g., at Gauss-points). One elegant solution is the ENO-like approach of Luo et al. (2005, 2008). However, in the presence of discontinuities this leads to a first-order discretization over the whole (large) element—exactly what is not desired. The solution has been h-refinement in the proximity of discontinuities, which for moving discontinuities implies an increased code complexity due to continuous mesh/DOF change and the required load rebalancing among parallel computing nodes when moving discontinuities are present.

Objection 2: Stencil Size and Shape

Consider a finite difference (FDM) stencil for a typical transport equation. Every degree of freedom inside the domain will have the same size and shape of stencil, i.e., will obtain information and use it to update its values from the same number of neighbors. For high-order FEMs/DGs a large variation of stencil size and shape from DOF to DOF is always present. Consider as an example the stencil size for continuous FEMs using Lagrange polynomials of order p in d dimensions. The resulting stencil sizes for different DOFs are given by

- Inner : $(p+1)^d$
- Corner: $2^d(p+1)^d$
- Edge : $2^{d-1}(p+1)^d$
- Face : $2^{d-2}(p+1)^d$

Consider next the stencil shape for the internal points of continuous FEMs or DGs using Lagrange polynomials of order p in one dimension. The DOF at p has only one DOF/point to the "right" (DOF/point $p+1$), but $p-1$ DOFs/points to the "left". The only way to alleviate this high variability in stencil shape is via isogeometric elements (Cottrell et al. 2009), which in turn have their own set of difficulties (non-locality, large bandwidth, etc.).

Objection 3: Real Order of Accuracy for Nonlinear Cases

Given the nonlinearity of the equations, a recurring question concerns the real order of accuracy achievable by high-order methods. Given a weighted residual approximation of a nonlinear flux F of the form:

$$r^i = \int W^i{}_{,j} F^j(u) d\Omega \ , \quad u = N^k \hat{u}_k \ ,$$

where W^i, N^k are shape functions and $\hat{u}_k$ the values at the degrees of freedom, two options to proceed are possible:

(a) integrate numerically "as is"; this is possible but expensive;
(b) approximate F with the same shape functions are u, i.e.,

$$r^i = \int W^i{}_{,j} N^k \hat{F}^j_k \ , \quad \hat{F}^j_k = F^j(\hat{u}_k) \ .$$

This implies that all geometric parameters can be precomputed, leading to a much faster overall procedure. On the other hand, this assumes $F(u)$ as being in the same polynomial space as u. What influence does that have on the overall order of accuracy? Is aliasing present?

One should remark that nonlinear fluxes are common. The compressible Euler/Navier–Stokes equations have many types of nonlinearities (nonlinear fluxes, nonlinear viscosities (e.g., Sutherland's law, k-e, power-laws, Casson model). Given the numerical integration scheme used, some errors are clearly unavoidable. Thus the concerns about the real order that is achievable for these nonlinear cases.

Objection 4: Accuracy When Butterfly Effects or Rogue Loads Are Present

As seen before, some of the more complex physical processes can exhibit butterfly effects or may lead to rogue loads. A number of open question still need to be answered for these cases:

- How important is spatial versus temporal accuracy?
- Are traditional accuracy measures still applicable?
- How accurate are high-order methods with respect to such measures as the mean, the Kurtosis, etc.?

Objection 5: Cost Versus Accuracy

As was stated before, engineering design and analysis can be viewed as an information gathering and creation process. But how accurate is the information provided to and obtained from the engineering models and physical approximations used? Even more: to what degree of accuracy can the required physical parameters, geometry, and boundary conditions be known? If for most of the engineering applications one can not know these to better than 1%, then this should be the relative accuracy needed. The question then becomes: Can high-order methods compete at 1% relative accuracy levels? Let us see if this is the case.

Work Estimates for High-Order Schemes

This section was copied almost verbatim from a previous paper by Löhner (2013). It is included here, because it shows why the fast high-order methods that are the subject of this book are required.

Basic Assumptions

All estimates, be they for work or errors, start from a set of assumptions. The present estimates are no exception. We briefly outline the most basic ones.

General, Nonlinear Physics: The **first assumption** is that we are interested in extendable, general-purpose codes that solve the compressible (or incompressible) Navier–Stokes equations. This implies that we have to assume general, nonlinear advective (equation of state for the pressure) and viscous (Sutherland's law or other empirical curvefits) fluxes that are not reducible to simple quadratic forms, implying the need for general quadrature rules to evaluate residuals.

Optimal Time Complexity: The **second assumption** (and perhaps the most contentious one) is that we have a way to solve for the unknown coefficients $\hat{u}$ of the resulting discrete system in **optimal time complexity**, i.e., it takes $O(n_m N)$ operations to do so, where N denotes the number of matrix entries and n_m the number of iterations/multigrid cycles. This would imply a fixed number of multigrid cycles and is certainly a lower bound, seldomly achieved in practice. For explicit time stepping (as would be the case for many LES problems), this same work estimate holds with $n_m = O(1)$.

Finite Element Shape Functions: The **third assumption** is that we have classic finite elements with C^0-continuous functions. This implies no loss of generality: similar, if not worse, estimates will ensure for discontinuous Galerkin schemes or isogeometric elements. Only elements that can be cast in tensor-product form result in less complexity and work. But they carry their own problems: explicit time-stepping constraints are severe, stability can be problematic, and tetrahedral elements (which are needed to mesh general domains) do not lend themselves easily to such shape functions.

(a) Lagrange Polynomials In Tensor Form:
Denoting by d the dimensionality of the problem, h the element size and p the order of the polynomial, the degrees of freedom per (quad/hex) element will be

$$DOF_{el} = (1+p)^d \ , \tag{1}$$

the number of matrix entries per element

$$n_{mat}^{FE} = (1+p)^{2d} \ , \tag{2}$$

and the number of elements

$$n_{el} = O(h^{-d}) \ . \tag{3}$$

As numerical integration is the usual way to build the matrix entries, this has to be taken into account. In order to integrate exactly the inner product of order $2p$ (mass matrix) or $2p-1$ (advection in flux form), at least $p+1$ Gauss-points are required in each dimension (tensor direction). Therefore, the work required to obtain the matrix entries will be

$$W_{el,g}^{FE} = (1+p)^{2d}(1+p)^d \quad . \tag{4}$$

This results in a work estimate for linear problems (i.e., only one matrix build) of

$$W_{d,p}^{FE} = c_1 h^{-d}(1+p)^{2d} \max\left(n_m, (p+1)^d\right) \quad . \tag{5}$$

The last term is difficult to estimate. For low-order elements the multigrid smoothers are well understood; this is not the case with higher order polynomials (Trottenberg et al. 2001). However, for nonlinear problems a new matrix needs to be built at every iteration. Given that many engineering problems of interest are nonlinear, we fold n_m into the constant c_1 and arrive at the following work estimate for nonlinear problems:

$$W_{d,p}^{FE} = c_1 h^{-d}(1+p)^{3d} \quad . \tag{6}$$

This same estimate holds for explicit time integration.

We remark that many authors have investigated specialized techniques to reduce the work of Gaussian integration, matrix formation, and matrix-vector multiplication, such as sum-factorization and elaborate integration rules (see, for example, Melenk et al. 1999, Vos et al. 2010, and Cantwell et al. 2011). The key idea is to use the tensor nature of the shape functions in order to reduce the general multi-dimensional shape/weight functions to a series of one-dimensional products. However, the conclusion of recent papers on this subject (Vos et al. 2010; Cantwell et al. 2011) was that even for the simple linear Helmholtz operator the element-based assembly or matrix-vector multiplication (used for the estimates above) was the fastest choice for $p < 10$. It is for this reason that the usual integration work estimates were used here.

(b) Polynomials For Simplex Elements:

The ratio of degrees of freedom for simplex (tria/tet) elements compared with tensor-product Lagrange (quad/hex) elements may be approximated by

$$\frac{DOF_{Si}}{DOF_{La}} = a + \frac{b}{(1+p)^c} \quad , \tag{7}$$

where the following values provide good fits (see Fig. 7.1):

- 2-D: $a = 0.51$, $b = 0.50$, $c = 1.10$;
- 3-D: $a = 0.18$, $b = 0.82$, $c = 1.30$;

implying for the degrees of freedom per element

$$DOF_{el} = (1+p)^d \left(a + \frac{b}{(1+p)^c}\right) \quad , \tag{8}$$

the number of matrix entries per element

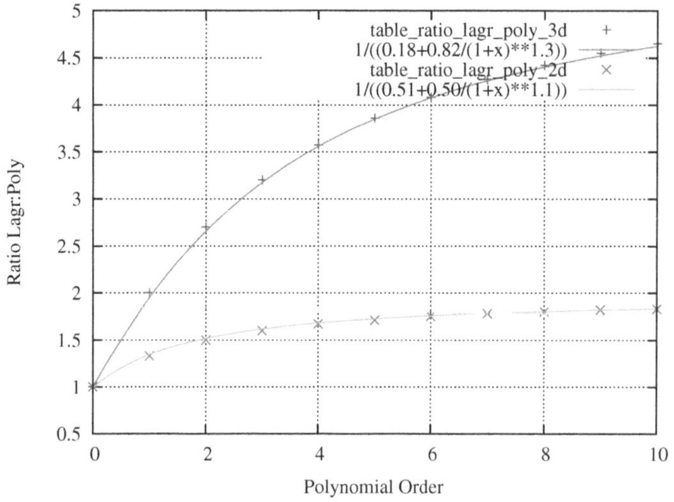

Fig. 7.1 Ratio of DOF tensor-product: simplex FE

$$n_{mat}^{FE} = (1+p)^{2d}\left(a + \frac{b}{(1+p)^c}\right)^2 , \tag{9}$$

and the number of elements

$$n_{el} = O(c_v h^{-d}) , \tag{10}$$

where $c_v = 2.0$ for 2-D and $c_v = 5.5$ for 3-D.

As before, the work required for numerical integration has to be taken into account. In reviewing the literature on cubature formulas (Jinyun 1984; Keast 1986; Geller and Hardbord 1991) one can approximate the number of Gauss-points required for a function of polynomial order p by

$$n_g = \left(\frac{p+1}{2}\right)^k , \tag{11}$$

where, surprisingly, $k = 2.5$ for 3-D (see Fig. 7.2).

Therefore, in order to integrate exactly the inner product of order $2p$ (mass matrix) or $2p-1$ (advection in flux form), at least

$$n_{el,g}^{FE} = \left(\frac{2p+1}{2}\right)^k . \tag{12}$$

Gauss-points are required, leading to a work estimate to obtain the matrix entries of

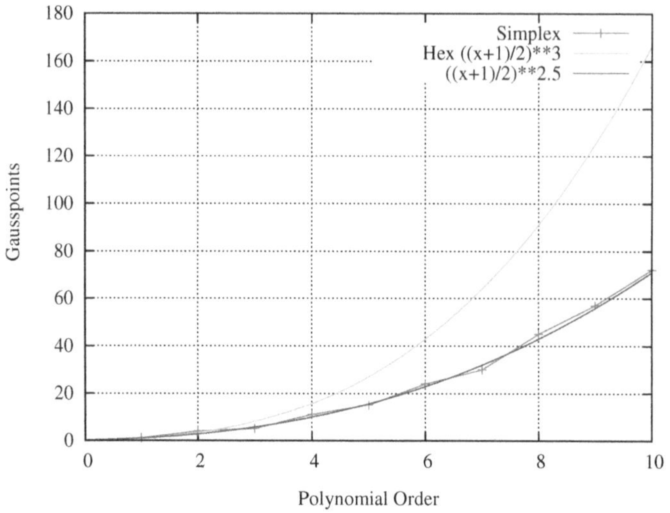

Fig. 7.2 Number of Gauss-points required

$$W_{el,g}^{FE} = (1+p)^{2d} \left(a + \frac{b}{(1+p)^c} \right)^2 \left(\frac{2p+1}{2} \right)^k . \tag{13}$$

This results in a work estimate for linear problems (i.e., only one matrix build) of

$$W_{d,p}^{FE} = c_1 c_v h^{-d} (1+p)^{2d} \left(a + \frac{b}{(1+p)^c} \right)^2 \max \left(n_m, \left(\frac{2p+1}{2} \right)^k \right) . \tag{14}$$

As before, we forego the speculation as to which is the dominant term in the last expression (max(...)) and concentrate on nonlinear problems and explicit time integration, which leads to

$$W_{d,p}^{FE} = c_1 c_v h^{-d} (1+p)^{2d} \left(a + \frac{b}{(1+p)^c} \right)^2 \left(\frac{2p+1}{2} \right)^k . \tag{15}$$

We remark that one can retain the tensor-structure of the quad/hex elements in the context of simplex elements by suitable mappings (Vos et al. 2010; Cantwell et al. 2011). This may offer some advantages for the numerical integration of very high-order elements, but also introduces unnecessary extra degrees of freedom with no measurable gain in accuracy. Therefore, only the typical simple shape functions were considered here.

(c) Finite Differences:
For a classic **finite difference** stencil on a Cartesian grid the number of neighbor points required increases linearly with the order of the approximation, i.e., the number

of matrix coefficients is given

$$n_{mat}^{FD} = 1 + 2dp \ , \tag{16}$$

yielding a work estimate of

$$W_{d,p}^{FD} = c_1 n_m h^{-d}(1 + 2dp) \ . \tag{17}$$

Mentioning finite differences in this context may produce mixed reactions in some readers. After all, finite elements offer the ability to treat complex geometries in a straightforward way. However, there will be many problems where the majority of the degrees of freedom is going to be in the field, and the scales to be resolved are such that a rather uniform distribution of mesh size is required. Acoustics, electromagnetics and LES runs fall under this category. And there is no reason not to use a finite difference (or any other structured grid method) in this region if a suitable linkage to the unstructured grid on the boundary can be found. See Darve and Löhner (1997) and Morgan et al. (2009) for examples of this kind.

Solution Smoothness: The **fourth assumption** is that the solution u is **smooth** (otherwise a high-order method would not make sense). The error ϵ_u is then given by

$$\epsilon_u = \|u - u^h\| < c_2 h^{p+1} |u|_{p+1} \ . \tag{18}$$

While this is the error estimate commonly given in textbooks, one can argue that the actual distance between degrees of freedom is not h but $h/\max(1, p-1)$. One then arrives at (see also Schwab 2004):

$$\epsilon_u = \|u - u^h\| < c_2 \frac{h^{p+1}}{\max(1, p-1)^{p+1}} |u|_{p+1} \ . \tag{19}$$

In both of these cases one can diminish the dependence of c_2 from p by invoking a generalized Taylor series and interpolation theory which would imply a division by $(p+1)!$. The error estimates then narrow to

$$\epsilon_u = \|u - u^h\| < c_2 \frac{h^{p+1}}{(p+1)!} |u|_{p+1} \ . \tag{20}$$

and

$$\epsilon_u = \|u - u^h\| < c_2 \frac{h^{p+1}}{\max(1, p-1)^{p+1} \cdot (p+1)!} |u|_{p+1} \ . \tag{21}$$

Clearly, the work estimates will directly depend on these error estimates. As they will be compared subsequently, we write them as

$$\epsilon_u = \|u - u^h\| < c_2 \frac{h^{p+1}}{g(p)} |u|_{p+1} \ . \tag{22}$$

Relative Error

The **fifth assumption** is that we are interested in the **relative error** of an unknown solution with a certain degree of spatial frequency of the form:

$$u = u_0 e^{i\omega x} \quad , \quad \omega = \frac{2\pi}{\lambda} \ , \tag{23}$$

where λ denotes the spatial wavelength corresponding to ω. In this case $u_{,p+1} = \omega^{p+1} u_0 e^{i\omega x}$ and

$$\epsilon_u = c_2 \frac{h^{p+1}}{g(p)} \omega^{p+1} |u|_0 = c_2 \frac{1}{g(p)} \left(\frac{2\pi h}{\lambda} \right)^{p+1} |u|_0 \ . \tag{24}$$

The relative error therefore becomes:

$$\epsilon = \frac{\epsilon_u}{|u|_0} = c_2 \frac{1}{g(p)} \left(\frac{2\pi h}{\lambda} \right)^{p+1} \ , \tag{25}$$

implying

$$h = \frac{\lambda}{2\pi} \left(\frac{\epsilon g(p)}{c_2} \right)^{\frac{1}{p+1}} \ . \tag{26}$$

Work Estimates

(a) Lagrange Polynomials In Tensor Form:
Combining Eqs. (6) and (26), one can see that for the **tensor-product elements** the work changes with the required relative error ϵ according to

$$W_{d,p}^{FE} = c_1 c_2^{\frac{d}{p+1}} \left(\frac{2\pi}{\lambda} \right)^d (\epsilon g(p))^{-\frac{d}{p+1}} (1+p)^{3d} \ . \tag{27}$$

Note that the constants c_1, c_2 depend on d, p, and are not always easy to estimate. In the remainder we will assume that the main effects of d, p are in the remaining terms, i.e., the desired relative error ϵ, the estimate function $g(p)$ and d, p. The "asymptotic estimate", obtained by removing from Eq. (27) the "coding-specific" constants c_1, c_2 and the problem specific factor λ are then given by

$$W_{d,p}^{FEa} = (\epsilon g(p))^{-\frac{d}{p+1}} (1+p)^{3d} \ . \tag{28}$$

(b) Simplex Elements:
For **simplex elements**, the "asymptotic estimate" is given by

$$W_{d,p}^{FEa} = c_v(\epsilon g(p))^{-\frac{d}{p+1}}(1+p)^{2d}\left(a + \frac{b}{(1+p)^c}\right)^2\left(\frac{2p+1}{2}\right)^k . \tag{29}$$

(c) Finite Differences:
For finite differences, the equivalent work estimate is given by

$$W_{d,p}^{FDa} = (\epsilon g(p))^{-\frac{d}{p+1}}(1+2dp) . \tag{30}$$

These estimates have been compared in Figs. 7.3, 7.4, and 7.5 for all four possible variations of $g(p)$. In these graphs, FE and FD denote Finite Element and Finite Difference, respectively. Interestingly, the simplex elements seem to be more efficient than the tensor-product (Lagrange) elements. This is in contrast to observations made by other authors for simplex elements that still use an underlying tensorial base for the shape functions (Vos et al. 2010; Cantwell et al. 2011). The tensor-simplex-elements have extra degrees of freedom which add extra work without a

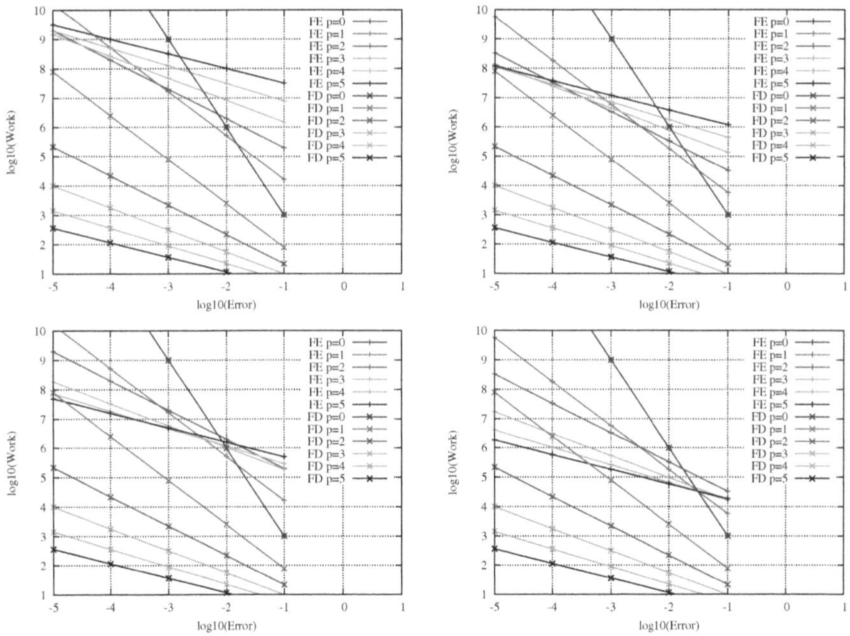

Fig. 7.3 Asymptotic work estimates (3-D, tensor-(Lagrange)-FE). Top left: $g(p) = 1$, top right: $g(p) = (p+1)!$, bottom left: $g(p) = \max(1, p-1)^{p+1}$, bottom right: $g(p) = (p+1)! \cdot \max(1, p-1)^{p+1}$

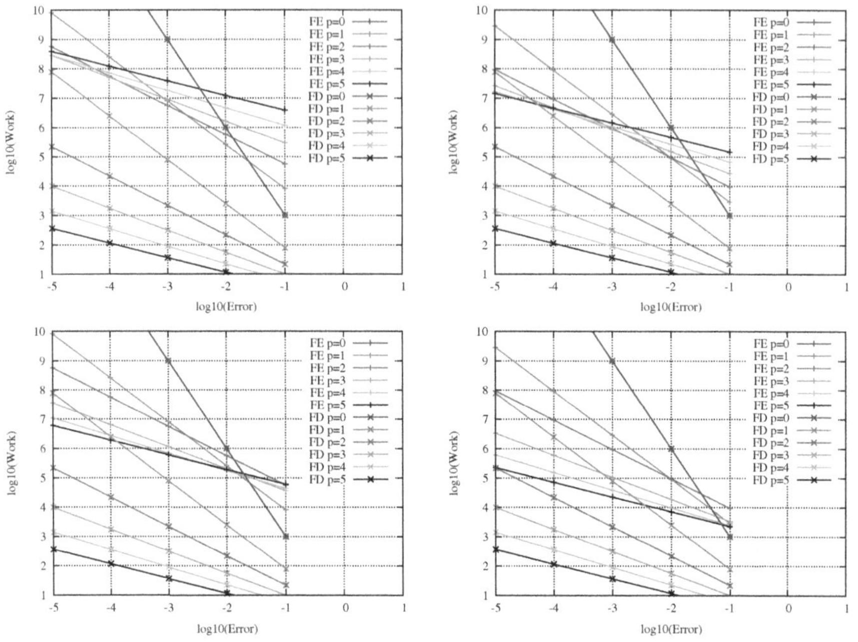

Fig. 7.4 Asymptotic work estimates (3-D, simplex-FE). Top left: $g(p) = 1$, top right: $g(p) = (p + 1)!$, bottom left: $g(p) = \max(1, p - 1)^{p+1}$, bottom right: $g(p) = (p + 1)! \cdot \max(1, p - 1)^{p+1}$

compensating reduction in errors. The ratios of work estimated for finite element and finite difference solvers have been summarized in Fig. 7.6.

Possible Objections

A number of possible objections may be raised for the estimates given. We list the most common ones:

(a) Cache and Local Memory
The most obvious objection to the estimates given is that the re-use of cache and local memory may not be the same for different element types. After all, higher order elements should lend themselves to a higher degree of cache and local memory re-use per multiplication if coded properly. This may be true, but it depends on coding. And if experience is any guide, properly coded low-order elements can achieve a very high degree of cache and local memory re-use as well.

(b) Derivatives
Another objection to the estimates given is that for many engineering applications one is interested in derivative information. A typical example may be shear stress on

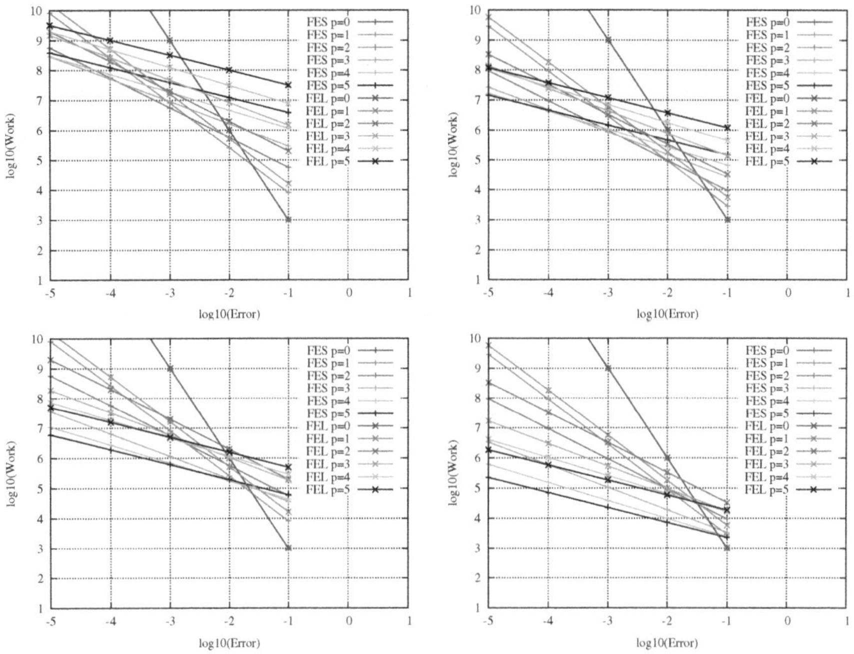

Fig. 7.5 Asymptotic work estimates (3-D, simplex-vs-lagrange-FE). Top left: $g(p) = 1$, top right: $g(p) = (p+1)!$, bottom left: $g(p) = \max(1, p-1)^{p+1}$, bottom right: $g(p) = (p+1)! \cdot \max(1, p-1)^{p+1}$

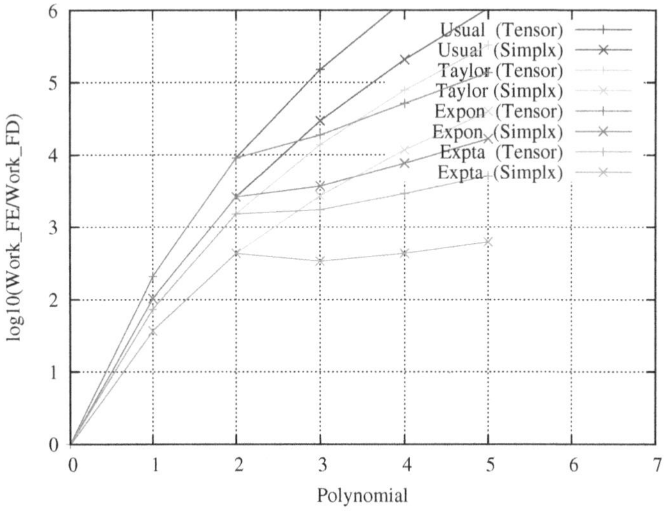

Fig. 7.6 Work ratios between finite elements and finite differences for 3-D

walls. Over the years, derivative "recovery" methods have been designed (Douglas and Dupont 1974; Levine 1985; Wheeler and Whiteman 1987; Mackinnon and Carey 1989; Thomée et al. 1989; Löhner et al. 2008), and superconvergence estimates have been obtained for both low- and high-order elements. From these, it appears that the quality of derivatives that can be obtained is similar to that of the unknowns. This would imply that the estimates derived above still hold.

(c) Limiting
A third objection is that the work estimated for finite differences may be too low. After all, if complex limiters (e.g., ENO, WENO) are used for high-order schemes, the algorithmic complexity increases nonlinearly with the order of the polynomial/finite difference scheme. This is certainly the case, but one can argue that the same will happen for finite element or discontinuous Galerkin methods when discontinuity capturing operators, stabilization, or limiting are used. As these are difficult to compare, they were left out of the present estimates.

(d) Gauss-Points
A fourth objection is that the number of Gauss-points used for the integration of the weighted residuals may be too low. After all, in order to integrate exactly the advection terms for shape functions of order p ones needs cubatures that are correct to order $2(p-1)$. This would imply that the work estimates given for the high-order FEM schemes **may be too low**. On the other hand, most codes use the same shape functions for the fluxes as for the unknowns. This introduces inconsistencies. One hopes that these inconsistencies are lower than the interpolation error of the scheme. This objection is valid, though, for elements with curved boundaries. For these, the Jacobian of the elements is not constant, requiring more Gauss-points than the ones considered here. Consider that for RANS grids and typical aerodynamic/automotive cases, a large portion of the volume mesh will have elements with curved boundaries (!). This same objection holds for isogeometric elements (Cottrell et al. 2009), were the rational nature of the shape functions requires further Gauss-points for integration (and furthermore the shape functions reach into neighboring elements).

(e) Work per Gauss-Point
A fifth objection is that the estimation of work proportional to the number of Gauss-points may be wrong. After all, the higher the polynomial/ complexity of the shape function, the higher the operation count for the evaluation of function values at Gauss-points. If one where to take this into consideration, the work estimates for high-order elements could potentially grow by another factor of $(1+p)^d$ for tensor-product Lagrange elements, and the corresponding factor given by Eq. (8) for simplex elements. These were not taken into consideration here, as one can find ways to minimize the number of operations to evaluate shape functions. However, the reduction factor is difficult to estimate, and one can argue that on modern microprocessors these terms are explicit, i.e., no transfer to memory is required, drastically reducing the CPU overhead.

(f) Convergence Rates
A sixth objection is that the work estimated for the solver may be too low for high-order methods. The implicit assumption of a multigrid solver was made. While p-multigrid has been shown to work well (Helenbrook et al. 2003; Luo et al. 2005; Kannan and Wang 2009; Liang et al. 2009), one should be aware that, as stated in Trottenberg et al. (2001) "in general, the efficient solution of problems discretized by high-order schemes is more difficult than that of low-order schemes since it becomes more difficult to find efficient smoothing schemes". Most p-multigrid papers only quote relative times and show independence of cycles to converge with respect to mesh size, but they give no comparison of the extra smoothing cycles needed as the order of the method is increased.

(g) Inadequacy of Low-Order Elements
Another objection is that one does not use high-order elements for speed, but for other properties, such as robustness, stability, and the removal of some of the numerical difficulties of linear/ low-order elements. An example in this last category is the deterioration of accuracy for very low Mach-numbers for compressible Navier–Stokes equation solvers. The present paper has only focused on the cases where convergence exists. However, the strong suspicion is that the same techniques that work for low-order elements (e.g., low Mach-number preconditioning) should also work advantageously for high-order elements. That this is indeed the case may be seen in Nigro et al. (2010).

Discussion

The immediate conclusion one draws from Figs. 7.3, 7.4 is that all conclusions will depend on the assumptions made for $g(p)$. Nevertheless, a surprising result is that linear elements (or the equivalent low-order schemes used in current production codes) do not compare as unfavorably as originally thought for engineering accuracy in 3-D. One can see from Figs. 7.3b, 7.4b that for a relative accuracy of $\epsilon = 10^{-2}$ (typical for engineering accuracy, given the uncertainties in geometry, boundary conditions, physics, etc.) the linear element is faster than any other element except for the most optimistic error estimate, i.e., $g(p) = (p+1)! \cdot \max(1, p-1)^{p+1}$. However, for linear elements the loops can be re-written over edges, saving considerable CPU and indirect addressing (empirical data for Intel Xeon chips: for heat conduction 1:7, for fluid dynamics 1:8, for linear elasticity 1:14). **This implies that the linear element will probably outperform all other element-based formulations even for** $\epsilon = 10^{-3}$.

The second conclusion is drawn from Fig. 7.6. As the order of the method increases, the ratio of work required for finite element, discontinuous Galerkin Methods, or isogeometric methods (i.e., micro-unstructured grids) as compared to finite difference methods grows to several orders of magnitude. Whereas for linear elements this could be reduced by changing data structures, no such option is presently available for high-order elements. The question then becomes whether the benefit of micro-unstructured grids is really worth 3–4 orders (!) of magnitude extra cost. Clearly, there are great opportunities for improvement here.

An immediate objection from many more mathematically inclined code developers will be that the 1% error bar is a moving scale. Indeed, over the years the levels of engineering accuracy have gone up when computations and models became more complex and ambitious. This is an objection that is hard to follow for any engineer performing CFD runs on a daily basis. In the vast majority of cases, material properties are not known to better than 1%. Only in rare cases, such as aerodynamics, are these levels of certainty reachable. Any one modeling melts, lipids, polymers, food processing, blood, combustion, explosions, and so many other liquids or gases knows how deficient our state of knowledge and models are as far as material properties.

LES Observations

Turbulent flows are characterized by vortices that need to be transported over large distances/times. The commonly accepted claim is that in order to accurately resolve turbulent flows, higher order methods are required (see, e.g., the NASA CFD Vision 2030 Study by Slotnick et al. 2014). An open question to date has been how to compare schemes of different spatial and temporal order. We add here recent empirical evidence in this respect.

Taylor Green Vortex

The Taylor Green Vortex (TGV) at $Re = 1600$ in 3-D is a canonical testcase where a simple initial flow condition causes a fully isotropic decay to a turbulence in time. This case is attractive as it is a "clean case" without any possible pollution from boundary conditions or mesh grading effects, where accuracy, levels of numerical dissipation, kinetic energy and physical dissipation can clearly be seen. It has been used repeatedly as a validation for high-order schemes, see, for example, (Gassner and Beck 2012; de Wiart et al. 2014; Bull and Jameson 2015). Furthermore it has been presented as a difficult and baseline type test case in the first, second, fourth, and fifth International Workshop on High-Order CFD Methods (High-Order CFD Methods 2015). The geometry of the problem consists in a triply periodic domain with size $0 \leq x, y, z \leq 2\pi$ with initial conditions,

$$u(t_0) = u_0 \sin\left(\frac{x}{L}\right) \cos\left(\frac{y}{L}\right) \cos\left(\frac{z}{L}\right) \ , \tag{31}$$

$$v(t_0) = -u_0 \cos\left(\frac{x}{L}\right) \sin\left(\frac{y}{L}\right) \cos\left(\frac{z}{L}\right) \ , \tag{32}$$

$$w(t_0) = 0 \ , \tag{33}$$

$$p(t_0) = p_0 + \frac{\rho_0}{16}\left(\cos\left(\frac{2x}{L}\right) + \cos\left(\frac{2y}{L}\right)\right)\left(\cos\left(\frac{2z}{L}\right) + 2\right) , \quad (34)$$

with constants $L = 1$, $u_0 = 1$, $\rho_0 = 1$ and $p_0 = 100$. The velocity of sound is $c_s = 10$, the viscosity is set to 6.2510^{-4} in order to obtain the desired Reynolds number $Re = 1600$ and the total simulation time is $t = 20$. As time advances the expected results consist of a series of key physical processes in turbulence: vortex roll-up, vortex stretch and interaction, and finally dissipation of the energy in the fluid leading to a steady state.

The simulations were carried out with FDFLO, a finite difference code that uses adaptive cartesian grids and allows for arbitrary (up to eight order) spatial and temporal discretization (Löhner et al. 2014b, 2019; Figueroa and Löhner 2019). The second- and fourth-order spatial discretizations were integrated with an explicit fourth-order Low-Storage Runge–Kutta (LSRK) scheme, while eighth-order spatial discretizations made use of an explicit eighth-order LSRK scheme.

Given the isotropic flowfield, the standard practice when comparing schemes for the TGV has been to monitor the following volume averaged quantities:

- Average Kinetic Energy (TKE):

$$E_k = \frac{1}{\rho_0 V}\int_V \frac{1}{2}\rho\mathbf{u}\cdot\mathbf{u}dV , \quad (35)$$

- Energy Dissipation Rate (DIS):

$$D_e = -\frac{dE_k}{dt} , \quad (36)$$

- Vorticity Dissipation Rate (NDI) (measures resolution of inertial range of turbulence):

$$D_\upsilon = \frac{2\mu}{\rho_0}\frac{1}{\rho_0 V}\int_V \frac{1}{2\rho}\omega\cdot\omega dV , \quad \omega_i = \varepsilon_{ijk}\frac{\partial u_k}{\partial x_j} , \quad (37)$$

where V is the total volume, ω is the vorticity and μ is the dynamic viscosity.

Three meshes were used in the simulations: 100^3 (coarse), 200^3 (medium) and 400^3 (fine) nodal points. As the time integration is explicit, the timestep size was proportional to the element/cell size.

Figure 7.7 shows the average kinetic energy with 2nd, 4th and 8th order FD obtained for the grids. These results are compared with DeBonis (2013), which uses a "Dispersion–Relation–Preserving" (DRP) scheme on a 512^3 element mesh, and Bull and Jameson (2015) which implements a Flux Reconstruction scheme that recovers a Spectral Difference method (FR-SD). Bull and Jameson (2015) noted that $D_e - D_\upsilon$ is an important error measure and corresponds to an estimation of the numerical dissipation introduced by the numerical method selected. Figure 7.8 compares the vorticity, kinetic energy, and numerical dissipation rates obtained. "FR-SD-63X4"

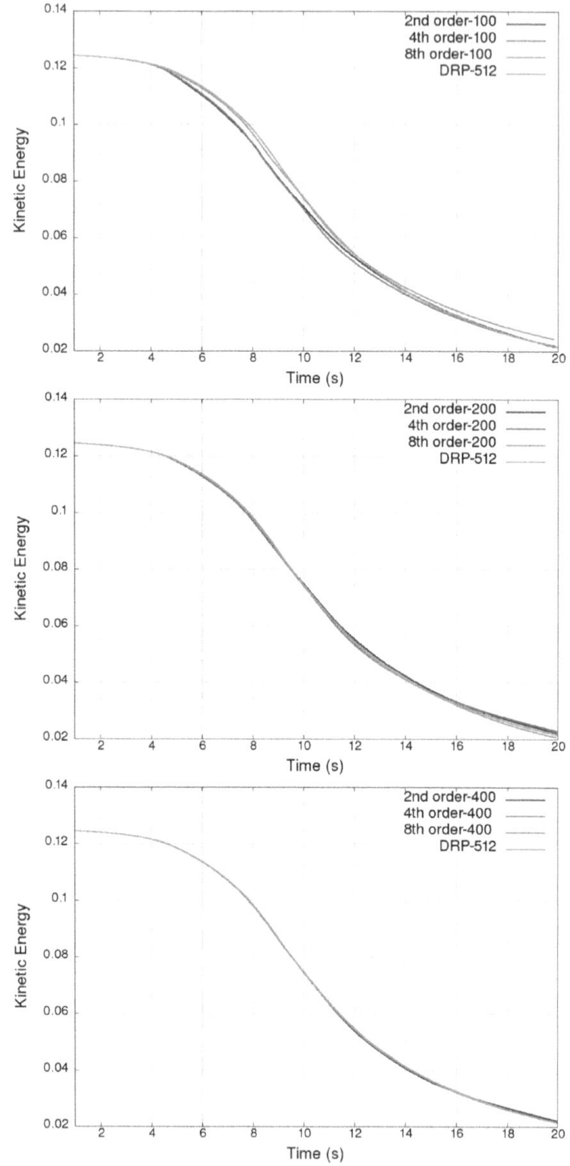

Fig. 7.7 TGV: kinetic energy versus time on grids of size 100^3, 200^3, 400^3

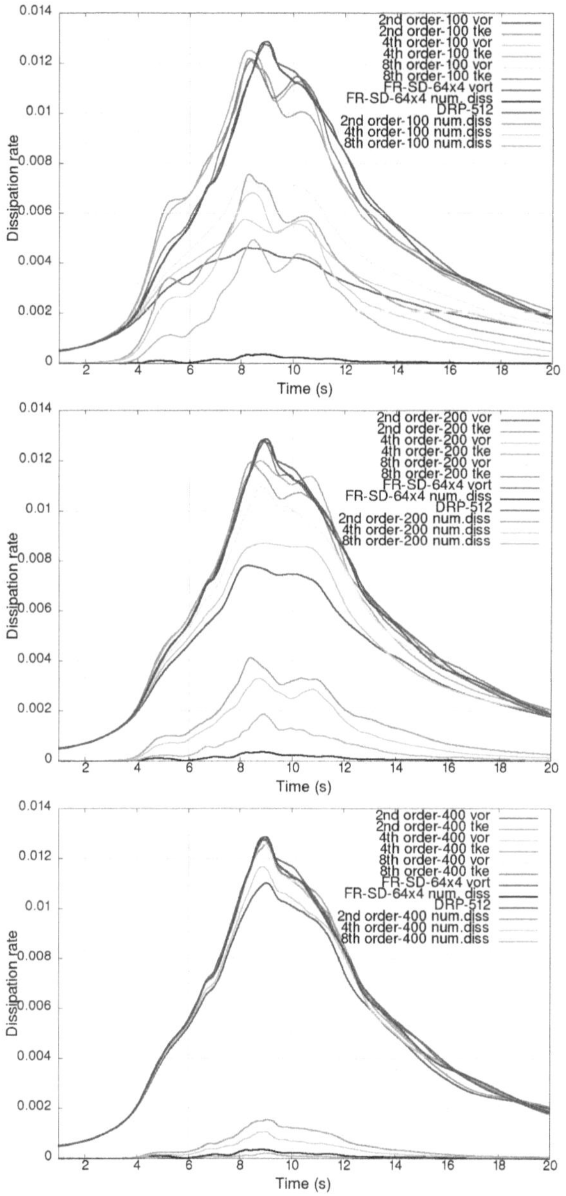

Fig. 7.8 TGV: TKE, VOR, NDI for different approximations. Orders and meshes (top: 100^3, middle: 200^3, bottom: 400^3)

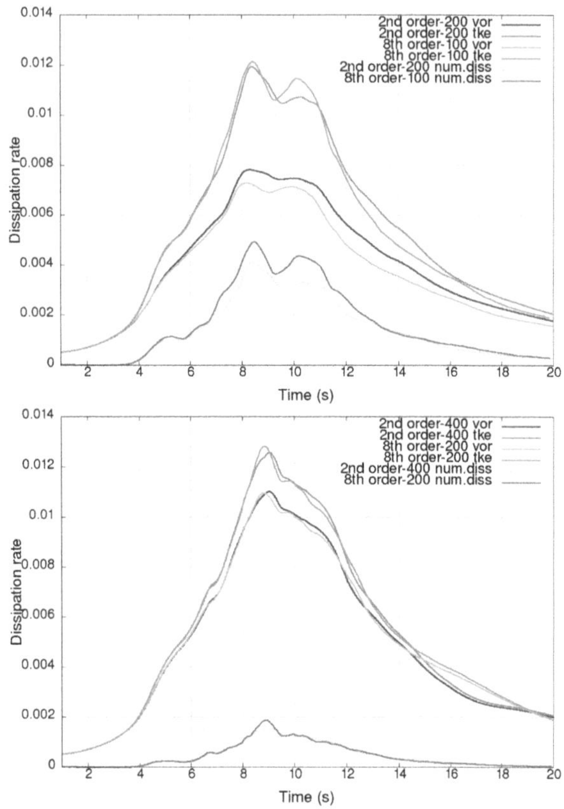

Fig. 7.9 TGV: TKE, VOR, NDI versus time. Top: comparison between 2nd order on 200^3 and 8th Order on 100^3 mesh. Bottom: comparison between 2nd Order on 400^3 and 8th order on 200^3 mesh

corresponds to a p_3 solution using the FR-SD method on a 64^3 mesh. As expected when the order of the approximation or the size of the mesh increase, the results approach the solution provided by the DRP scheme on the 512^3 mesh. Furthermore, the numerical dissipation decreases and, as can be seen for the 400^3 mesh, is similar to the numerical dissipation introduced by the FR-SD method in a 64^3 mesh with a p_3 solution (i.e., corresponding to a mesh of 256^3 degrees of freedom).

Another important observation contained in Fig. 7.8 **is that the results of** 8th **order FD approximations on a grid of size** $2h$ **are similar to** 2nd **order FD approximations on a grid of size** h. Figure 7.9 substantiates the last statement by comparing kinetic energy, vorticity and numerical dissipation rate using the 2nd and 8th order FD method and varying the mesh size. For further comparisons, see Figueroa (2020).

These results were obtained using finite difference solvers. It remains to be seen if similar behaviors are also observed for other numerical schemes, such as high-order DG schemes. But if they are, the case for high-order methods weakens considerably.

Conclusions

In the present chapter an attempt was made to place numerical methods in an engineering context. Starting with the observation that engineers create new things and hence always have to deal with incomplete information, a critical review was made of the accuracy of the available physical and modeling parameters. This showed that in many cases key physical and modeling parameters such as viscosities, boundary conditions, geometry, and even physics are not known to within 1%. This raises the question as to whether any numerical method needs to be of even higher quality than this threshold or a fraction thereof.

The attention then turned to work estimates for classic high-order elements. The work estimates for the so-called "fast high-order methods" can be found in the other chapters of this book. All of these estimates are naturally based on many assumptions, many of which will be incorrect for particular implementation or codes. They do, however, yield an overall picture of the relative merit high-order elements can provide.

The comparison of error and work estimates shows that even for relative accuracy in the 0.1% range, which is one order below the typical accuracy of engineering interest (1% range), linear elements may outperform all higher order elements. It is interesting to note that the empirical evidence in this respect corroborates the estimates derived. Note that a very different conclusion would have been reached for 1-D and 2-D problems. Here the high-order elements have a clear advantage, even for relatively high relative errors. This may, in part, explain the marked prevalence of publications that only consider 1-D and 2-D problems for high-order elements.

As expected, the estimates also show that the optimal order of elements in terms of work and storage demands depends on the desired relative accuracy.

Finally, the comparison of work estimates for high-order elements and their finite difference counterparts reveal a work-ratio of several orders of magnitude. It thus becomes questionable if general geometric flexibility via micro-unstructured grids is worth such a high cost.

An open question that has haunted the CFD community at large is how to compare schemes of different order for LES applications. A partial answer to this question was attempted by solving the classic Taylor–Green vortex decay. This case is attractive as it is a "clean case" without any possible pollution from boundary conditions or mesh grading effects, where accuracy, levels of numerical dissipation, kinetic energy and physical dissipation can clearly be seen. The results showed that the results of 8th order FD approximations on a grid of size $2h$ are similar to 2nd order FD approximations on a grid of size h.

Even though these conclusions and observations may dampen the expectations placed on high-order methods, in no way do they imply that high-order methods are not useful and should not be explored further. There are many applications where they have been proven essential to achieve good results, such as in wave phenomena, where essential aspects of the physics are lost if there is too much numerical dissipation.

Acknowledgements It is a pleasure to acknowledge many fruitful discussions on high-order methods with Profs. Ramon Codina (UPC, Barcelona), Antonio Huerta (UPC, Barcelona), Dominique Pelletier (Montreal), as well as Drs. Romain Aubry (GMU, NRL), Adrien Loseille (INRIA, Rocquencourt), Frederic Alauzet (INRIA, Rocquencourt) and Javier Principe (CIMNE, Barcelona) throughout the decades.

References

Atkins, H. L., & Shu. C.-W. (1998). Quadrature free implementation of discontinuous Galerkin method for hyperbolic equations. *AIAA Journal, 36*(5).

Bassi, F., & Rebay, S. (1997). A high-order accurate discontinuous finite element method for the numerical solution of the compressible Navier-Stokes equations. *Journal of Computational Physics, 131*(2), 267–279.

Bey, K. S., Oden, J. T., & Patra, A. (1996). A parallel hp-adaptive discontinuous Galerkin method for hyperbolic conservation laws. *Applied Numerical Mathematics, 20*, 321–336.

Brown, J. (2010). Efficient nonlinear solvers for nodal high-order finite elements in 3-D. *Journal of Scientific Computing, 45*, 48–63.

Bull, J., & Jameson, A. (2015). Simulation of the Taylor-Green vortex using high-order flux reconstruction schemes. *AIAA Journal, 53*(9), 2750–2761.

Camelli, F., & Löhner, R. (2004). Assessing maximum possible damage for contaminant release events. *Engineering Computations, 21*(7), 748–760.

Camelli, F., & Löhner, R. (2006). VLES study of flow and dispersion patterns in heterogeneous urban areas. *AIAA*-06-1419.

Cantwell, C. D., Sherwin, S. J., Kirby, R. M., & Kelly, P. H. J. (2011). From h to p efficiently: Strategy selection for operator evaluation on hexahedral and tetrahedral elements. *Computers and Fluids, 43*(1), 23–28.

Castro, M. A., Putman, C. M., & Cebral, J. R. (2006). Computational fluid dynamics modeling of intracranial aneurysms: Effects of parent artery segmentation on intra-aneurysmal hemodynamcis. *American Journal of Neuroradiology, 27*, 1703–1709.

Cockburn, B., Hou, S., & Shu, C.-W. (1990). TVD Runge-Kutta local projection discontinuous Galerkin finite element method for conservation laws IV: The multidimensional case. *Mathematics of Computation, 55*, 545–581.

Cockburn, B., Karniadakis, G. E., & Shu, C.-W. (Eds.). (2000). *Discontinuous Galerkin methods, theory, computation, and applications. Springer lecture notes in computational science and engineering (Vol. 11)*. Berlin: Springer.

Cottrell, J. A., Hughes, T. J. R., & Bazilevs, Y. (2009). *Isogeometric analysis*. New Jersey: Wiley.

Darve, E., & Löhner, R. (1997). Advanced structured-unstructured solver for electromagnetic scattering from multimaterial objects. *AIAA*-97-0863.

DeBonis, J. R. (2013). Solutions of the Taylor–Green vortex problem using high-resolution explicit finite difference methods. *NASA-TM-2013-217850*, also *AIAA*-13-0382.

de Wiart, C., Hillewaert, K., Duponcheel, M., & Winckelmans, G. (2014). Assessment of a discontinuous Galerkin method for the simulation of vortical flows at high Reynolds number. *International Journal for Numerical Methods in Fluids, 74*(7), 469–493.

Douglas, J., & Dupont, T. (1974). Galerkin approximation for the two-point boundary-value problem using continuous piecewise polynomial spaces. *Num. Math., 22*, 99–109.

Draper, L., (1964). 'Freak' ocean waves. *Oceanus* (X:4).

Draper, L. (1971). Severe wave conditions at sea. *J. Inst. Navig., 24*(3), 273–277.

Duncan, B., Fischer, A., & Kandasamy, S. (2010). Validation of Lattice-Boltzmann aerodynamics simulation for vehicle lift prediction. *ASME 2010 3rd Joint US-European Fluids Engineering Summer Meeting*, (pp. 2705–2716).

Figueroa, A., & Löhner, R. (2019). Postprocessing-based interpolation schemes for nested Cartesian finite difference grids of different size. *International Journal for Numerical Methods in Fluids*, *89*(6), 196–215.
Figueroa, A. (2020). Improvement of nested Cartesian finite difference grid solvers. *PhD Thesis*, George Mason University.
Gassner, G., & Beck, A. (2013). On the accuracy of high-order discretizations for underresolved turbulence simulations. *Theoretical and Computational Fluid Dynamics*, *27*, 221–237.
Geier, M., Pasquali, A., & Schönherr, M. (2017). Parametrization of the cumulant Lattice Boltzmann method for fourth order accurate diffusion part II: Application to flow around a sphere at drag crisis. *Journal of Computational Physics*, *348*(1), 889–898.
Geller, M., & Harbord, R. (1991). Moderate degree cubature formulas for 3-D tetrahedral finite element approximations. *Comm. Appl. Num. Meth.*, *7*(6), 487–495.
Hafez, M. (2003). Non-uniqueness problems in transonic flows. In H. Sobieczky (Ed.), *IUTAM Symposium Transsonicum IV, Fluid Mechanics and its Applications* (Vol. 73). Dordrecht: Springer.
Hartmann, R., & Houston, P. (2008). An optimal order interior penalty discontinuous Galerkin discretization of the compressible Navier-Stokes equations. *Journal of Computational Physics*, *227*(22), 9670–9685.
Haver, S. (2003). *Freak wave event at Draupner Jacket January 1 1995*. PTT-KU-MA: Statoil Technical Report.
Helenbrook, B. T., Mavriplis, D., & Atkins, H. L. (2003). Analysis of p-multigrid for continuous and discontinuous finite element discretizations. *AIAA*-03-3989.
Huerta, A., Angeloski, A., Roca, X., & Peraire, J. (2013). Efficiency of high-order elements for continuous and discontinuous Galerkin methods. *International Journal for Numerical Methods in Engineering*, *96*, 529–560.
International Workshop on High-Order CFD Methods, https://how5.cenaero.be
Jameson. A. (1991). Nonunique solutions to the Euler equations. *AIAA*-91-1625.
Jinyun, Y. (1984). Symmetric Gaussian quadrature formulae for tetrahedronal regions. *Computer Methods in Applied Mechanics and Engineering*, *43*, 348–353.
Kannan, R., & Wang, Z. J. (2009). A study of viscous flux formulations for a p-multigrid spectral volume Navier-Stokes solver. *Journal of Scientific Computing*, *41*(2), 165–199.
Karniadakis, G. E., & Sherwin, S. J. (2005). *Spectral/hp element methods for computational fluid dynamics* (2nd ed.). Oxford: Oxford University Press.
Keast, P. (1986). Moderate-degree tetrahedral quadrature formulas. *Computer Methods in Applied Mechanics and Engineering*, *55*, 339–348.
Klaij, C. M., van Raalte, M. H., van der Ven, H., & van der Vegt, J. J. W. (2007). h-multigrid for space-time discontinuous Galerkin discretizations of the compressible Navier-Stokes equations. *Journal of Computational Physics*, *227*(2), 1024–1045.
Kroll, N. (2006). ADIGMA—A European project on the development of adaptive higher order variational methods for aerospace applications. In P. Wesseling, E. Oñate & J. Périaux (Eds.), *Proceedings of the ECCOMAS CFD 2006*, TU Delft.
Laflin et al. K., (2004). Summary of data from the second AIAA CFD drag prediction workshop. *AIAA*-2004-0555.
Levine, N. D. (1985). Superconvergent recovery of the gradient from piecewise linear finite-element approximations. *IMA Journal on Numerical Analysis*, *5*, 407–427.
Liang, C., Kannan, R., & Wang, Z. J. (2009). A p-multigrid spectral difference method with explicit and implicit smoothers on unstructured triangular grids. *Computers and Fluids*, *38*(2), 254–265.
Lin, Y., & Chin, Y. S. (1993). Discontinuous Galerkin finite element method for Euler and Navier-Stokes equations. *AIAA Journal*, *31*, 2016–2023.
Löhner, R. (2011). Error and work estimates for high order elements. *AIAA*-11-0211.
Löhner, R. (2013). Improved error and work estimates for high order elements. *International Journal for Numerical Methods in Fluids*, *72*(11), 1207–1218.
Löhner, R., & Camelli, F. (2005). Optimal placement of sensors for contaminant detection based on detailed 3-D CFD simulations. *Engineering Computations*, *22*(3), 260–273.

Löhner, R., Appanaboyina, S., & Cebral, J. R. (2008). Parabolic recovery of boundary gradients. *Communications in Numerical Methods in Engineering, 24*, 1611–1615.
Löhner, R., Britto, D., Michalski, A., & Haug, E. (2014a). Butterfly-effect for massively separated flows. *Engineering Computations, 31*(4), 742–757.
Löhner, R., Corrigan, A., Wichmann, K. R., & Wall, W. A. (2014b). Comparison of Lattice-Boltzmann and finite difference solvers. *AIAA*-2014-1439.
Löhner, R., Haug, E., Michalski, A., Britto, D., Degro, A., Nanjundaiah, R., et al. (2015). Recent advances in computational wind engineering and fluid-structure interaction. *Journal of Wind Engineering and Industrial Aerodynamics, 144*, 14–23.
Löhner, R., Figueroa, A., & Degro, A. (2019). Recent advances in a Cartesian solver for industrial LES, *AIAA*-2019-2328.
Luo, H., Baum, J. D., & Löhner, R. (2005). A p-multigrid discontinuous Galerkin method for the compressible Euler equations on unstructured grids. *Journal of Computational Physics, 211*, 767–783.
Luo, H., Baum, J. D., & Löhner, R. (2008). On the computation of steady-state compressible flows using a discontinuous Galerkin method. *International Journal for Numerical Methods in Engineering, 73*, 597–623.
Mackinnon, R. J., & Carey, G. F. (1989). Superconvergent derivatives: A Taylor series analysis. *International Journal for Numerical Methods in Engineering, 28*, 489–509.
Melenk, J. M., Gerdes, K., & Schwab, C. (1999). Fully discrete hp-finite elements: Fast quadrature. *ETH Report*, 99-15.
Michalski, A., Kermel, P. D., Haug, E., Löhner, R., Wüchner, R., & Bletzinger, K.-U. (2011). Validation of the computational fluid-structure interaction simulation at real-scale tests of a flexible 29 m umbrella in natural wind flow. *Journal of Wind Engineering and Industrial Aerodynamics, 99*, 400–413.
Morgan, K., Xie, Z. Q., & Hassan, O. (2009). A parallel hybrid time domain method for large scale electromagnetic simulations. In B. H. V. Topping & P. Iványi (Eds.), *Parallel, distributed and grid computing for engineering, Chapter 14* (pp. 309–328). Stirlingshire, UK: Saxe-Coburg Publications.
Nastase, C. R., & Mavriplis, D. J. (2006). High-order discontinuous Galerkin methods using an hp-multigrid approach. *Journal of Computational Physics, 213*(1), 330–357.
Nigro, A., De Bartolo, C., Hartmann, R., & Bassi, F. (2010). Discontinuous Galerkin solution of preconditioned Euler equations for very low Mach number flows. *International Journal for Numerical Methods in Fluids, 63*(4), 449–467.
Persson, P.-O., Bonet, J., & Peraire, J. (2009). Discontinuous Galerkin solution of the Navier-Stokes equations on deformable domains. *Computer Methods in Applied Mechanics and Engineering, 198*, 1585–1595.
Salas, M. D., Gumbert, C. R., & Turkel, E. (1984). Nonunique solutions to the transonic potential flow equation. *AIAA Journal, 22*(1), 145–146.
Schwab, C. (2004). *p- and hp- finite element methods*. Oxford: Oxford Science Publications.
Slotnick, J., Khodadoust, A., Alonso, J., Darmofal, D., Gropp, W., Lurie, E., & Mavriplis, D. (2014). CFD vision 2030 study: A path to revolutionary computational aerosciences. *NASA/CR-2014-218178, NF1676L-18332*.
Steinhoff, J., & Jameson, A. (1981). Multiple solutions of the transonic potential flow equation. *AIAA*-81-1019.
Thomée, V., Xu, J., & Zhang, N. Y. (1989). Superconvergence of the gradient in piecewise linear finite-element approximation to a parabolic problem. *SIAM Journal on Numerical Analysis, 26*(3), 553–573.
Trottenberg, U., Oosterlee, C., & Schüller, A. (2001). *Multigrid*. London: Elsevier Academic Press.
Visbal, M. R., & Gaitonde, D. V. (2002). On the use of higher-order finite-difference schemes on curvilinear and deforming meshes. *Journal of Computational Physics, 181*(1), 155–185.

Vos, P. E. J., Sherwin, S. J., & Kirby, R. M. (2010). From h to p efficiently: Implementing finite and spectral/hp element discretisations to achieve optimal performance at low and high order approximations. *Journal of Computational Physics*, *229*, 5161–5181.

Wang, Z. J. (2007). High-order methods for the Euler and Navier-Stokes equations on unstructured grids. *Progress in Aerospace Sciences*, *43*, 1–41.

Wheeler, M. F., & Whiteman, J. (1987). Superconvergent recovery of gradients on subdomain from piecewise linear finite element approximations. *Numerical Methods P.D.E.'s*, *3*, 357–374.

Index

M. Kronbichler and P.-O. Persson (eds.), *Efficient High-Order Discretizations for Computational Fluid Dynamics*, CISM International Centre for Mechanical Sciences 602, https://doi.org/10.1007/978-3-030-60610-7

The manufacturer's authorised representative in the EU is Springer Nature Customer Service Centre GmbH, Europaplatz 3, 69115 Heidelberg, Germany. If you have any concerns regarding our products, please contact ProductSafety@springernature.com

Printed and bound by CPI Group (UK) Ltd, Croydon, CR0 4YY
30/09/2026
02225690-0014